Lecture Notes in Electrical Engineering 1131

The book series *Lecture Notes in Electrical Engineering* (LNEE) publishes the latest developments in Electrical Engineering—quickly, informally and in high quality. While original research reported in proceedings and monographs has traditionally formed the core of LNEE, we also encourage authors to submit books devoted to supporting student education and professional training in the various fields and applications areas of electrical engineering. The series cover classical and emerging topics concerning:

- Communication Engineering, Information Theory and Networks
- Electronics Engineering and Microelectronics
- Signal, Image and Speech Processing
- Wireless and Mobile Communication
- Circuits and Systems
- Energy Systems, Power Electronics and Electrical Machines
- Electro-optical Engineering
- Instrumentation Engineering
- Avionics Engineering
- Control Systems
- Internet-of-Things and Cybersecurity
- Biomedical Devices, MEMS and NEMS

For general information about this book series, comments or suggestions, please contact leontina.dicecco@springer.com.

To submit a proposal or request further information, please contact the Publishing Editor in your country:

China

Jasmine Dou, Editor (jasmine.dou@springer.com)

India, Japan, Rest of Asia

Swati Meherishi, Editorial Director (Swati.Meherishi@springer.com)

Southeast Asia, Australia, New Zealand

Ramesh Nath Premnath, Editor (ramesh.premnath@springernature.com)

USA, Canada

Michael Luby, Senior Editor (michael.luby@springer.com)

All other Countries

Leontina Di Cecco, Senior Editor (leontina.dicecco@springer.com)

**** This series is indexed by EI Compendex and Scopus databases. ****

Jason C. Hung · Neil Yen · Jia-Wei Chang
Editors

Frontier Computing on Industrial Applications Volume 1

Proceedings of Theory, Technologies and Applications (FC 2023)

Springer

Editors
Jason C. Hung
Department of Computer Science
and Information Engineering
National Taichung University of Science
and Technology
Taichung City, Taiwan

Neil Yen
School of Computer Science and Engineering
University of Aizu
Aizuwakamatsu, Japan

Jia-Wei Chang
Department of Computer Science
and Information Engineering
National Taichung University of Science
and Technology
Taichung City, Taiwan

ISSN 1876-1100 ISSN 1876-1119 (electronic)
Lecture Notes in Electrical Engineering
ISBN 978-981-99-9298-0 ISBN 978-981-99-9299-7 (eBook)
https://doi.org/10.1007/978-981-99-9299-7

This Springer imprint is published by the registered company Springer Nature Singapore Pte Ltd.
The registered company address is: 152 Beach Road, #21-01/04 Gateway East, Singapore 189721, Singapore

Paper in this product is recyclable.

Contents

Surface Defect Detection of Frozen Dumplings Based on Improved U-Net Network

Zhigen Fei, Peiting Li(✉), Xiaoxiao Song, Xinchang Zhao, and Yanqiu Xiao

Mechanical and Electrical Engineering Institute, Zhengzhou University of Light Industry, Zhengzhou, Henan, China
1778702840@qq.com

Abstract. In order to realize automatic detection of surface defects of frozen dumplings, a U-shaped semantic segmentation ResNet_Unet based on U-Net is proposed in this paper. ResNet50 is used as the encoder of the semantic segmentation network to enhance the feature extraction capability. In order to enhance the recognition accuracy of small targets, expanding the receptive field of the multi-scale output feature map of the encoder is proposed. And embedding BN normalizes the data before decoding module convolution, which can speed up training and improve generalization of the model. In addition, the label is smoothed when the loss is calculated to prevent the model from placing too much faith in the predictions. The experimental result shows that the mPA and mIoU of the ResNet_Unet network are 83.20% and 80.44%, respectively. This study provides a reference for intelligent segmentation of surface defects in frozen dumplings.

Keywords: Frozen Dumpling Surface Defect · Resnet_Unet · Dilated Convolution · BN

1 Introduction

Businesses and consumers steadily raised their expectations for the quality of product manufacture as society and The Times developed. Surface defects are simple to develop in the production, processing, and freezing operations for frozen dumplings [1]. According to the visual and sensory requirements of frozen food in GB 19295–2021 "National Standard for Food Safety Frozen Noodle Rice and Prepared Food," as well as the defect identification method provided by Henan Sanquan Food Co., LTD., this paper divides the defects into six types, including leak, broken belly, broken ends, crack, adhesion, and special-shaped. Manual sorting and conventional visual inspection procedures have a tough time meeting the inspection standards because of the range of surface flaws present in quickly frozen dumplings.

The convolutional neural network of deep learning can extract features of different dimensions in the image and strengthen the network's study of various defect categories through the learning of multi-dimensional features [2]. Wang et al. [3] used the DenseNet121 convolutional neural network as the backbone network of YOLOv4 to detect the defects of apple leaves. Wang et al. [4] used the improved U-Net network

J. C. Hung et al. (Eds.): FC 2023, LNEE 1131, pp. 1–8, 2024.
https://doi.org/10.1007/978-981-99-9299-7_1

and two twin dense branch networks to realize the detection of the four defects of the plate steel, including cobwebbing, inclusion, plaque and scratch.. Zhou et al. [5] established WideResNet50-AdamW-Wce model on the basis of WideResNet to classify five categories of green plum including rot, crack, scar, rain spot and normal. Fu et al. [6] proposed that YOLOv4 algorithm was used to detect potato surface sprouting, mechanical damage, rot, insect attack and scab.

In conclusion, multi-class defect detection tasks benefit from the convolutional neural network's strong detection effect. There is, however, no effective method to detect and classify the surface defects of frozen dumplings. Therefore, to achieve the segmentation and classification of frozen dumpling surface defects, this research develops a U-shaped semantic segmentation network called ResNet_Unet based on Unet.

2 U-Net Algorithm

U-Net [7] is an end-to-end U-shaped semantic segmentation network, which is mainly composed of encoders, decoders and skip connections, as shown in Fig. 1. The encoder is mainly used for feature extraction. Skip connection is responsible for copying and clipping the encoder multi-scale output feature map, and then Concat that merges channels of the up-sampled feature map in the decoder with channels of the processed encoder multi-scale output feature map to achieve the fusion of shallow detailed features. The decoder receives the feature map of the same scale in the multi-scale output feature map of the encoder and realizes the upsampling operation through deconvolution, so that the feature map is gradually restored to the input resolution. Finally, pixel-by-pixel classification is achieved through 1×1 convolution to obtain the final segmentation effect map.

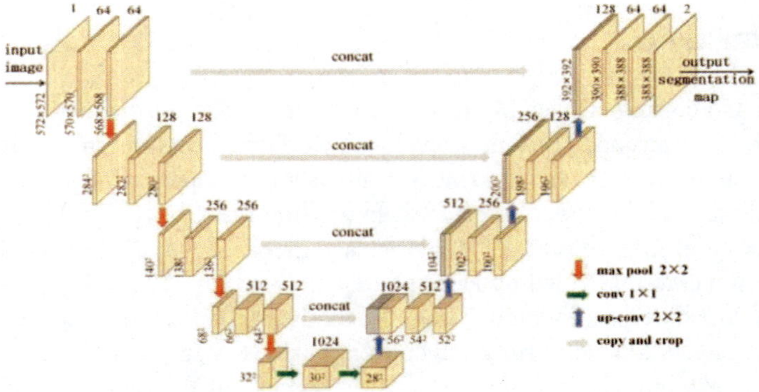

Fig. 1. U-Net architecture

3 ResNet_Unet Network Architecture

ResNet_Unet is a U-shaped architecture, and its encoder is composed of ResNet50 [8]. The part of skip connection is embedded into the dilated convolution with a dilated rate of 2 to expand the receptive field, and the BN operation is added before each convolution of the decoder. The overall network structure is shown in Fig. 2.

Among them, ResNet50 is used as an encoder network for U-net feature extraction [9], which follows the U-shaped structure of Unet, and U-ResNet is constructed. The expansive convolution is to expand the convolution kernel size, which can add zeros between kernel elements of conventional convolution [10]. Compared with conventional convolution, dilated convolution can expand the receptive field of feature map without changing the spatial resolution. Batch Normalization (BN) is a method of data processing. It solves the problem of uneven distribution of deep layers's data by normalizing input [11]. By smoothing the input distribution of hidden layers, it helps with random gradient descent and reduces the negative impact of weight updates on subsequent layers [12].

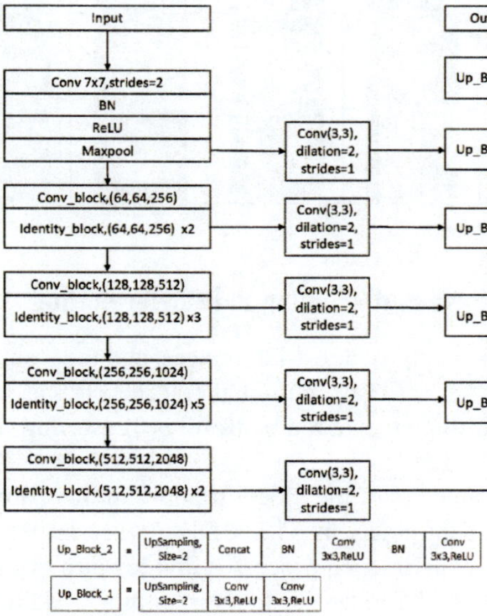

Fig. 2. ResNet_Unet architecture

4 Experiment

4.1 Dataset Introduction

The defective frozen dumplings in this paper were provided by Henan Zhengzhou Sanquan Food Co., LTD., and were shot with the camera and lens of IRAYPLE, as shown in Fig. 3, with a resolution of 2592 × 2048. Defects are divided into six types. Table 1 shows images of normal and defective frozen dumplings.

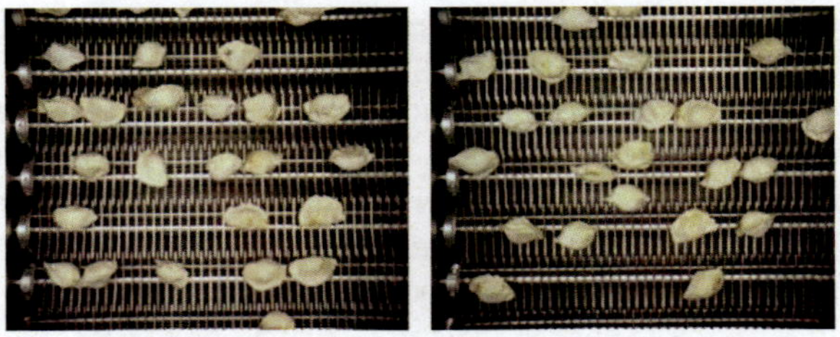

Fig. 3. The original image of the collection

Table 1. Comparison of frozen dumplings with different defects

Normal	Leak	Broken_ belly	Broken_ ends	Crack	Adhesion	Special-shaped
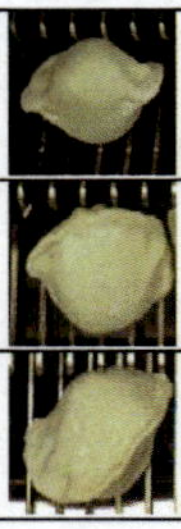				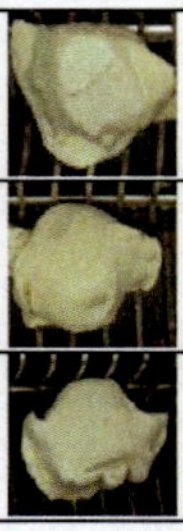		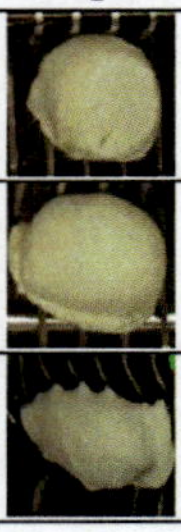

4.2 Experimental Platform and Training Parameter Setting

The dataset in this paper is composed of 1426 images, and the resolution of each image is 2592 × 2048. Among them, 143 were used as test sets, 129 as verification sets, and 1154 as training sets. Our implementation platform is Tensorflow 1.31.1, CUDA 11.2, CUDNN 8.0.1, and Keras2.1.5.

The experimental parameters of the model training in this paper are configured as follows: The input size of the image is 512 × 512. Based on the pre-training model of ImageNet training, the experimental process is divided into two stages: freezing and non-freezing. In the training phase, we use label smoothing, and batch_size is set to 2. The Adam optimizer is used, the initial learning rate is set to 1e-4 and 1e-3, respectively, and the cosine annealing strategy is adopted. We adapt the dice_loss loss function for backpropagation. In addition, 10 epochs are trained in the freezing phase, and 90 epochs are trained in the non-freezing phase.

4.3 Evaluation Indicator

In this paper, four evaluation indices are used for the semantic segmentation network, which are mIoU (mean Intersection over Union), FPS (Frames Per Second), mPA (mean Pixel Accuracy), and Dice Coefficient. IoU is the intersection and union ratio between the category prediction result and the real label, and the calculation formula is Formula (1).

TP represents the number of pixels predicted into a certain category and predicted correctly; FP represents the number of pixels predicted into a certain category but predicted incorrectly; and FN represents the unpredicted number of pixels in this class.

$$IoU = \frac{X \cap Y}{X \cup Y} = \frac{TP}{TP + FP + FN} \tag{1}$$

FPS is the number of images model processes per second as a measure of how fast the model detects images. CPA (Class Pixel Accuracy): represent the pixel accuracy of a class, representing the probability of all pixels predicted to be of a certain class actually belonging to that class, as shown in Eq. (2).

$$CPA = \frac{TP}{TP + FP} \tag{2}$$

mPA: average pixel accuracy for each category of pixel accuracy, as shown in Eq. (3).

$$mPA = \frac{1}{n}\sum_{i=1}^{n} CPA_i \tag{3}$$

Dice Coefficient: used to calculate the similarity between two sets, and its value range is between [0,1], as shown in Eq. (4):

$$Dice = \frac{2TP}{2TP + FP + FN} \tag{4}$$

4.4 Experimental Results and Analysis

4.4.1 Method Performance Comparison

Figure 4 shows the CPA comparison results of the three models. As can be seen from the figure, compared with U-ResNet, the dilated convolution network'CPA improves by 8.66% in crack, by 0.85% in broken ends, and by 8.98% broken belly pixels.

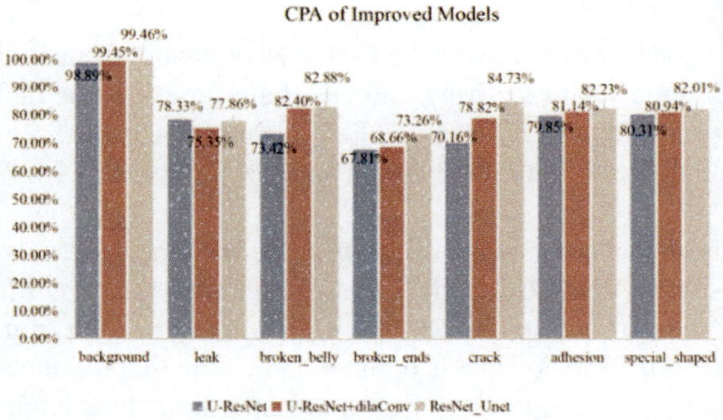

Fig. 4. CPA comparison of the optimization methods corresponding to the models

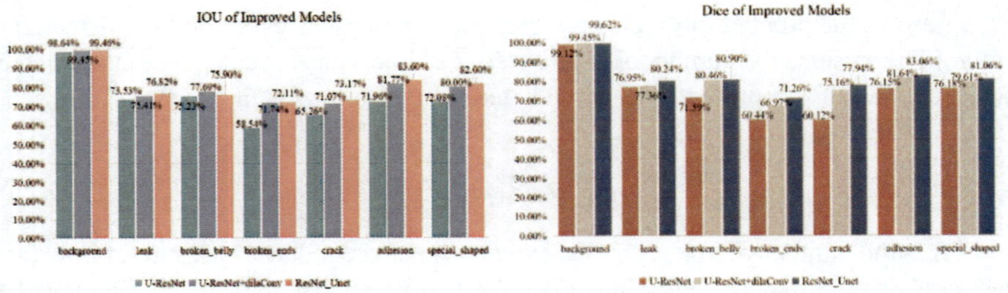

Fig. 5. IoU and Dice comparison of the optimization methods corresponding to the models

ResNet_Unet'CPA also improved by 5.91% and 4.6% in cracks and broken ends respectively.

Figure 5 and Fig. 6 respectively shows the IoU and Dice comparison results of the three models.

Table 2 shows the optimization method comparison result by four evaluation indexes. For the optimized semantic segmentation ResNet_Unet model, although the segmentation speed decreases to a certain extent, ability of the model to identify defects and correctly identify pixel categories is improved.

Table 2. Comparison of improved models

Models	U-ResNet	Dilated	BN	Size (MB)	mIOU (%)	mPA (%)	FPS (Frames/s)
U-ResNet	√			165	73.61	78.40	29.07
U-ResNet+ Dilated	√	√		186	79.59	80.97	26.16
ResNet_Unet	√	√	√	186	80.44	83.20	26.41

4.4.2 Contrast Experiment

The proposed method is compared with the U-shaped semantic segmentation constructed by VGG16 and MobileNet respectively, and the results are shown in Table 3. Compared with VGG_Unet and MobileNet_Unet, the FPS of ResNet_Unet is 5.91 lower than that of MobileNet_Unet, but the mIOU of resnet_unet is 10.32% and 14.47% higher than those.

Figure 6 is the visual comparison of the test results of the three networks. (a) is the segmentation result by VGG_Unet, (b) is the segmentation result by MobileNet_Unet, and (c) is the segmentation result by ResNet_Unet. As can be seen from the figure, the segmentation effect of ResNet_Unet is optimal, indicating that the semantic segmentation model proposed in this paper is an update and extension of the intelligent automatic detection of frozen dumpling surface defects.

Table 3. Comparison of different networks

Models	Size (MB)	mIOU (%)	FPS (Frames/s)
VGG_Unet	96	70.12	25.35
MobileNet_Unet	60.6	65.97	32.32
ResNet_Unet	186	80.44	26.41

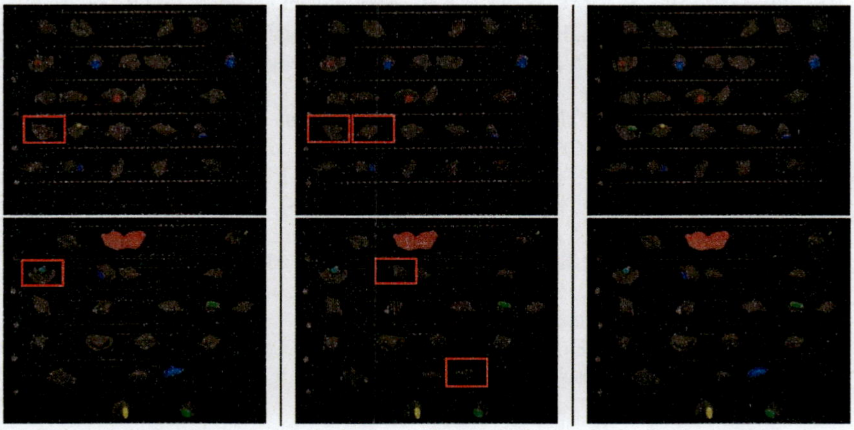

(a) (b) (c)

Fig. 6. Defect segmentation result of frozen dumplings. (a) Result of VGG_Unet, (b) result of MobileNet_Unet, (c) result of ResNet_Unet

5 Conclusion

In this paper, a U-shaped semantic segmentation network ResNet_Unet based on U-Net is proposed to realize the intelligent detection of surface defects of frozen dumplings. The self-made data set of surface defects of frozen dumplings was used to evaluate the model through mIoU, mPA, Dice Coefficient, and FPS. The experimental results showed that the method in this paper was more suitable for the detection of surface defects in frozen dumplings, with 80.44% mAP and 83.20% mPA. It is of great significance for the automatic detection of surface defects in frozen dumplings. The experimental results show that the FPS of the proposed method is not very high, so the next step is to improve the detection speed of the network.

References

1. Wang, N.: Don't eat cracked dumplings. News Digest (02) (2014)
2. Li, W.: Research on fine-grained image classification technology based on deep learning. Inf. Comput. **34**(11), 65–68 (2022)

3. Wang, Q., Lü, L., Huang, D., Fu, S., Yu, H.: Research of apple leaf disease defect detection based on improved YOLOv4 algorithm. J. Chin. Agricult. Mechaniz. **43**(11), 182–187 (2022)
4. Daolei, W., Liu Yiteng, D., Wenbin, Z.R.: A cascaded twin dense network based metallic surface defect detection algorithm. J. Comput.-Aided Des. Comput. Graph. **34**(06), 946–952 (2022)
5. Zhou, C., Wang, H., Liu, Y., et al.: Green plums surface defect detection based on deep learning methods. IEEE Access **10**, 100397–100407 (2022)
6. Yunlong, F., Dan, L., Dongtai, L., Ning, F., Xianer, C.: Potato surface defect detection based on machine vision and YOLO algorithm. Machinery **59**(08), 82–87 (2021)
7. Ronneberger, O., Fischer, P., Brox, T.: U-Net: convolutional networks for biomedical image segmentation. In: Navab, N., Hornegger, J., Wells, W.M., Frangi, A.F. (eds.) MICCAI 2015. LNCS, vol. 9351, pp. 234–241. Springer, Cham (2015). https://doi.org/10.1007/978-3-319-24574-4_28
8. He, K., Zhang, X., Ren, S., et al.: Deep residual learning for image recognition. In: Proceedings of the IEEE Conference on Computer Vision and Pattern Recognition, pp. 770–778 (2016)
9. Li, Y., Li, Y., Liu, J., et al.: Research on segmentation of steel surface defect images based on improved res-UNet network. J. Electron. Inf. Technol. **44**(5), 1513–1520 (2022)
10. Khalaf, M., Dhannoon, B.N.: MSRD-Unet: multiscale residual dilated U-Net for medical image segmentation. Baghdad Sci. J. **19**(6 Supplement), 1603 (2022)
11. Dimitriou, N., Arandjelović, O.: Sequential normalization: embracing smaller sample sizes for normalization. Information **13**(7), 337 (2022)
12. Ioffe, S., Szegedy, C.: Batch normalization: accelerating deep network training by reducing internal covariate shift//International conference on machine learning. In: PMLR, pp. 448–456 (2015)

Application of Distributed Database System in Financial Management

Baifang Liu(✉), Yajing Ji, and Chenxi Zhou

School of Business, Beijing Language and Culture University, Beijing, China
liubaifang@blcu.edu.cn

Abstract. With the rapid development of information technology, information technology has gradually been incorporated into financial management, and more and more enterprises are using financial management in their operations. The financial management level in today's enterprises is relatively low and cannot meet the needs of the enterprise, so it is necessary to further upgrade financial management. This article studied the application of distributed database systems in financial management, aiming to further improve the level of financial management through distributed management systems. This article tested the efficiency improvement of financial management after using a distributed database system through experiments. The experimental data showed that the efficiency has improved by at least 13% and the highest by 19%. Through this experimental data, it can be proven that distributed database systems can indeed have good results in financial management.

Keywords: Financial Management · Distributed · Database Systems · Information Technology

1 Introduction

Many scholars have conducted research on financial management. Atmadja A T studied the quantitative asset pricing implications of financial intermediaries facing leverage constraints. He used recursive methods to construct global solutions to explain occasionally bound constraints [1]. Cumming D J found that during financial crises, the predictive power of all traditional financial distress prediction models has decreased [2]. Asaff R believed that financing decisions are one of the important areas in financial management to increase shareholder wealth, and companies can use debt or equity capital to finance their assets [3]. Although there is a lot of research on financial management, there are still many shortcomings in financial management.

2 Use of Distributed Database Systems in Financial Management

2.1 Current Financial Management Issues

The problems in financial management are shown in Fig. 1.

J. C. Hung et al. (Eds.): FC 2023, LNEE 1131, pp. 9–16, 2024.
https://doi.org/10.1007/978-981-99-9299-7_2

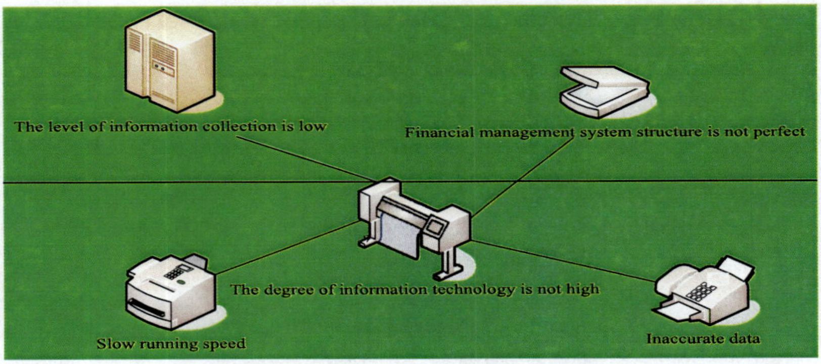

Fig. 1. Problems in financial management

The level of financial information collection for enterprises is relatively low. Whether the work of enterprise management accounting can be successfully carried out largely depends on the level of financial information collection of the enterprise [4, 5]. However, the current level of financial informatization in most enterprises is only at the level of purchasing a set of financial software to conduct accounting and supervision, and cannot achieve the goal of establishing a data center for enterprises. Management accounting can choose appropriate methods and models, and obtain corresponding data results. In the process of comparing these data, various risks that exist in the development process of the enterprise can be identified in a timely manner. While doing internal control well, targeted improvement measures can also be proposed to help managers make decisions. Therefore, the low level of financial informatization in enterprises has become one of the factors restricting the development of management accounting [6, 7].

Only in this way can the efficiency and quality of financial management work be improved, thus promoting the healthy and stable development of the company. However, so far, the financial management of many state-owned enterprises still remains in the traditional management mode, and the level of informatization is not high, which has a significant impact on the efficiency and quality of enterprise financial management [8, 9].

The system and structure of enterprise information financial management are not perfect. At present, most enterprises have not yet formed a unified processing system for financial big data [10]. Even some enterprises have not yet established a unified software platform for managing financial big data, making it difficult to integrate these data.

2.2 Construction of Financial Management Big Data Platform

In recent years, with the rise of the "Internet", many enterprises have begun to turn to this model. The biggest change brought about by enterprise transformation is that the business activities of enterprises have gradually shifted partially or completely from offline to online. The change in business activities has led to significant changes in the way companies obtain financial information. The process of building a big data platform is shown in Fig. 2.

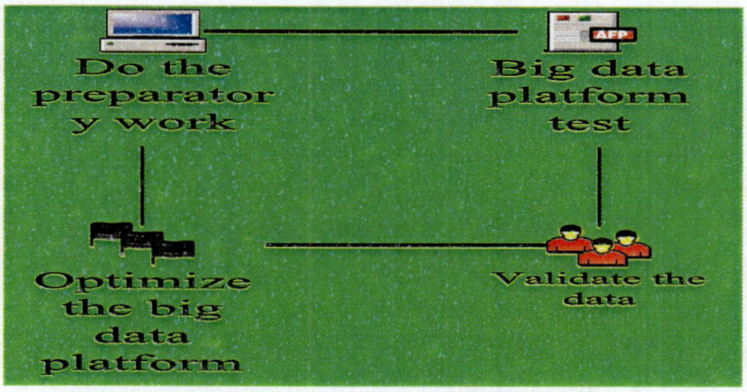

Fig. 2. The process of building a big data platform

Firstly, enterprises need to make the most basic preparations. Enterprises should collect necessary information and materials during the process of building a big data platform based on their industry and business requirements, in order to ensure the efficiency of building a big data platform. Secondly, testing of big data platforms is required. Referring to relevant materials, a system platform that integrates cash flow management, profit distribution management, and fund financing management can be established. On this basis, combined with the actual situation after industry integration, the role of the financial management big data platform was verified. Thirdly, it is necessary to optimize the functionality of the financial big data platform.

2.3 Overview of Distributed Database

Distributed database technology is a product that combines database technology with distributed technology. One database technology was specifically mentioned. Different databases are geographically dispersed, but logically, they belong to the same system. The system does not attach importance to centralized control of the system, but rather emphasizes the autonomy of each database node. In addition, in order to reduce the workload of programmers when writing programs and the possibility of system errors, no consideration is usually given to the distribution of data, so that the distribution of system data is always transparent.

2.4 Application of Distributed Database in Financial Management

Distributed databases have a wide range of applications in daily life. For example, the familiar shopping website Taobao has countless people logging into it every day. The reason why people can see their shopping records and shopping related information is because they have a powerful database. The method of using databases for financial management is as follows: Firstly, the application of distributed databases in financial management greatly reduces people's work intensity, eliminating the need to spend a lot of time and effort on detailed registration of each member's personal information and economic status. The financial management system is completed by computer programmers from inputting data to printing the final results, without manual intervention, thus freeing financial managers from tedious accounting and statistical work.

Under the requirements of modern enterprise management, databases are an effective means to achieve modernization of enterprise financial management. Nowadays, financial information is becoming increasingly rich and important. As a company, they need to understand and analyze these data in the first place, and then make scientific predictions and decisions. In this sense, database technology can effectively solve the contradiction between the increasing amount of information and the lagging management of human resources and funds. The more truthful the financial information of a company, the greater its profits. The use of database technology for refined management of enterprise operations can effectively ensure the financial security of enterprises and promote their sustainable development.

Secondly, it has put an end to the history of handmade tables and greatly improved work efficiency. In work with the same amount of data, the processing time of computers is definitely much shorter than that of manual labor, and the accuracy and efficiency of database processing data are also unmatched by manual labor. This application gives people the ability to process information in real-time that manual operations do not possess. Using a database for financial management has transformed manual management from multiple individuals to machine management for one person, saving both manpower and material resources.

On the other hand, the application of database technology has also improved the quality of reports. Originally, manually summarizing financial statements was not conducive to accountants conducting amateur assessments. After using a database to summarize financial statements, some non-standard units filled in the reports can be directly displayed in the database, thereby improving the qualification rate of the reports.

2.5 Financial Management Database System

Analyzing the operational performance of basic data: For ordinary enterprises, information system operators are not information technology workers and cannot handle the complexity of information systems. For financial management information systems, the main task they need to do is to input some system data into the financial management information system, and the rest of the analysis work is basically completed automatically by the financial management information system. In financial management work, a lot of financial data and information are involved. Therefore, in order to manage these data information, it is necessary to fully utilize information technology and network technology, in order to improve the efficiency of financial management.

2.6 Innovative Financial Management Work

In the context of big data, innovation drives development. At the same time, the financial management of enterprises should also take reform as the driving force for development, and strengthen it at every link. In the process of data collection, the authenticity and reliability of data quantification should be taken as the basis to ensure that all financial accounting information in the system has authenticity, comprehensiveness, and objectivity. In data storage work, it is necessary to organically combine information technology and big data technology, and continuously innovate data storage technology to enhance data security while ensuring the integrity of the data chain. Therefore, in specific management work, it is necessary to continue to improve the construction of databases, so that they can better utilize the value of accounting information resources, classify data reasonably, and accelerate data query and integration.

2.7 Use of Distributed Database Algorithms

There is a naive Bayesian algorithm in distributed database algorithms. The naive Bayesian algorithm provides a series of training samples and instances related to the objective function when applied, and then predicts the target values of new instances for classification. The goal of the naive Bayesian algorithm for new instance classification is to describe the attribute value $<c_1, c_2, c_3, ...c_n>$ of the instance. P represents probability, and the obtained target value is shown in Formula (1):

$$A = \arg\max P(c_1, c_2, c_3, ..., c_n) \tag{1}$$

The naive Bayesian algorithm classification is based on a basic assumption that the target value attribute values are independent of each other. Based on this assumption, Formula (1) can be transformed into Formula (2):

$$B = \arg\max \mathrm{P}(\mathrm{c}_1) \prod \mathrm{P}(\mathrm{c}_\mathrm{n}) \tag{2}$$

It is assumed that R is a random discrete variable, the local probability of each variable in the Bayesian algorithm is represented by P(R I D), and the calculation method for P (R I D) is shown in Formula (3):

$$P(R|D) = zP(R) \prod P(D_i|U_i) \tag{3}$$

In Formula (3), z is a generalization factor; U_i represents the set of parent nodes of variable D_i; ; P (R) represents a prior probability; $P(D_i|U_i)$ represents the local probability distribution of D_i.

3 Simulation Experiment on the Application of Distributed Database System in Financial Management

Distributed database systems can bring convenience to financial management. This article tested the efficiency improvement of financial management after using a distributed database system through experiments. If the efficiency improvement is good, it indicates that distributed data systems can indeed achieve good results in financial management. This article selected five groups of enterprises that used a distributed database system for financial management, and the improvement in financial management efficiency is shown in Fig. 3.

From the experimental results in Fig. 3, it can be seen that the efficiency of enterprise financial management using a distributed database system was improved by at least 13% and at most by 19%. From this experimental data, it can be seen that distributed database

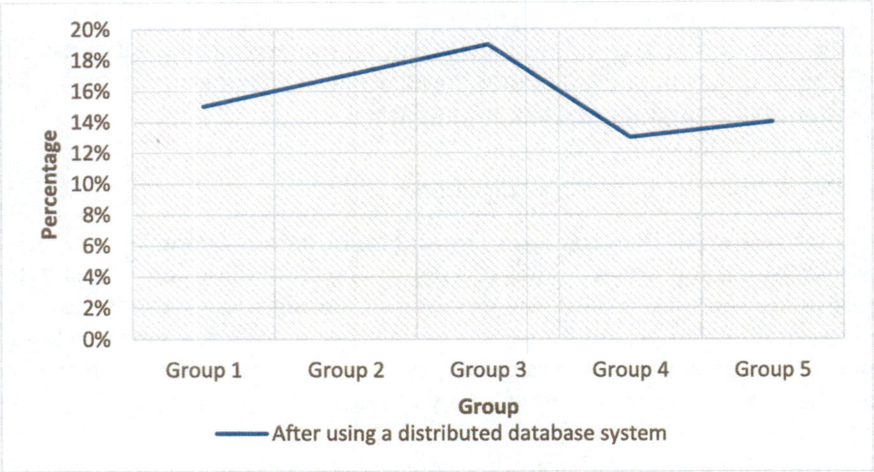

Fig. 3. Financial management efficiency improvement of enterprises using distributed database system

systems can achieve good results in financial management, as they can improve the efficiency of enterprise financial management.

This article once again tested the amount of unnecessary loss of financial assets within a month using a distributed database system for financial management, and then compared it with financial management without using a distributed database system. The comparison results are shown in Fig. 4.

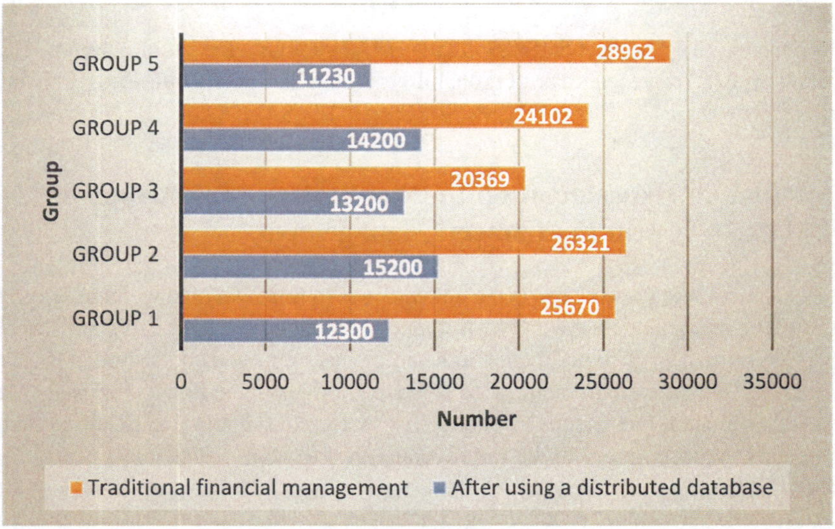

Fig. 4. Two kinds of financial management property loss situation comparison

According to the experimental results in Fig. 4, it can be seen that the maximum amount of unnecessary loss of property within one month after using distributed database financial management was 15200 yuan, and the minimum was 11230 yuan; the maximum amount of unnecessary loss of property in traditional financial management within a month was 28962 yuan, and the minimum was 20369 yuan. From this experimental data, it can be seen that distributed databases have the effect of reducing unnecessary property consumption in companies, which is of great significance for enterprises. Through the above two experiments, it can be seen that distributed databases and financial management have a good fit.

4 Conclusions

Financial management is very important for enterprises. Having good financial management in an enterprise can greatly improve the efficiency of handling affairs and save a lot of manpower and resources. So, nowadays many enterprises choose to use financial management, but in fact, using financial management requires the use of distributed databases to achieve better results. Distributed databases can store a lot of personnel information data, and can be saved for a long time, so that personnel information data is not prone to errors and omissions. This article focused on the application of distributed database systems in financial management. This article tested the efficiency improvement of financial management after using a distributed database system through experiments, and found that the data is considerable, indicating that the distributed database system can achieve good results in financial management. Due to space limitations, the experiment conducted in this article is not bad and would be improved in the future. Finally, it is hoped that financial management would become more and more in place.

References

1. Atmadja, A.T., Saputra, K.A.K.: Determinant factors influencing the accountability of village financial management. Acad. Strateg. Manag. J. **17**(1), 1–9 (2018)
2. Cumming, D.J., Leboeuf, G., Schwienbacher, A.: Crowdfunding models: Keep-it-all vs. all-or-nothing. Financ. Manag. **49**(2), 331–360 (2020)
3. Asaff, R., Suryati, S., Rahmayani, R.: Pengaruh financial attitude dan financial knowledge terhadap financial management behavior. JEMMA (J. Econ. Manag. Account.) **2**(2), 9–22 (2019)
4. Azar, J., Raina, S., Schmalz, M.: Ultimate ownership and bank competition. Financ. Manag. **51**(1), 227–269 (2022)
5. Jagtiani, J., Lemieux, C.: The roles of alternative data and machine learning in fintech lending: evidence from the LendingClub consumer platform. Financ. Manag. **48**(4), 1009–1029 (2019)
6. Da, Z., Shive, S.: Exchange traded funds and asset return correlations. Eur. Financ. Manag. **24**(1), 136–168 (2018)
7. Nawn, S., Banerjee, A.: Do proprietary algorithmic traders withdraw liquidity during market stress? Financ. Manag. **48**(2), 641–676 (2019)
8. Carter, T.D., Walker, W.D., III.: Retaining, cultivating, and investing: strategies for leading millennial managers. J. Govern. Financ. Manag. **67**(2), 20–25 (2018)

9. Das, C.P., Swain, R.K.: Influence of capital structure on financial performance. Parikalpana: KIIT J. Manag. **14**(1), 161–171 (2018)
10. Crespi, F., Giacomini, E., Mascia, D.V.: Bail-in rules and the pricing of Italian bank bonds. Eur. Financ. Manag. **25**(5), 1321–1347 (2019)

Research of the Influencing Factors of Hotel Customers' Green Behavior Based on the DEMATEL-ISM Model Under the "Dual Carbon" Target

Zirui Qiu and Bin Zhao(✉)

School of Management, Tianjin University of Technology, Tianjin, China
zhaobin_tjut@hotmail.com

Abstract. Under the influence of the "dual carbon" strategy, China's hotel industry has gradually begun to transform into green environmental protection. The research was based on literature and interview research, 13 influencing factors of hotel customers' green consumption behavior were identified by the Delphi method. The decision laboratory method (DEMATEL) was used to aggregate the 13 forming factors into four-factor gathers: strong cause, weak cause, strong effect, and weak effect. Furthermore, the multilevel hierarchical model of the factors that affect hotel guests' green consumption habits is adopted to explain the structural model (ISM). To research the elements that influence hotel guests' green consumption habits and systematically assess the relative relevance of the many influencing factors, it is of great significance for hotels to attract customers to produce green consumption behavior. The results indicate that the green consumption behavior of hotel customers is the outcome of direct factors, intermediate factors and deep factors. Dual carbon strategy, green certification, green fiscal, and corporate green marketing are the primary factors affecting the formation of customers' green consumption behavior, which should be paid attention to.

Keywords: Dual Carbon Strategy · Green Hotel · Green Consumption · Decision Laboratory Method · Structure Interpretation Model Introduction

1 Introduction

China officially proposed the "dual carbon" target of reaching a carbon peak by 2030 and carbon neutrality by 2060 during the 75th United Nations General Assembly in September 2020. The "dual carbon" strategy advocates the formation of a green, environmentally friendly, and low-carbon lifestyle. As important guidance for China's economic development and green transformation, it has had a significant impact on various industries in China. Among them, the hotel industry is a vital part of the development of China's tourism industry, and the green and low-carbon transformation of the hotel has also become a hot topic of social attention. The hotel industry as the "Large energy consumption household", the traditional production-consumption patterns lead to the

J. C. Hung et al. (Eds.): FC 2023, LNEE 1131, pp. 17–24, 2024.
https://doi.org/10.1007/978-981-99-9299-7_3

consumption of resources and wastewater volume increasing, the hotel itself is also aware of the problem of serious, formed by actively building and managing the "green hotel" as the goal of the green transformation strategy, hotel management has begun an increasing trend of "green practice", And pay further attention to environmental issues. But the true sense of the green hotel does not only need the hotel's green low-carbon business philosophy, advanced environmental protection technology, and equipment but also needs the support and cooperation of consumers themselves, gradually developing consumption habits.

In the current situation of increasingly depleted ecological resources, Green consumerism is viewed as a key approach for addressing natural resource issues and promoting long-term socioeconomic growth. In January 2022, the National Development and Reform Commission and other seven ministries and administrations jointly issued the Implementation Plan for Promoting Green Consumption [1]. Therefore, to study the impacting factors of hotel customers' green consumption conduct, for the hotel to improve the green management, and publicity system; Undertake the responsibility of social green environmental protection; It is of great significance to strengthen the active selection of green hotels by clients.

Different scholars do not completely agree on the attribution of green consumption behavior of hotel customers, which can be roughly distinguished from three aspects: consumer characteristics, hotel behavior, and policy guidance. From the perspective of consumer characteristics, customers' personal characteristics have different impacts on their green hotel cognition and green consumption standpoint, and their green hotel cognition will also affect their green consumption willingness and the promotion of environmental protection measures [2].In addition, The environmental ethics of customers will also have a significant impact on their willingness to consume green products [3]; Environmental responsibility is an important intrinsic driver of green purchasing behavior [4]. In terms of external environmental impact, group pressure has a momentous impact on individual green consumption behavior [5]; As most people begin to believe in green consumption, group pressure is changing others' attitudes towards green consumption, which is seen as a form of consistent consumer behavior [6]. Hotel behavior is also the object of exploration and research by many scholars. In the external behavior of the hotel, green hotel enterprises should make the value of green hotels explicit by using various technologies and marketing means so that consumers can quickly and accurately perceive the value of the green hotel, so as to stimulate their green hotel consumption intention [7]; In the internal management of the hotel, Its employees guide customers through their own green behavior and play a core role in the realization of the hotel environment practice. In addition, the customer's visual experience is also an important part [8], the decorative style plays a crucial role in explaining the purchase intention of Chinese customers. In recent years, More academics are becoming aware of how important it is for the government to encourage green consumer behavior. On the one hand, the government can promote green consumption by rationally using supporting tools such as preferential tax and normative tools such as green certification [9]. On the other hand, the government actively educates citizens and popularizes green environmental protection knowledge through public welfare publicity activities, which can also positively stimulate citizens' green awareness and behavior [10].

Based on the above literature research, it is not difficult to find that although a large number of scholars analyze and discuss the influencing factors of the green consumption behavior of hotel customers, they do not point out the main influencing factors and the relationship between the influencing factors. The resources of society and enterprises are limited. Identity most critical factor among many influencing factors that can help enterprises to accurately play the role of key factors to stimulate customers' green consumption, and to better fulfill the corporate social responsibility of green and low-carbon hotels. In view of this, this research will combine the decision laboratory analysis method and the explanation structure model method to analyze the influencing factors of hotel customers' green consumption behavior, reveal the hierarchical structure relationship among the influencing factors, and identify the key factors, so as to supply preliminary countermeasures for improving the formation of hotel customers' green consumption behavior.

2 Identification of Impacted Factors of Hotel Customers' Green Consumption Conduct

To identify the influencing factors of hotel customers' green consumption conduct, firstly, literature was used, Based on the research, the semi-structured interview outline was compiled, and the elements that affect hotel guests' green consumption habits were preliminarily identified through the interview research of Long-term stay in the hotel consumers. After that, Delphi method was adopted to invite five experts (two professors, two associate professors and one hotel practitioner) engaged in consumer behavior research and corporate green marketing to discuss the representativeness and accuracy of the description of influencing factors. Based on the purpose of simplifying indicators and ensuring that the meaning of indicators is not repeated as much as possible, 13 influencing factors of hotel customers' green consumption behavior are finally determined from three perspectives: consumer characteristics, hotel behavior, and policy guidance, as exposed in Table 1.

Table 1. Influencing factors of hotel customers' green consumption behavior

Serial number	The affecting Factors	Serial number	The affecting Factors
W_1	Age stage	W_2	Educated level
W_3	Green consumption cognition	W_4	Environmental responsibility
W_5	Group pressure	W_6	Green marketing
W_7	Green management	W_8	Green decoration
W_9	Green services	W_{10}	Green certification
W_{11}	Green fiscal policy	W_{12}	Dual carbon strategy
W_{13}	Public welfare publicity		

3 Model Construction and Analysis

3.1 DEMATEL-ISM Method Model Construction

According to the influencing factors of hotel customers' green consumption behavior constructed in Table 1, 10 experts engaged in consumer behavior and psychology research (5 from universities and 5 from the management of hotel enterprises) were requested to judge the influence of the factors. The values are assigned on a scale of 0 to 4. To get rid of the unique variations in expert evaluations, the 10 initial direct influence matrices are taken as the integers of the average to construct the direct influence matrix of impacted factors of hotel customers' green consumption behavior, as displayed in Table 2.

By calculation,the influence degree, influenced degree, centre degree, and cause degree of each forming factor can be obtained.

According to the centre and cause degree values of each forming factor calculated in Table 2, the cause-and-effect diagram of the factors affecting guests' green consumption behavior is constructed using Matlab software as shown in Fig. 1.

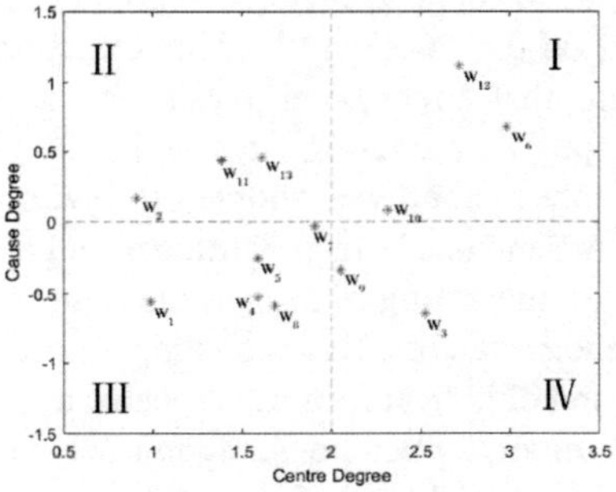

Fig. 1. Central-causality distribution of the influencing factors of hotel customers' green consumption behavior

It is necessary to establish a threshold in order to eliminate the relation with the least effect and produce the reachability matrix. Therefore, Select a threshold of 0.12 and plot the nodality change curve, as shown in Fig. 2. From the analysis of the node degree change diagram, it is found that when λ is 0.12, the corresponding node degree is moderate, and the coincidence rate of the causative factors with large node degree and the key causative factors calculated from the central degree is high the adjacency influence matrix A is obtained, and raising the adjacency influence matrix to the identity matrix I results in the acquisition of the holistic influence matrix B. I'm going to multiply the matrix B by the maximin operator until it stays the same. The results of the overall influence matrix B can obtain the reachable matrix.

The reachability matrix represents the matrix of factors influencing other factors through a certain length, indicating whether there is an influencing relationship between system factors. The skeleton matrix can be obtained by shrinking the reduced point and making the minimalist loop.

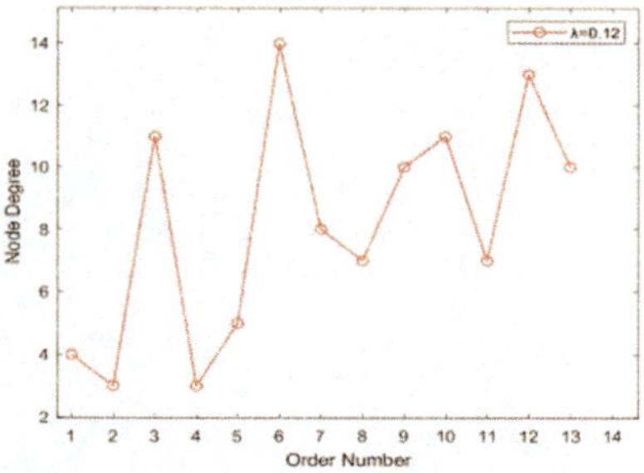

Fig. 2. Node degree change at fixed threshold

Table 2. Direct influence matrix

factors	W_1	W_2	W_3	W_4	W_5	W_6	W_7	W_8	W_9	W_{10}	W_{11}	W_{12}	W_{13}
W_1	0	0	2	1	1	0	2	0	1	0	0	0	0
W_2	0	0	3	3	2	0	1	0	1	0	0	0	0
W_3	0	0	0	2	3	2	0	1	2	1	0	2	2
W_4	0	0	3	0	3	2	0	0	0	2	0	2	0
W_5	1	0	3	1	0	2	0	0	0	0	0	1	1
W_6	2	3	3	2	2	0	4	3	3	3	2	3	2
W_7	1	2	1	1	0	2	0	3	4	2	1	1	1
W_8	2	0	1	1	0	1	0	0	0	3	2	0	0
W_9	3	0	2	2	2	1	1	0	0	3	0	1	0
W_{10}	1	2	2	0	1	3	4	4	4	0	0	0	0
W_{11}	0	0	1	0	0	2	3	2	1	4	0	1	0
W_{12}	2	0	4	3	3	4	3	3	3	2	3	0	3
W_{13}	3	1	3	2	2	0	1	2	2	1	0	2	0

According to the matrix R', the reachable set $R(W_i)$ of the impacting factors of hotel customers' green consumption conduct can be obtained and the antecedent set $Q(W_i)$, it is verified that when i = 1, 3, 5, 8, 9, meet $Q_i(W_i) \cap R_i(W_i) = R(X_i)$, W_1, W_3, W_5, W_8 and W_9 is the first influencing factor; Cross out the corresponding rows and columns of these factors in the matrix and repeat the above steps to obtain the second layer of forming factor W_2, W_4, W_7, and W_{13}; Similarly, the third layer formation factor can be obtained W_6, W_{10}, W_{11}, and W_{12}. Therefore, the influencing aspects of hotel guests' green consumption patterns may be categorized into three tiers, as publicized in Fig. 3.

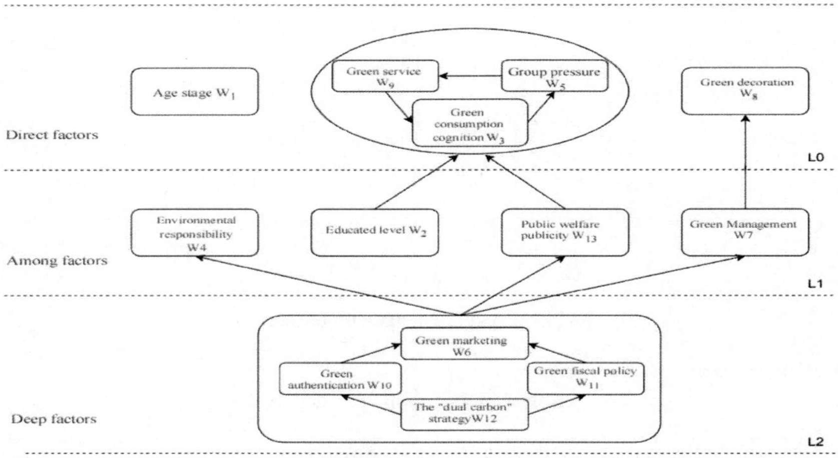

Fig. 3. Hierarchical explanatory model of influencing factors of hotel customers' green consumption behavior

3.2 Analysis of Results

Results of the DEMATEL calculations of the elements that influence hotel guests' green consumption behavior (Fig. 1) show that four categories are used to group the aspects that affect hotel guests' green consumption habits: Category 1 is a strong cause factor subset (zone I), these elements significantly affect how clients of hotels develop their green consumption habits, And it has a large effect on other outcome-type factors, Its cause degree distribution is between 0–1, Centrality degree is distributed at 2.4–3.0; Category 2 is a subset of weak cause factors (zone II), and Its centrality is smaller than the type 1 factor, Cause degree is basically flat; category 3 is a weak outcome factor subset (zone III), These factors are the outcome of the interaction of several additional potential causes, It contains the most elements, Reason degree is negative, Low centrality; category 4 is a strong outcome factor set (zone IV), Such factors are also the result of the combination of other causal-type factors, Its low cause value, But the central values are between 2.1–2.5, It has a very important influence on the formation of customers' green consumption behavior. Were calculated using the DEMATEL factor, the greater the center the greater the degree of influence.

4 Conclusion

4.1 Research Conclusion

1) Analysis of the deep factors

Dual carbon strategy (W_{12}), green certification (W_{10}), green fiscal (W_{11}), and enterprise green marketing (W_6) is the deep factor affecting the shape of customers' green expenditure conduct. There is a unique phenomenon in China: the consumer market and consumers have a strong desire to respond to national policies and guidance. Green certification and green fiscal (tax incentives, etc.) as the government departments to approve the development of industry standards and incentives, will inevitably be positively responded to the hotel enterprises, therefore, hotel green marketing is deeply affected by green certification and green finance; Green marketing is the hotel to environmental protection, green culture as the center of the marketing model, which by undertaking corporate social responsibility (CSR) to establish the hotel image, therefore consumers have a positive impression on the hotel. The above factors lead to the formation of other factors, which is the deep cause of the formation of customers' green consumption behavior.

2) Analysis of the among factors

Environmental responsibility (W_4), educated level (W_2), public welfare publicity (W_{13}) and green management (W_7) belong to the second layer of the hierarchical model, which is the intermediate factor in the formation of customers' green consumption behavior. Among them, environmental responsibility is mainly reflected in the individual's recognition of responsibility for maintaining the environment due to moral influence, which is mainly affected by the dual carbon strategy and government public service advertising; Public welfare publicity (public service advertising, etc.) is mainly manifested as the government's statement and promotion of green environmental protection concepts that contribute to the public interest in various ways, which has a great impact on consumers' green cognition and behavior. Green management is mainly reflected in the hotel integration of the concept of sustainable resources into the management, including green human resources management so that the hotel staff has behavior in line with green and sustainable development; These factors are partly affected by the deep factors, and at the same time affect other factors, so that hotel customers produce green consumption behavior.

3) Analysis of the direct factors

Age stage (W_1), green services (W_9), group pressure (W_5), green consumption cognition (W_3) and hotel green decoration (W_8) is the direct factor for the formation of customers' green consumption behavior. Green service is mainly reflected in the hotel through improving staff's awareness of green service and service quality to meet the needs of customers to enjoy a high-quality green service experience; Group pressure is reflected in that individual green consumption behavior is influenced by the combination of social and personal standardization due to conformity psychology. For example, consumers will consume because of the guidance of experts and authorities. Green consumption cognition is embodied in the personal subjective consciousness of green consumption held views and attitudes, which has a profound impact on personal behavior; Hotel green decoration is reflected in the hotel through simple, environmentally friendly materials on the hotel environment theme decoration, so as to create a green and low-carbon environment atmosphere. These factors will have a positive or negative impact on the green consumption behavior of hotel customers.

4.2 Management Inspiration

The factors affecting the green consumption behavior of hotel customers are more complex, and there is obvious hierarchy among the various factors. The government must strengthen the implementation of dual carbon strategy, green certification, and green fiscal, only by making the transformation of green environmental protection into a new direction of the hotel, and making the concept of green environmental protection deeply rooted in the customer level, can the generation of customer green consumption behavior be promoted. The hotel should take effective measures to improve the green marketing model, and innovate the channels for customers to join the hotel green model so that customers can actively participate in green consumption. In daily life, the hotel should pay close attention to the green consumption trend of customers, and timely make changes to the business publicity strategy to meet the consumption needs of customers.

4.3 Limitations and Prospects

Due to the influencing factors of hotel customers producing green expenditure conduct being mixed and massive, the degree of factors interaction and factors identification are evaluated by experts, in setting threshold lambda is obtained through the expert, and has certain subjectivity. In the follow-up research, the index factors will be further improved, combined with the green practices of hotels and consumers, and the green consumption behaviors of hotel customers will be further studied.

References

1. National Development and Reform Commission. To promote green consumption and implementation plan (2022). http://www.gov.cn/zhengce/zhengceku/2022/01/21/content_5669785.htm
2. Zhu, W.: A Study on the Influence of customers' green hotel cognition on consumption intention. Market. World **26**, 49–51 (2020)
3. Akhtar, R., Sultana, S., Mehedi Masud, M., Jafrin, N, Al-Mamun, A.: Consumers' environmental ethics, willingness, between lower and higher income groups. Resour. Conserv. Recycling **168**, 105274 (2021)
4. Du, J., Duan, S.: The Impact of environmental responsibility on consumers' green purchase behavior: the chain multiple mediating effects of green self-efficacy and green perceived value. J. Nanjing Univ. Technol. (Soc. Sci. Edn.) **21**(03), 48–60+115–116 (2022)
5. Yu, W.: Formation mechanism analysis of consumers' green consumption behavior: based on group pressure and environmental cognition. Consum. Econ. **25**(04), 75–77+96 (2009)
6. Yeh, S.-S., Guan, X., Chiang, T.-Y., Ho, J.-L., Huan, T.-C.: Reinterpreting the theory of planned behavior and its application to green hotel consumption intention. Int. J. Hosp. Manag. **94**, 102827 (2021)
7. Xu, Y.: Research on the Impact of Green Perceived Value on Green Hotel Consumption Intention. Wuhan University of Technology (2019). https://doi.org/10.27381/dcnki.Gwlgu.2019.001345
8. Bitner, M.J.: Servicescapes: the impact of physical surroundings on customers and employees. J. Mark. **56**, 57 (1992)
9. Wu, H., Hao, Y., Ren, S.: How do environmental regulation and environmental decentralization affect green total factor energy efficiency: evidence from China. Energy. Econ. **91**, 104880 (2020)
10. Yeşiltaş, M., Gürlek, M., Kenar, G.: Organizational green culture and green employee behavior: differences between green and non-green hotels. J. Clean. Product. **343**, 131051 (2022)

IAST-A Model for Teaching Chinese as a Foreign Language Based on Fuzzy Neural Network Algorithm and Cloud Computing

Yan Yang(✉)

School of Humanities, The College of Arts and Sciences, Kunming, Yunnan, China
YangYankm@21cn.com

Abstract. Considering the proportion of stylistic skills in language communication, this article combines the characteristics of Chinese and needs to establish a correct positioning in teaching Chinese as a foreign language. The IAST-A (Interaction Audio-Video Share Text Activities) model constructed in this article uses social networks as a medium, and its core element is "activity". This project requires mastering Chinese listening, speaking, reading and writing skills, but also emphasizes the ability to communicate in Chinese. The Chinese as a Foreign Language Education System based on cloud computing in this article refers to the combination of English education and English education based on cloud technology, making English education more intelligent, conducting listening teaching through cloud platforms, and conducting multi-channel interactive learning. This paper uses fuzzy neural network method to evaluate the quality of teaching, including the evaluation of teachers, students, and teaching managers. In this article, the proportion of students in the first 80–90 score range using the IAST-A model is 8.4%, while the proportion of students in the 80–90 score range after using the model is 43.5%. This reflects the effectiveness of the IAST-A model for teaching Chinese as a foreign language in this article.

Keywords: Teaching Chinese as a Foreign Language · IAST-A Model · Fuzzy Neural Networks · Cloud Computing

1 Introduction

Nowadays, in foreign language education, people's overall understanding of language ability was vague. On the one hand, there were great differences in their cognition, and on the other hand, they did not give a reasonable answer to the problems that occurred. With the development of modern information technology, oral Chinese was widely used in the modern society. In the modern society, oral Chinese was an indispensable teaching method. The Chinese class of cloud calculation was a brand new Chinese class. It created a more realistic situation for the students in the form of sound, pictures and words, which was of great significance to improve the quality of Chinese class.

During the development of foreign language education in the past few years, the key point of it was to cultivate the language communication ability, not to teach knowledge

J. C. Hung et al. (Eds.): FC 2023, LNEE 1131, pp. 25–32, 2024.
https://doi.org/10.1007/978-981-99-9299-7_4

and theory. It opened four skill courses, namely listening, speaking, reading and writing, and adopted a more direct teaching method, as well as the concept of "Teaching more and practicing more" in the class. All these had changed from the concept of "Professor" to the concept of "ability". Ke E's study examined the promotion of multiple word reading for adult foreign language students in the environment of English form consciousness [1]. Yu A thought that the high drop out rate of Chinese students had always been a concern for foreign language teachers. He had discussed the theoretical foundation and meaning of Chinese characters teaching from the very beginning [2]. In order to promote remote collaboration and communication in the field of foreign language teaching, Luo H focused on five important subjects: mode, task, challenge, technology and new trend of remote collaboration [3]. At present, the teaching of English for senior students was focused on the teaching of English, which inevitably involved the problem of language strategy. It was obvious that their research did not involve this point.

This article made a detailed and comprehensive investigation and analysis of the application methods of all kinds of information technology in the current teaching of Chinese as a foreign language, which provided a solid support for the combination of information technology and the international education of Chinese. At the same time, based on the study of Chinese teaching, the problems that foreign students had to face when they were learning Chinese by voice teaching and network technology were analyzed and summarized, and tried to give some solutions to "seek advantages and avoid disadvantages".

2 Inquiry Method of IAST-A Model in Teaching Chinese as a Foreign Language

2.1 Teaching Chinese as Foreign Language

With the rapid development of China's economy, her status in the world was becoming more and more prominent. In this case, there was a "Chinese fever" in the world, which had already formed a common social phenomenon [4, 5]. Experts analyzed and said that there was a "Chinese heat" in the world, mainly because there was a huge demand for Chinese talents all over the world [6]. In addition, many advanced countries began to pay attention to studying Chinese, such as America and England. Besides, many students from other countries came to China to study every year. This further increased the worldwide enthusiasm for Chinese [7, 8]. But because of the difference between China and other countries, foreign students encountered many difficulties in learning Chinese, mainly listening, speaking, reading and writing [9].

With the continuous development of the Chinese class technology, the teacher's demand for teaching resources is also increasing. Although there are many websites of the University and corresponding data has been established for the Chinese class resources, there is no corresponding data for a specific type of class and its corresponding teachers [10, 11]. At the same time, the current teaching resources have a large number, a variety of, scattered memory and other characteristics, which has brought a lot of difficulties to teachers' search and use [12]. In some areas, there was an urgent problem to be solved, that was, how to use modern information technology like network and media, and how to provide a high-quality multimedia teaching resource library for the current spoken Chinese education, and how to establish a continuous and updated multimedia teaching resource system.

2.2 Comprehensive Evaluation of Foreign Language Based on the Fuzzy Nerve Network

In this paper, a method based on fuzzy neural network is proposed for the quality of foreign Chinese teaching. The construction of the evaluation model of the blurry nerve network was a math tool to make a comprehensive decision with the help of the blurry line transformation and the blurry reflection, which could effectively solve the problems in the current teaching quality evaluation system of the school. Classroom teaching relied on the exchange of information between teachers and students. With its own dynamic, multiple factors, slow effect and blurry characteristics, the traditional teaching quality assessment was facing some difficulties.

To confirm the set Y_S of factors for teaching Chinese as a foreign language:

$$Y_S = (U_1, U_2, \cdots, U_n) \tag{1}$$

P_J for the assessment of foreign language teaching:

$$P_J = (V_1, V_2, \cdots, V_m) \tag{2}$$

To make sure if she succeeded in learning Chinese, the root mean deviation R_C was:

$$R_C = \sqrt{\sum (T_p - O_p)/n_p} \tag{3}$$

3 Construction of IAST-A Model for Teaching Chinese as Foreign Language

3.1 Design of the Foreign Language Teaching System

With the rapid development of internet technology, online education has increasingly become a mainstream learning mode. To better adapt to this internet trend, cloud computing technology has become an important technical support. Based on cloud computing technology, we have developed an efficient and convenient Chinese teaching system for foreign students. The system is primarily designed to improve teaching efficiency and students' learning outcomes, while also providing students with more learning options.

The cloud-based system facilitates online Chinese language teaching for students via various devices. It comprises several modules: the Student Information Module records basic student data and tracks progress; the Learning Resources Module offers diverse materials for language learning; the Teaching Information Module provides teaching schedules and content; the Live Teaching Module delivers online courseware and videos, including advanced scenario simulations; the Testing Module assesses students' language proficiency; the Learning Report Module generates personalized reports to guide learning strategies; the Extended Learning Module enriches the learning experience with cultural content; and the Communication and Discussion Module encourages interaction among students and teachers. This system enhances the efficiency and quality of Chinese language learning by offering rich, personalized, and diverse resources and methods.

3.2 Construction of IAST-A Model

The IAST-A model is a Chinese language teaching model based on social networks. The model mainly promotes students' Chinese language abilities through activities. In this model, social networks are a key media element, and students can participate in learning activities through social network platforms to experience and learn Chinese language knowledge and skills. Of course, the design and implementation of the IAST-A model need to consider students' personalized requirements, so it has strong flexibility and customizability. In daily life and usage, the application scope of this model is also very wide, not only suitable for traditional teaching modes but also can be widely used in online education and blended education.

4 Construction of IAST-A Model

The teacher used the multimedia teaching materials to focus on teaching, and the teaching was based on the book. Only by an objective understanding of the book could they better discover and create new books. Therefore, before the design of the spoken Chinese teaching, the assessment of the spoken Chinese books used by the current teachers was carried out a survey, and the results were as shown in Table 1.

The purpose of spoken English textbooks was to cultivate students' ability to use Chinese correctly and take the initiative to communicate. It should have a strong communication nature, and its content should meet the needs of students in real life. However, according to the investigation, 65% of teachers thought that the current spoken English textbooks lacked real language materials. From this point, it could be seen that the English teacher had different evaluation of the current spoken English textbooks. However, most of the teachers thought that it was necessary to use the teaching resources of cloud advertising. If they only relied on the knowledge in the book, they couldn't meet the needs of students. Therefore, when the teachers were preparing the course, they had to add some expanding materials related to spoken Chinese and many kinds of media materials related to the course content, only in this way could they adapt to the needs of students of different levels. Therefore, he set up a resource library of cloud advertising that matched the teaching material, so that it could better provide the teaching materials for the teachers.

The teacher's approval of the use of the foreign language teaching system was shown in Fig. 1 (creating a scene to arouse the students' interest, expand their learning scope and broaden their horizons). But it can help students correct their pronunciation and tone as shown in Fig. 1 (b). In Chinese class, the teachers agreed with the positive effect of the multimedia teaching resources. Among them, the choices of creating a scene to arouse students' interest occupied 74%; the choices of expanding students' learning

Table 1. Evaluation of the Chinese textbooks used by the teachers

Identification situation	Lack of authentic language materials (%)	Have an impact on students' actual oral communication ability (%)	Cloud Computing is necessary for teaching (%)
Agree	40	15	50
Comparative agreement	25	20	20
Disagree	15	25	15
Unclear	20	40	15

range and broaden their horizons occupied 62%; the choices of helping students correct their pronunciation and tone occupied 51%. It could also be seen that if the foreign language teaching system was applied to the spoken Chinese, it could indeed attract the students from many aspects and angles, so that the spoken Chinese class would be closer to the real life. Through the personal experience, the students could cultivate their understanding of the language communication ability.

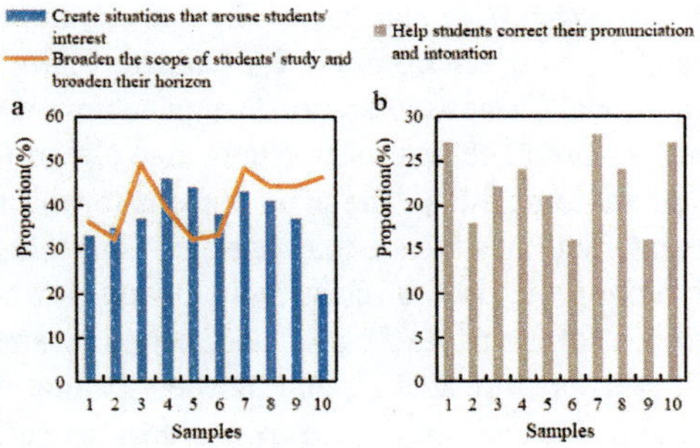

Fig. 1. The teacher's approval of the use of the foreign language teaching system

On this basis, it was necessary to have a preliminary discussion about the interaction teaching mode and its use in the online spoken Chinese class. Based on this situation, it was obvious superiority to combine the characteristics of the spoken English course with the local characteristics. However, because of the lack of interaction in class, there were still some problems in the application of this mode. Therefore, this paper planned to integrate this mode into the Chinese teaching, and design a foreign language teaching course based on interaction, in order to learn Chinese on the Internet.

The comparison of students' scores before and after the IAST-A model was shown in Fig. 2. The former was shown in Fig. 2(a), and the latter was shown in Fig. 2(b). The ratio of students in the top 80–90 scores of the IAST-A model was 8.4%, and the ratio of students in the top 80–90 scores of the IAST-A model was 43.5%.

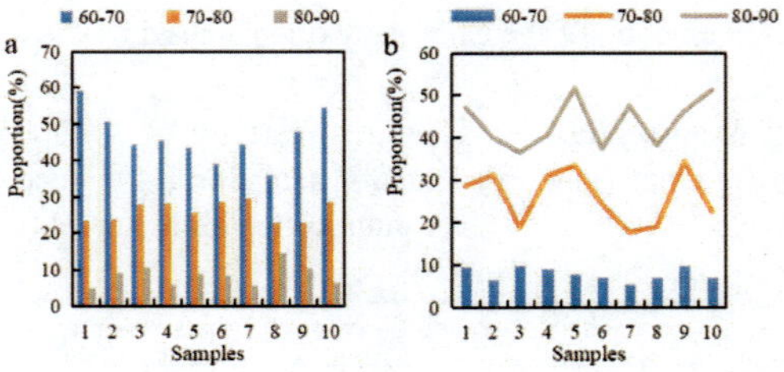

Fig. 2. Comparison of students' scores before and after using the IAST-A model for teaching Chinese as a foreign language

IAST-A teaching model was a two-way interaction process, so teachers should propose some inspiring questions and give guidance to students, so that they could actively participate in the class and increase their frequency of speaking. In the teaching of spoken Chinese, the teacher was the leader and the student was the main body. If students could stimulate their enthusiasm to participate in the class and improve their motivation of learning.

With the rapid development of China and the enhancement of national strength, the development of internationalization of Chinese language has increasingly influenced the world, and has also stimulated many people's enthusiasm for learning Chinese as a foreign language. In this era of rapid technological development, and there are many online teaching websites and various learning platforms. The rise of online education platforms not only provides a good learning environment for Chinese learners, but also achieves teaching effects that traditional Chinese education cannot achieve through various forms.

The level of teacher performance in a foreign language classroom is directly related to students' language interaction ability. The main reasons for this are: First, in the process of teacher-student interaction, students may ask questions that are not related to the classroom content, disrupting the pace of class and slowing down teaching, making it difficult for teachers to control the pace of class; The second reason is that teachers and students are located in two separate spaces, and teachers cannot monitor and manage the classroom in real time. Instead, they need to use devices such as mobile phones and microphones to monitor the classroom. Third, attention to students has become distracted. The above problems can cause difficulties for teachers to control the classroom, thereby affecting the interaction between students. Although in the teaching process, various types of interaction can be used to attract students' attention. However, if students' focus is not within the scope of the course content, it is difficult to achieve true interactive effects.

After using the IAST-A model, students' enthusiasm for learning Chinese is shown in Fig. 3 (relatively positive and general, as shown in Fig. 3(a), with no effect, as shown in Fig. 3(b)). The proportion considered to be relatively positive was 62.4%. The rate of no effect was only 13.9%.

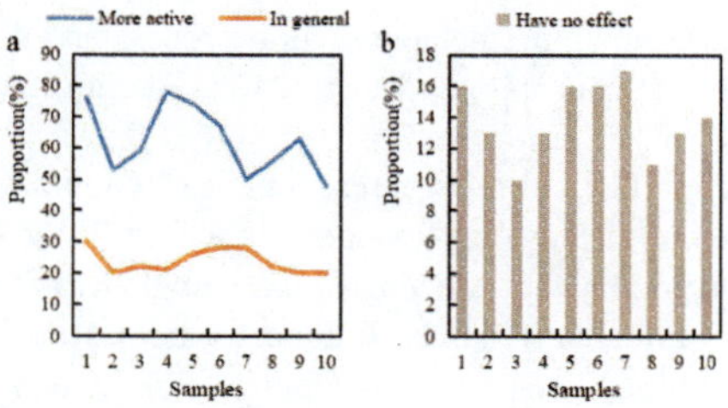

Fig. 3. Students' enthusiasm for learning Chinese after adopting the IAST-A model

5 Conclusions

This article briefly reviews the emerging "cloud" technology in recent years, compares the application and popularity of fuzzy neural networks and cloud computing in other fields, explores new ideas for the cloud computing and Chinese language education, and gives reasonable design assumptions based on this. This article also proposes countermeasures to "seek advantages and avoid disadvantages" in order to better use cloud computing technology for teaching Chinese as a foreign language. At the same time, this paper briefly reviews the two new technologies, "cloud computing" and "fuzzy neural network", which have emerged in recent years, as well as their application and popularization in other fields, explores the integration approaches of new technologies and Chinese language teaching, and proposes corresponding design schemes. This article analyzes and summarizes some theories and examples used in a large number of literature searches in order to make a more comprehensive and correct discussion of this issue.

References

1. Ke, E., Koda, K.: Transfer facilitation effects of morphological awareness on multicharacter word reading in Chinese as a foreign language. Appl. Psycholinguist. **42**(5), 1263–1286 (2021)
2. Yu, A., Geng, G.: Introducing Chinese Hanzi at the beginning of teaching Chinese as a Foreign Language (CFL). Global Chinese **6**(2), 359–380 (2020)
3. Luo, H., Yang, C.: Twenty years of telecollaborative practice: Implications for teaching Chinese as a foreign language. Comput. Assist. Lang. Learn. **31**(5–6), 546–571 (2018)
4. Han, Y., Gao, X., Xia, J.: Problematising recent developments in non-English foreign language education in Chinese universities. J. Multiling. Multicult. Dev. **40**(7), 562–575 (2019)
5. Gong, Y., Lyu, B., Gao, X.: Research on teaching Chinese as a second or foreign language in and outside mainland China: a bibliometric analysis. Asia Pac. Educ. Res. **27**(4), 277–289 (2018)
6. Dolce, F.M., van Compernolle, R.A.: Topic management and student initiation in an advanced Chinese-as-a-foreign-language classroom. Classroom Discourse **11**(1), 80–98 (2020)
7. Wang, D.: Translanguaging in Chinese foreign language classrooms: Students and teachers' attitudes and practices. Int. J. Biling. Educ. Biling. **22**(2), 138–149 (2019)
8. Gong, Y.F., Gao, X.A., Lyu, B.: Teaching Chinese as a second or foreign language to non-Chinese learners in mainland China (2014–2018). Lang. Teach. **53**(1), 44–62 (2020)

9. Liu, S., Hsieh, C.Y.C.: Developing metaphorical awareness and competence in Chinese as a foreign language through concept-based instruction. Foreign Lang. Ann. **53**(3), 478–504 (2020)
10. Hua, X., Hu, G., Liu, F.: The cultural representation of Chinese-speaking groups in US-produced Chinese as a foreign language textbooks. Asia Pac. J. Educ. **43**(1), 126–143 (2023)
11. Huang, F., Teo, T., Zhou, M.: Factors affecting Chinese English as a foreign language teachers' technology acceptance: a qualitative study. J. Educ. Comput. Res. **57**(1), 83–105 (2019)
12. Li, C.: A positive psychology perspective on Chinese EFL students' trait emotional intelligence, foreign language enjoyment and EFL learning achievement. J. Multiling. Multicult. Dev. **41**(3), 246–263 (2020)

Development and Application of Russian MOOC System Based on Neural Network Algorithm

Bingqing Li(✉) and Peng Sun

Sanya Aviation and Tourism College, Sanya, China
libingqing83@163.com

Abstract. Many countries list Russian as their second or third foreign language, and the demand for learning Russian is constantly increasing. People have higher requirements for intelligent information processing, automated management, and digital communication, and they are more enthusiastic about the development of the Russian Massive Open Online Course (MOOC) system. This article used neural network algorithms to study the MOOC system, which was helpful for the teaching of Russian majors. This article analyzed the performance of the Russian MOOC system through experimental design and algorithm comparison. Experimental data showed that the convolutional neural network algorithm had the highest accuracy among the four algorithms, reaching 95%. Therefore, convolutional neural network algorithms can be used to develop Russian MOOC systems.

Keywords: Neural Network Algorithm · Russian Learning · MOOC System · Development and Application

1 Introduction

Traditional French teaching methods often have problems such as low efficiency and lack of teaching resources, which cannot meet the needs of students. Therefore, the MOOC system based on neural network algorithms can provide learners with a more flexible and personalized learning experience, improving learning efficiency and quality. This article mainly introduces the development of a Russian MOOC system based on neural networks. The advantage of neural network algorithms lies in their ability to learn and train from a large amount of data, enabling the processing and decision-making of complex problems.

In Russian learning, neural network algorithm can help students better master Russian language knowledge by analyzing the language features and rules in a large number of Russian corpora. For this reason, many scholars have studied neural network algorithms and methods related to Russian learning. For example, some scholars have proposed a digital language generation strategy based on neural networks [1, 2]. Some scholars have also used questionnaire surveys to train and evaluate the oral skills of Russian major students [3, 4]. In addition, some scholars have established the relationship between text language and the implemented functions when studying language [5, 6]. Therefore, this article can utilize neural networks, combine traditional methods

J. C. Hung et al. (Eds.): FC 2023, LNEE 1131, pp. 33–41, 2024.
https://doi.org/10.1007/978-981-99-9299-7_5

with modern technology, and explore a predictive simulation system based on learning models and adaptive training algorithms.

The Russian MOOC system based on neural network algorithms adopts an intelligent recommendation and evaluation mechanism based on deep learning, which can provide accurate diagnosis and feedback based on students' performance and individual differences, and provide targeted suggestions and learning strategies for students. The use of natural speech processing technology and large format corpus training algorithms has effectively improved the recognition and understanding capabilities of Russian MOOC systems. These innovations have greatly improved the teaching quality and learning efficiency of the Russian language learning system, providing learners with more practical, effective, and personalized learning methods.

2 Russian MOOC System Based on Neural Network Algorithm

2.1 Neural Network Algorithm

Neural network algorithms are computer algorithms that simulate the working principles of neural networks in the human brain. It allows for data processing and analysis by transmitting information between the network and multiple nodes (also known as neurons) [7]. Neural network algorithms typically include three levels: input level, hidden level, and output level. The input layer is used to receive data input, the output layer is used to generate results, and the hidden layer performs complex calculations and processing based on the input data [8]. The main characteristics of neural network algorithms are self-learning and adaptability. It can continuously optimize the model, improve processing power and data accuracy by adjusting parameters such as weights and thresholds [9].

Neural network algorithms are widely used in fields such as machine learning, data mining, image recognition, natural speech processing, and have achieved good results in speech recognition, image recognition, and autonomous driving [10]. At present, neural network algorithms mainly include pre neural networks, recurrent neural networks, convolutional neural networks, etc. [11]. Each type of neural network algorithm has its own application scenarios, advantages and disadvantages. When using neural network algorithms, it is necessary to select, optimize, and adjust appropriate algorithms based on different needs and actual situations to obtain the best results [12].

2.2 Russian MOOC System

The Russian MOOC school system is a teaching platform based on the Internet and modern educational technology, providing comprehensive and personalized learning services and support for Russian learners. This system combines many functions, such as online video courses, learning resource libraries, intelligent learning aids, etc., to help students learn Russian more effectively and improve their Russian skills [13].

The main characteristics of the Russian MOOC system are: rich and colorful learning content, including Russian pronunciation, vocabulary, grammar, reading, writing, culture, history, geography, and other knowledge [14]. Personalized learning, this system can provide personalized learning services and support for students based on their learning situation and needs, such as recommending appropriate learning resources and providing intelligent learning assistance tools [15]. A powerful learning evaluation system allows for comprehensive evaluation and analysis of students' learning, including progress, learning outcomes, learning difficulties, and other aspects, to help students better understand their learning and progress. An excellent teaching team, the teachers of this system have rich teaching experience and professional teaching skills, and can provide high-quality teaching and support services for students. Advanced teaching technology, this system adopts advanced teaching technologies such as online video and intelligent teaching assistants, making it easier and faster for students to learn Russian [16]. In short, the Russian MOOC system is an advanced teaching platform based on the Internet and modern educational technology, which can provide personalized and effective learning services and support for Russian learners to improve their knowledge and efficiency in Russian learning.

The modules typically included in Russian MOOC systems are shown in Fig. 1.

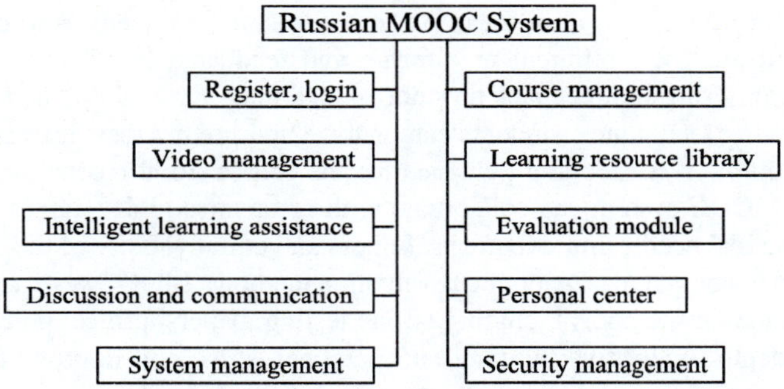

Fig. 1. Russian MOOC System

The system comprises multiple modules: the Registration/Login Module manages account information; the Course Module administers course content; the Video Module provides online video courses [17]; the Learning Resource Library Module contains all learning resources; the Intelligent Learning Assistance Module integrates various learning tools; the Assessment Module offers comprehensive learning evaluation and analysis; the Discussion and Communication Module promotes interaction between students and teachers [18]; the Personal Center Module includes student personal information and learning records; the System Management Module is responsible for system configuration and operations; the Security Management Module ensures system security and reliability. These modules ensure the system's operation and performance meet the expected goals, with specific modules varying depending on the system.

2.3 MOOC System Improvement Measures

The following optimization and improvement measures can be taken to solve some problems in the MOOC system.

Improving user experience: The course system should pay more attention to students' personal needs and learning experience. By using intelligent technology, suggestions can be made on students' learning paths, interests, and hobbies to improve their satisfaction. Optimizing content quality: The course content should comply with course standards and teaching requirements, while taking into account students' interests and learning needs. By inviting excellent teachers and industry experts as lecturers and developing evaluation mechanisms, the rigor and effectiveness of the content can be ensured. Strengthening interactive communication: Excessive self-learning can make students feel lonely and lost, so it is necessary to strengthen interactive communication between teachers and students. Online Q&A, classroom discussion, real-time interaction, and other functions can be set up to improve teaching efficiency and student engagement. Customized course design: Based on factors such as age, interest, and educational background, customized course design can better meet students' needs, improve learning efficiency and quality. Intelligent monitoring and feedback: MOOC systems should strengthen monitoring and feedback on students' learning. By applying data mining and machine learning techniques, students can analyze and predict their learning progress, providing accurate feedback and guidance for the course. Joint business development course: The MOOC system can collaborate with companies to develop online courses that meet practical needs, improve the usefulness and employability of the courses, and support talented companies. Diversified evaluation methods: MOOC systems must adopt different evaluation methods, including online testing, papers, project practice, etc., to promote students' innovative thinking and practical skills, and improve their overall quality. The MOOC system needs to be continuously optimized and improved to adapt to the constantly changing learning needs and social development requirements, and to provide high-quality online learning services for more students.

3 Development and Application of Russian MOOC System

3.1 System Development

Russian MOOC modeling can be divided into the following steps:

In order to determine the needs, it is first necessary to clarify the needs and objectives of the Russian language learning system, including curriculum, learning resources, teaching methods, and how to evaluate students' learning and progress. Data collection: According to system requirements, relevant data should be provided, such as course materials, exercise questions, student information, etc. Data purification and processing: The collected data is purified and processed, duplicate data is deleted, and missing data is filled in. Data analysis and modeling: Based on data cleaning, data analysis and modeling can be carried out using machine learning algorithms to predict and evaluate students' learning. Model optimization and customization: Based on the actual situation and feedback results, the model is optimized and customized to improve its accuracy and efficiency. In actual systems, the established Russian MOOC system model has been implemented and tested, and validated to ensure the stability and reliability of the system. In summary, Russian MOOC modeling is a multi-step process that requires comprehensive consideration of system requirements, data collection and processing, model creation and optimization to ensure that system functionality and performance meet expected goals.

4 Development Environment

The development environment of the Russian MOOC system is shown in Table 1.

Table 1. Development Environment of the Russian MOOC System

Exploitation environment	Examples
Operating system	Windows, Linux
Development tool	Eclipse, IDEA, Visual Studio
Data base	MySQL、Oracle、PostgreSQL
Front end technology	JavaScript
Backend frame	Spring, Spring Boot, Django and Flask

Operating system: Depending on the developer and system functionality, the operating system is selected. Tools are developed and various programming tools are used. The development languages include Java, Python, etc., and the specific selection depends on the project requirements and the skills of the developers. Relational or non relational databases are selected. Version control: Version control tools have been selected for code management, which facilitates collaboration and code backup among multiple people. In summary, the development environment of the Russian MOOC system needs to comprehensively consider operating system, development tools, development language, database, front-end technology, back-end technology, version control, etc., to ensure the efficiency and quality of development.

Neural networks are a very popular machine learning algorithm used to solve problems such as model classification, regression, and grouping. The following are commonly used performance indicators in neural network algorithms:

Accuracy is the ratio of the number of correctly classified samples to the total number of samples. The calculation formula should be expressed as follows:

$$A = \frac{Z}{T_N} \tag{1}$$

Z is the correct quantity, A is the accuracy. Recall rate: Recall rate refers to the proportion of successful predictions among all positive cases, which can be used to evaluate the number of positive cases detected by the model. The formula is:

$$\mathrm{R} = \frac{\mathrm{S}}{\mathrm{Z} + \mathrm{E}} \tag{2}$$

Accuracy refers to the ratio between the number of correctly predicted samples and the total number of samples in the prediction results.

$$P = \frac{Z_Y}{T_N} \tag{3}$$

The F-value is the weighted average of accuracy and recall, used to measure the overall effectiveness of the model. The formula is:

$$F = 2 * P * \frac{R}{P + R} \tag{4}$$

The loss function is used to measure the performance of the neural network during training. Methods such as cross entropy or average absolute error are usually used to calculate the loss function. The lower the value of the loss function, the more accurate the model is. Training time is an indicator of neural network efficiency. If the training time is too long, it means more resources and costs. This article selects the scores of Russian major students as a sample to evaluate the ability of the MOOC system.

5 Comparison of MOOC Systems Based on Neural Networks

MOOC systems based on neural networks can be compared in various ways, including:

Algorithm effects: The effects of different neural network algorithms in MOOC systems vary greatly. The best algorithm can be found by comparing multiple algorithms horizontally and comparing their performance in predicting learning outcomes and ranking students. Datasets: Different MOOC systems collect different datasets. The size, quality, and characteristics of different datasets can affect the performance of neural network algorithms. Therefore, longitudinal comparisons are needed to compare the differences in algorithm efficiency and applicability between different datasets. Experimental parameters: When applying neural network algorithms to MOOC system analysis, it may be necessary to adjust algorithm parameters, different training strategies, etc. The differences in experimental parameters affect the reliability of the results, so it is necessary to conduct experimental comparisons to verify the rationality and effectiveness of the strategy. Application scenario: The characteristics and requirements of different systems also vary. Some systems place greater emphasis on providing personalized services for learners, while others place greater emphasis on improving course quality. Therefore, it is necessary to compare the neural network analysis methods in different MOOC system application scenarios to find the most suitable method for this scenario.

6 Russian MOOC Experimental Results

The MLP model is simple in structure and easy to implement and train, but it needs to manually select parameters such as the number of hidden layers, nodes and regularization. In the Russian MOOC system, text data can be converted into image format and then processed by CNN. CNN requires more computing resources and training time than

MLP, but can achieve better prediction results. In the Russian MOOC system, LSTM can be used to recommend course content and predict learning outcomes. LSTM can automatically learn long-term correlations in sequence data and can handle long sequence data better than traditional recurrent neural networks. In the Russian MOOC system, students' learning history can be modeled using DBN and their current learning status and future performance can be predicted. DBN has advantages in unmanned pre training, but its model structure is complex and requires a lot of resources and training time. The specific time and accuracy are shown in Fig. 2.

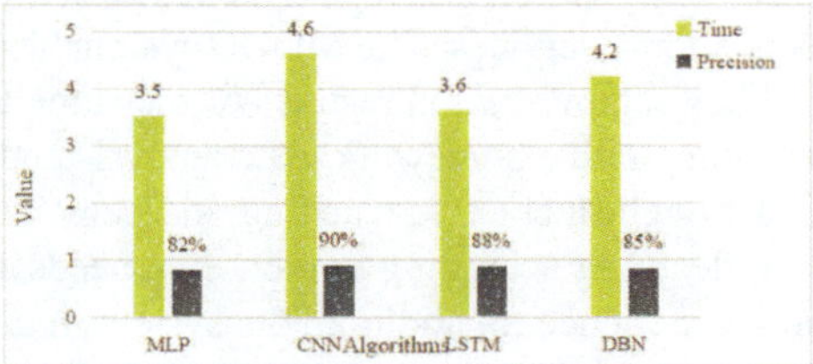

Fig. 2. Time-consuming and Precision of the Different Algorithms

As shown in Fig. 2, this article can find that the CNN algorithm takes the longest time among the four algorithms, while the MLP algorithm takes the shortest time. The CNN algorithm has the highest accuracy, while the MLP algorithm has the lowest accuracy. The LSTM algorithm takes 3.6 s and has an accuracy of 88%.

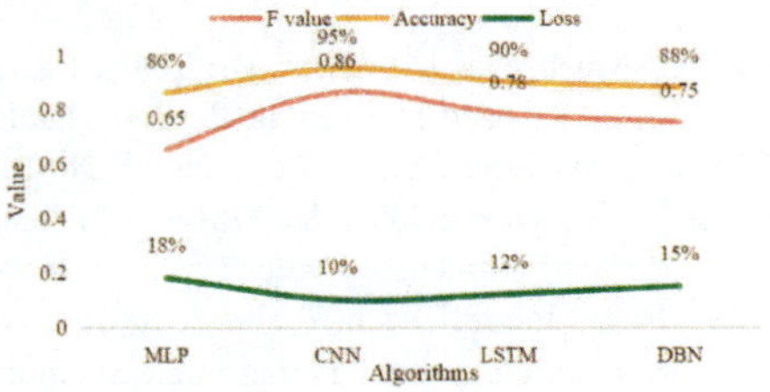

Fig. 3. Accuracy, Loss Function Value and F Value of the Different Algorithms

As shown in Fig. 3, it can be observed that MLP has the lowest accuracy, at 86%. The CNN algorithm has the highest F-value, while the MLP algorithm has the lowest F-value. The accuracy of LSTM algorithm is 90%, the loss function value is 12%, and the F value is 0.78. The accuracy of the DBN algorithm is 88%, the loss function value is 15%, and the F value is 0.75.

7 Conclusions

This article described the development and application of a Russian MOOC system based on neural network algorithms. The system uses deep learning and natural language processing technology, through in-depth corpus training and screening, can improve the efficiency and quality of Russian learning. Through this system, students can learn online, including listening, speaking, reading and writing. The system can provide intelligent advice and scores for students' performance, helping them better master Russian skills. The system also has data analysis and user feedback functions, which can be customized and optimized based on students' performance and feedback to provide more personalized and high-quality educational services. The MOOC system has been widely applied in practical teaching and has achieved good results. At the same time, the system also supports multimedia teaching, online homework, data analysis and other functions, and has many applications. Through interactive teaching and personalized learning paths, students can participate in learning according to their own needs and interests, thereby obtaining a good learning experience. In addition, the system in this article has a social evaluation and feedback mechanism, encouraging students to share and cooperate in learning.

References

1. Jalilian, A., Mateu, J.: Assessing similarities between spatial point patterns with a Siamese neural network discriminant model. Adv. Data Anal. Classif. **17**(1), 21–42 (2023)
2. Raphaël, M.J.I.: Larsen, marc-oliver pahl, gouenou coatrieux: multipath neural networks for anomaly detection in cyber-physical systems. Ann. des Télécommun. **78**(3–4), 149–167 (2023)
3. Drugova, E., Zhuravleva, I., Aiusheeva, M., Grits, D.: Toward a model of learning innovation integration: TPACK-SAMR based analysis of the introduction of a digital learning environment in three Russian universities. Educ. Inf. Technol. **26**(4), 4925–4942 (2021)
4. Zemlyanova, M., Muravyeva, N., Masterskikh, S., Shilova, L., Shevtsova, A.: Advancing English language learners' speaking skills using VoiceThread in mobile learning for Russian tertiary context. Int. J. Web Based Learn. Teach. Technol. **16**(6), 1–11 (2021)
5. Elangovan, R., et al.: Static and dynamic isolated Indian and Russian sign language recognition with spatial and temporal feature detection using hybrid neural network. ACM Trans. Asian Low Resour. Lang. Inf. Process. **22**(1), 26:1-26:23 (2023)
6. Novikova, O., Kalugina, Y., Mustafina, A., Edeleva, I.: Linguistic analysis of insta, twit posts and lj blogs in the context of their functions (based on the Russian Language). Int. J. Interact. Mob. Technol. **15**(5), 66–86 (2021)
7. Gomathi, P., Muniraj, C., Periasamy, P.S.: Micro calcification detection in mammogram images using contiguous convolutional neural network algorithm. Comput. Syst. Sci. Eng. **45**(2), 1887–1899 (2023)
8. Sudha, S., Srinivasan, A., Devi, T.G.: Cross-validation convolution neural network-based algorithm for automated detection of diabetic retinopathy. Comput. Syst. Sci. Eng. **45**(2), 1985–2000 (2023)
9. Falahzadeh, M.R., Farokhi, F., Harimi, A., Sabbaghi-Nadooshan, R.: Deep convolutional neural network and gray wolf optimization algorithm for speech emotion recognition. Circ. Syst. Signal Process. **42**(1), 449–492 (2023)

10. Mohanty, S., Dash, R.: A flower pollination algorithm based Chebyshev polynomial neural network for net asset value prediction. Evol. Intell. **16**(1), 115–131 (2023)
11. Alajas, O.J.Y., et al.: Grape pseudocercospora leaf specked area estimation using hybrid genetic algorithm and recurrent neural network. J. Adv. Comput. Intell. Intell. Informatics **27**(1), 35–43 (2023)
12. Bawankule, K.L., Dewang, R.K., Singh, A.K.: Early straggler tasks detection by recurrent neural network in a heterogeneous environment. Appl. Intell. **53**(7), 7369–7389 (2023)
13. Lebedeva, M.: Instructional design of skill-balanced LMOOC: a case of the Russian language MOOC for beginners. J. Univ. Comput. Sci. **27**(5), 485–497 (2021)
14. Kelih, E.: Quantitative approaches to the Russian language. J. Quant. Linguist. **27**(1), 80–83 (2020)
15. Akhmetov, I., Krassovitskiy, A., Ualiyeva, I., Gelbukh, A.F., Mussabayev, R.: An open-source lemmatizer for Russian language based on tree regression models. Res. Comput. Sci. **149**(3), 147–153 (2020)
16. Lagutina, N.S., Lagutina, K.V., Adrianov, A.S., Paramonov, I.V.: Russian-language thesauri: automatic construction and application for natural language processing tasks. Autom. Control. Comput. Sci. **53**(7), 705–718 (2019)
17. Cobos, R., Ruiz-Garcia, J.-C.: Improving learner engagement in MOOCs using a learning intervention system: a research study in engineering education. Comput. Appl. Eng. Educ. **29**(4), 733–749 (2021)
18. Sezgin, S., Sevim-Cirak, N.: The role of MOOCs in engineering education: an exploratory systematic review of peer-reviewed literature. Comput. Appl. Eng. Educ. **29**(4), 950–968 (2021)

A Vector Autoregressive Model-Based Numerical Measurement Approach for Coordinating Relation Between Tourism Development and Financial Support

Changyan Cai[1,2], Beatrice Lim Fui Yee[1(✉)], and Kasim Mansur[1]

[1] Faculty of Business, Economics and Accountancy, Universiti Malaysia Sabah, 88400 Kota Kinabalu, Sabah, Malaysia
beatrice@ums.edu.my
[2] Xichang University, Xichang 615000, China

Abstract. Tourism industry, as one of the important driving forces for economic growth, can greatly promote regional economic development and accelerate the positive externalities of resource flow among regions. Finance is a vital component of modern economies, playing a critical role in allocating resources and driving economic development. Firstly, the study will conduct data analysis on the current situation of tourism development and financial support in liangshan prefecture. Secondly, an explanation of the vector autoregressive model will be provided; then, the vector autoregression model will be used to evaluate the coordinating relation between tourism development and financial support.

Keywords: Vector Autoregression Model · Financial Support · Tourism Development · China

1 Introduction

Tourism is an important force for the development of the national economy.As an important tool for resource allocation and macroeconomic regulation, finance is becoming increasingly important in the diversified tourism. With the rapid development of the tourism industry, the government has gradually increased its financial support for the tourism industry, issuing financial support policies. These policies actively guide all parties to provide financial assistance to support the tourism sector, providing financial guarantees for the structural upgrading of the tourism industry, the cultivation of advantageous industries, and the development of emerging industries.

LS Yi Autonomous Prefecture (refer to LS) is situated in the southwest region of China. LS tourism industry is a significant force for economic development, and poverty alleviation. However, in 2020, due to the impact of COVID-19, the tourism industry experienced a decline. In order to revitalize and advance tourism in LS, government have ramped up their support for the industry. This study will use the Vector Autoregression model to evaluate the coordinating relation between tourism development and financial

J. C. Hung et al. (Eds.): FC 2023, LNEE 1131, pp. 42–49, 2024.
https://doi.org/10.1007/978-981-99-9299-7_6

support. In order to facilitate the harmonious development of both the tourism and financial industries in LS, it is crucial to implement effective financial support measures that can foster the growth of the tourism.

2 Literature Review

Lately, finance plays an important role in the national economy, leading to a growth in scholarly research on the coordination between tourism development and financial support.

Markandya et al. (2005) proposed that World Bank projects should provide more help and support to the tourism industry, and that the expected benefits of the tourism industry must be carefully evaluated [1]. Pericet et al. (2011) emphasized that the economic impact of the tourism industry is significant, and international financial institutions should pay more attention to investing in the tourism industry [2], which is consistent with the research conclusions of Markandya et al. (2005). Dar and Mehta (2014) pointed out that the credit support of financial institutions has a positive economic impact on the development of the tourism industry. Financial institutions can provide various types of tourism loan programs to tourism development [3]. Katircio and Altinay (2017) pointed out that there is a long-term interactive relationship between Turkey's tourism development and financial development, and finance has a positive impact on growth of Turkey's tourism economy [4]. Cannonier and Burke (2017) indicated that tourism expenditure has positive impact on financial development in Caribbean countries, and increasing support for the tourism industry can also lead to a higher level of financial system depth [5]. Ohlan (2017) analyzed India's tourism industry, economic growth and financial development from 1960 to 2014. It was found that the relationship is stable [6]. Cozorici et al. (2019) found through their research that the main problem faced by tourism companies is a shortage of funds, and proposed the path of using the capital market to expand tourism company financing [7]. AYH Saif and Saha (2021) studied the data of 17,077 banks and found that there was an interaction between tourism and finance, with high-income countries having the least negative impact of banks on the tourism industry compared with developing countries [8].

The previous studies findings have revealed a strong correlation between finance and industry. However, as each country has unique national conditions and different levels of development, the significance of the tourism industry in the local economy may vary. Thus, this study will use vector autoregression model to study the financial support for tourism development in Liangshan Prefecture.

3 The Financial Support for the Development of Tourism in LS

The government of LS has introduced multiple policies and measures to support tourism, mainly including the following aspects. Increase financial support, optimize financial services, introduce differentiated and specialized financial products and services, and improve the financing capacity of tourism enterprises. Strengthening financial regulation and risk management, improving the financial regulatory system, and preventing and controlling financial risks. Strengthen tourism marketing, increase the influence of the

LS tourism brand, expand the market share of LS tourism, and increase tourism revenue. Optimizing tax policies to reduce the burden on tourism enterprises and improve their profitability.

There are still some shortcomings, lack of targeted financial products and services for the needs of tourism enterprises; lack of professional financial institutions and personnel support; lack of sound financial regulation and risk management systems; and lack of financial guidance and training for tourism enterprises.

4 Methodology

Vector Autoregression model is based on the statistical nature of the data model, the general mathematical expression of the Vector Autoregression model is[9]:

$$Y_t = \alpha + \sum_{i=1}^{p} A_i Y_{t-i} + \sum\nolimits_{j=1}^{r} B_j X_{t-j} + \varepsilon_t \tag{1}$$

In the formula (1), Y_t is the vector of endogenous variables; X_t is the vector of exogenous variables; p is the lag order of endogenous variables; r is the lag order of exogenous variables; A_i and B_j are the parameter matrix to be estimated, and ε_t is the random perturbation variable.

(1) **Stability test principle**

Before testing, we must ensure that the time series data are stable, otherwise the results will not be ideal. The equation is as follows:

$$\Delta \mathrm{y_t} = \alpha + \beta_\mathrm{t} + (\gamma - 1)\mathrm{y_{t-1}} + \sum\nolimits_{\mathrm{j-1}}^{\mathrm{p}} \delta \Delta \mathrm{y_{t-1}} + \varepsilon_\mathrm{t} \tag{2}$$

In the formula (2), α, β, γ, δ are the parameters, ε_t is the random error term, and the value of formula p is the minimum p value that guarantees ε_t.

(2) **Cointegration test principle**

The specific operation of co-integration inspection are as follows:

First, test the CI (d, b) order cointegration relationship of sequence Y_t and X_t. .First, the unit root test was conducted on each variable, and each variable was an I (d) sequence, and then the variable Y_t was selected to perform OLS regression on X_t, that is, the co-integration regression equation as formula 3:

$$Y_t = \alpha + \beta X_t + \varepsilon_t \tag{3}$$

Using $\hat{\alpha}$ and $\hat{\beta}$ to express the estimate of the regression coefficient, the model residual estimate as formula(4):

$$\hat{\varepsilon} = Y_t - \hat{\alpha} - \hat{\beta} X_t \tag{4}$$

Second, the unit root test was performed with ADF for the residual term ε_t in Eq. (4.4). If the results indicate that ε_t is an I (0) sequence, ε_t~I(0), then ε_t is a stationary sequence, and it can be concluded that Y_t and X_t are CI (d, b) order, and their cointegration vector is (I, -$\hat{\beta}$).

(3) **The Granger causality test principle**

The formula is as followsin Eq. (5):

$$X \to Y \leftrightarrow \sigma^2(y_t/y_{t-h}, h > 0) > \sigma^2(y_t/y_{t-h}, X_{t-h}, h > 0) \tag{5}$$

The Granger causality test model is as follows:

$$y_t = a_{10} + \sum\nolimits_{i-1}^{m} a_{1i}y_{t-i} + e_{1t} \tag{6}$$

$$y_t = a_{20} + \sum\nolimits_{i-1}^{m} a_{2i}y_{t-i} + \sum\nolimits_{i-1}^{m} b_{2i}x_{t-i} + e_{2t} \tag{7}$$

To test the single-term Granger causality from Xt to Yt, testing the null hypothesis H0: $b_{2j} = 0(j = 1, 2, \ldots n)$ test statistic:

$$F = \frac{(ESS_1 - ESS_2)/n}{ESS_1/[T - (M + N + 1)]} \sim F_{\{m, T-(m+n+1)\}} \tag{8}$$

In the formula, ESS_1 and ESS_2 are the sum of the residual squares of the formulas (6) and (7), respectively, and T is the total number of observations of Y. With α as confidence, in formulas (8), if F > Fa, H0 is rejected, that is, X has Granger causality for Y with the probability of 1- α; otherwise, accept the null hypothesis, X does not have Granger causality for Y [10].

5 Analysis of Financial Support for Tourism in LS: An Empirical Study

In this chapter, we conduct analysis of the relationship between tourism and financial support in LS.

5.1 Variable Selection

This paper selects the tourism income (recorded as TI) as the index of the tourism development in LS. And selects the following three indicators: the proportion of financial deposits and loans in GDP (recorded as FS) to present the financial development, the per capita deposit (recorded as SDP) to present financial developmen and the loan-to-deposit ratio (recorded as FE) to present the financial development.

5.2 Data Preprocessing

The data of this paper are mainly derived from the Statistical Bulletin of National Economic and Social Development of LS from 2002 to 2020. To avoid issues with data heteroscedasticity, the paper first takes the natural logarithm of the variables. Thus, the natural logarithm of each variable is denoted as LnTI1, LnFS1, LnSDP1, and LnFE1, respectively.

Table 1. ADF test results for the related variables

Variables	Inspection type	ADF	prob	stable or not
Lnti1	(C,N,0)	−0.836813	0.7839	not
Lnfs1	(C,N,0)	−1.612969	0.4559	not
Lnsdp1	(C,N,0)	−1.338658	0.5864	not
Lnfe1	(N,N,0)	−0.506905	0.4825	not
△Lnti1	(C,N,0)	−4.753345	0.0018	stable
△Lnfs1	(C,N,1)	−5.070829	0.0011	stable
△Lnsdp1	(N,N,2)	−3.096727	0.0486	stable
△Lnfe1	(C,N,3)	−6.914860	0.0001	stable

5.3 Empirical Analysis

(1) ADF inspection

The testing framework is as follows, See Table 1.

Based on the results of the ADF test presented in the table, it can be observed that the time series data for LnTI1, LnFS1, LnSDP1, and LnFE1 are non-stationary. However, after performing first-order differencing on the series (△LnTI1, △LnFS1, △LnSDP1, and △LnFE1), they become stationary, meeting the requirements for the Engle-Granger two-step co-integration test.

(2) E-G two-step co-integration test

1 Cointegrating Equation(s): Log likelihood 75.48640

Normalized cointegrating coefficients (standard error in parentheses)

LNTI1	LNFS1	LNSDP1	LNFE1
1.000000	-6.021335	-1.081794	-3.644806
	(0.71749)	(0.16327)	(0.52983)

Fig. 1. Co-integration test result

According to the results of the cointegration test in Fig. 1, we can get the consolidation equation:

$$\mathbf{LnTI1 = -6.021335LnFS1 - 1.081794LnSDP1 - 3.644806LnFE1}$$

From the results of the consolidation test, there is a long-term balanced relationship between financial support and tourism revenue.

(3) Granger test of causality

The Granger test of causality results are presented below:

VAR Granger Causality/Block Exogeneity Wald Tests
Date: 06/07/22 Time: 09:54
Sample: 2002 2020
Included observations: 17

Dependent variable: LNTI1

Excluded	Chi-sq	df	Prob.
LNSDP1	7.032008954...	2	0.029717936...
LNFS1	8.528123264...	2	0.014065059...
LNFE1	6.259714638...	2	0.043724035...
All	12.50132393...	6	0.051675021...

Dependent variable: LNSDP1

Excluded	Chi-sq	df	Prob.
LNTI1	1.440153757...	2	0.486714836...
LNFS1	0.273555770...	2	0.872163925...
LNFE1	3.561218564...	2	0.168535430...
All	6.417994232...	6	0.378029189...

Dependent variable: LNFS1

Excluded	Chi-sq	df	Prob.
LNTI1	1.170589951...	2	0.556941553...
LNSDP1	1.526188187...	2	0.466221659...
LNFE1	4.719457655...	2	0.094445830...
All	8.577486417...	6	0.198771014...

Dependent variable: LNFE1

Excluded	Chi-sq	df	Prob.
LNTI1	1.789230292...	2	0.408764883...
LNSDP1	1.134334481...	2	0.567129707...
LNFS1	0.036073604...	2	0.982124887...
All	6.940725693...	6	0.326363684...

Fig. 2. Granger causality test result

According to the results of Fig. 2: LnTI1, LnFS1, LnSDP1, LnFE1 have a one-way causal relationship, LnFS1, LnSDP1, LnFE1 are Granger reasons for LnTI1, LnTI1 is not the Granger reason for LnFS1, LnSDP1, LnFE1.

(4) Stability test of the Vector Autoregression model

To ensure the reliability of the analysis, this paper examines the stability of the Vector Autoregression model as Fig. 3 shown:

Root	Modulus
0.534984 - 0.760776i	0.930047201701601...
0.534984 + 0.760776i	0.930047201701601...
0.835964 - 0.122021i	0.844822061002333...
0.835964 + 0.122021i	0.844822061002333...
-0.504599 - 0.486508i	0.700935528162967...
-0.504599 + 0.486508i	0.700935528162967...
0.274424 - 0.610909i	0.669715089674758...
0.274424 + 0.610909i	0.669715089674758...

No root lies outside the unit circle.
VAR satisfies the stability condition.

Fig. 3. Stability test result

(5) Variance analysis

The results in Fig. 4 indicate that over time, the contribution of financial development to tourism income in LS will gradually increase.

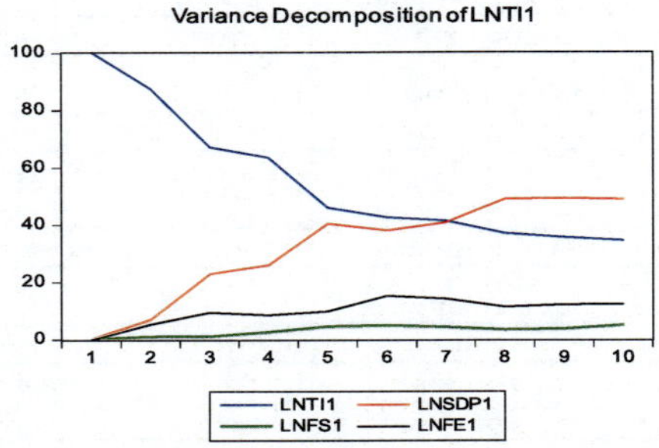

Fig. 4. Variance analysis result

(6) Regression analysis

In order to analyze the impact of these three indicators on the tourism development more specifically, the stepwise forwards regression analysis is conducted, and the regression results are as Fig. 5 shown:

Variable	Coefficient	Std. Error	t-Statistic	Prob.*
C	4.221790	0.343825	12.27888	0.0000
LNSDP1	1.697460	0.096193	17.64632	0.0000
LNFE1	0.900700	0.397609	2.265293	0.0377

R-squared	0.951846	Mean dependent var	4.291845
Adjusted R-squared	0.945827	S.D. dependent var	1.426064
S.E. of regression	0.331919	Akaike info criterion	0.776088
Sum squared resid	1.762724	Schwarz criterion	0.925210
Log likelihood	-4.372835	Hannan-Quinn criter.	0.801325
F-statistic	158.1331	Durbin-Watson stat	1.409865
Prob(F-statistic)	0.000000		

Fig. 5. Regression results

The following regression equations can be obtained from the above regression results:

$$\text{LnTI1} = 1.69746\text{LnSDP1} + 0.9007\text{LnFE1} + 4.22179 \tag{9}$$

According to the regression results in formula (9), the adjusted R squared is 0.945827, and the model had a high goodness of fit. The F statistic is 158.1331, and the concomitant probability is 0.000, indicating that the model overall passed the significance test. According to the regression coefficient, the financial development level and financial development efficiency have a positive relationship with the development of tourism. When the level of SDP and FE increase by 1% respectively, the TI will increase by 1.69% and 0.90%. In general, the above empirical results show that increasing the financial support of LS can promote the further development of LS tourism.

Acknowledgement. Fudings: Sichuan Province Social Science Key Research Base - China Panxi Health Industry Research Center, Project Name: Study on Tourist Satisfaction of Kangyang Tourism in LS, Project Number: PXKY-YB-202113.

References

1. Markandya, A., Taylor, T., Pedroso, S., et al.: Tourism and sustainable development: lessons from recent world bank experience. Econ. Tour. Sustain. Dev. 225–251 (2005)
2. Peric, J., Mujacevié, E., Simunic, M.: International financial institution investments in tourism and hospitality. J. Int. Bus. Cult. Stud. 1–17 (2011)
3. Dar, A.A., Mehta, S.K.: Assessment of the role of financial institutions in tourism development of Kashmir: a field study from demand side of market. J. Kashmir Tour. Cater. Technol. **1**(2), 14–23 (2014)
4. Katircio, Lu.S., Altinay, M.: Interactions between tourism and financial sector development: evidence from Turkey. Serv. Ind. J. **38**, 519–542 (2017)
5. Cannonier, C., Burke, M.G.: Tourism and financial development in small states: evidence from Caribbean countries. Tour. Econ. Bus. Finan. Tour. Recreat. **23**(6), 1369–1377 (2017)
6. Ohlan, R.: The relationship between tourism, financial development and economic growth in India. Future Bus. J. **3**(1), 9–22 (2017)
7. Cozorici, A.N., Prelipcean, G., Scutaru, L.: Financing tourism companies through the capital market. In: Springer Proceedings in Business and Economics, pp. 111–119. Springer, Heidelberg (2019). https://doi.org/10.1007/978-3-319-89468-3_9
8. Saif-Alyousfi, A., Saha, A.: Do tourism receipts affect bank profitability? analytical evidence from 85 tourism economies. Res. Int. Bus. Financ. **4**, 101437 (2021)
9. Saori, O., Takashi, S.: Re-examining the effects of official development assistance on foreign direct investment applying the VAR model. Economies **10**(10), 236 (2022)
10. Yu, J., Zhang, M., Liu, R., Wang, G.: Dynamic effects of climate policy uncertainty on green bond volatility: an empirical investigation based on TVP-VAR models. Sustainability **15**(2), 1692 (2023)

Application of Artificial Intelligence Technology in Text Recognition and Detection Algorithms

Junxia Liang[1] and Yongjun Qi[1,2](✉)

[1] Faculty of Megadata and Computing, Guangdong Baiyun University, Guangzhou 510450, Guangdong, China
qyj200702022@baiyunu.edu.cn

[2] School of Information and Communication Technology, Mongolian University of Science and Technology, Bayanzurkh district 13341, Ulaanbaatar, Mongolia

Abstract. With the development of the times, words play an indispensable role in daily life. Therefore, text recognition detection is very important. In recent years, although there have been many research achievements and progress in text recognition detection, there are still situations of ambiguity and distortion in text recognition. Therefore, it is necessary to further upgrade the text recognition detection algorithm. This paper studied the application of artificial intelligence (AI) technology in text recognition and detection algorithms, aiming to further improve the accuracy of text recognition and detection algorithms through AI technology. This article tested the improvement of text recognition detection accuracy using AI technology through experiments. The experimental data showed an improvement of at least 11% and at most 19%, indicating that AI technology can achieve good results in text recognition and detection algorithms.

Keywords: Text Recognition · Detection Algorithms · Artificial Intelligence · Algorithm Applications

1 Introduction

Text recognition is currently a popular research topic. Text recognition can scan documents and images, which can bring great convenience for people to obtain this article. Text recognition is constrained by the clarity and color difference of the image. Therefore, text recognition detection needs further improvement, and studying how to upgrade text recognition detection algorithms is of great significance.

Many scholars have conducted some research on text recognition. Scholar Chen X introduced an image database for handwritten text recognition research, which included digital images of approximately 5000 city names, 5000 state names, 10000 postal codes, and 50000 alphanumeric characters [1]. The architecture of the scene text recognition system proposed by Lin H had two unique characteristics. (1) It is end-to-end trainable, and most existing algorithm components are trained and tuned separately. (2) It naturally processes sequences of any length without involving character segmentation or horizontal scale normalization [2]. Petrova O believed that text recognition in natural scenes

J. C. Hung et al. (Eds.): FC 2023, LNEE 1131, pp. 50–58, 2024.
https://doi.org/10.1007/978-981-99-9299-7_7

had always been an active research topic in the fields of computer vision and pattern recognition [3]. Although the research on text recognition is deep, there are still some shortcomings that need to be addressed.

The improvement of text recognition detection algorithms is now a very urgent problem to be solved. This article studied text recognition detection algorithms using AI technology. This article tested the user satisfaction of text recognition detection algorithms using AI technology through experiments, and found that the satisfaction was very good. The improvement of text recognition detection accuracy has been tested again, and it has been found that the accuracy improvement is good, which proves that AI technology can play a good role in text recognition detection algorithms.

2 Text Recognition and Detection Based on AI

2.1 Overview of Text Recognition

Optical character recognition, also known as text recognition, converts images of typed, handwritten, or printed text into machine encoded text electronically or mechanically, whether these characters come from scanned documents, document photos, or scene photos (such as text on signs and billboards in landscape photos) or subtitle text superimposed on images (such as from television broadcasts). Optical text recognition is a discipline that integrates pattern recognition, AI, and computer vision [4, 5]. Optical character recognition allows for easy viewing, searching, and recognition of text in images and labels. Before AI was applied to text recognition, people attempted to solve the machine learning problem of text recognition using traditional computer vision technology. These traditional methods focus on structured text in text recognition. In a standard background, using appropriate lines and standard fonts, most of the text is relatively dense and can be well recognized [6, 7]. The problems with text recognition are shown in Fig. 1:

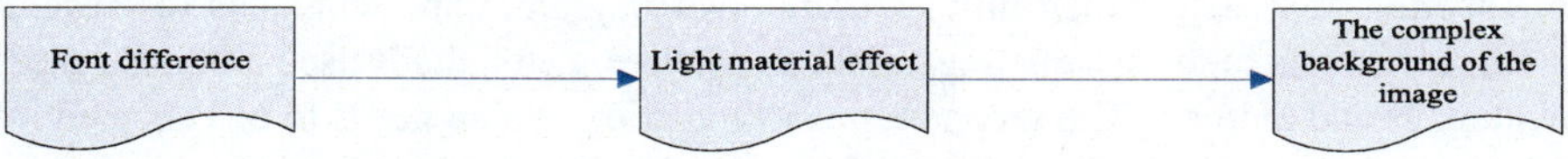

Fig. 1. Problems with text recognition

(1) Significant differences in handwriting and font (such as handwriting and font). (2) Due to factors such as light and material, the same text can have significant differences in different scenes [8, 9]. (3) The complex background, noise, flash, multi word, geometric distortion, etc., of an image. It is in this context that text recognition based AI methods are widely studied and applied, and are widely applied in various industries. The use of AI optical character recognition technology can: (a) Reduce costs. (b) Accelerate workflow speed. (b) Automation of file routing and content processing. (d) Centralization and security of data. (e) Obtain the latest and most accurate information on improving services [10, 11].

Automatic text detection and reading in natural scenes is an important part, which can be used for several challenging tasks, such as image based machine translation, autonomous vehicle or picture, video indexing. In recent years, the task of detecting and recognizing text in natural scenes has received great attention from computer vision and document analysis. In addition, in other fields of AI, recent breakthroughs have made it possible to create better scene text detection and recognition systems.

2.2 Steps for Text Recognition

In general, text recognition in images involves two stages. One is to detect the text in the image, and the other is to use recognition technology to obtain information about the text. Text detection is a prerequisite for text recognition, while traditional text detection techniques mainly focus on document images. In these images, there is a significant difference between text and background, and background information does not cause significant interference with the text. In general, in these images, the text carrying important information is black, while the non important background is white. Therefore, it is easy to separate the required text itself from the image, and then recognize the separated text through text recognition technology to obtain text content information. However, in natural scenes, there are significant differences between text and background, such as excessive contrast between text and background, or exposure to light. In addition, in natural scene images, there are differences in text size, font, color, and other aspects. For example, large and small characters can appear simultaneously in images, and text can be non horizontal or even circular. A character can have many colors, which makes text recognition very difficult.

2.3 Overview of AI

AI, abbreviated as AI, belongs to a branch of computer science and is a discipline used to simulate, extend, expand, and learn the theories, methods, technologies, and application systems of human intelligence. AI products include intelligent systems, intelligent machines, robots, and so on [12, 13]. There are two ways to implement AI, one is the Engineering approach, which utilizes traditional programming techniques to present intelligent effects to the system, regardless of whether the methods used are those used by humans and animals. The second type is simulation, which needs to be executed in the same way as humans and other organisms [14, 15]. The process of using AI is shown in Fig. 2:

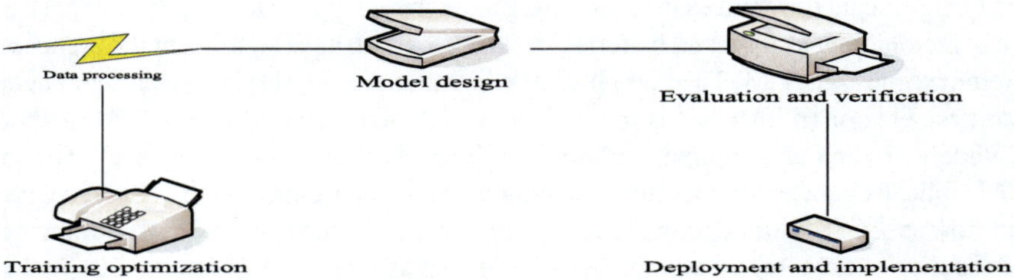

Fig. 2. Processes used by AI

2.4 Main Application Fields of AI

Natural language processing: In the application of AI, the most critical is natural language processing, which enables efficient interaction between people. Deep learning algorithms based on big data and parallel computing have made great progress in natural language processing [16, 17]. At present, the technologies in this field include language engineering, data processing, linguistics, etc. Among them, the most representative ones are customer service, robot chat, etc. In 2017, Google, Apple, Amazon, and other companies all released audio and home appliances controlled by sound [18, 19].

In the field of computer vision: AI is mainly applied in the field of computer vision, which imitates the human visual system and enables computers to determine the position, motion status, and recognize objects. In the process of application in the field of vision, there are a total of three stages: target detection, target recognition, and behavior recognition. At present, the most mature ones are facial recognition systems, pupil recognition systems, and fingerprint recognition systems. The most advanced machine vision technology can automatically extract the features of targets from a large amount of data and recognize them based on these features, greatly improving the accuracy of recognition [20].

2.5 Technical Support for AI

The technical support of AI includes algorithms: the so-called algorithms describe the strategies and mechanisms for solving problems in a systematic way, so that they can obtain the desired results in a limited time based on specific rules. In recent years, with the continuous development of new algorithms, the ability of machine learning has been greatly improved, especially in the increasingly mature situation of deep learning theory. Many companies provide advanced technology to the industry through cloud computing or open source methods, and package advanced algorithms into easy-to-use products, greatly promoting the development of AI technology.

Big Data: With the rapid development of mobile internet, massive amounts of data such as social media, mobile terminals, and inexpensive sensors are being quickly collected. With the continuous discovery of the value of data, various data management and analysis technologies are also constantly developing. In the field of AI, there are many machine learning algorithms, such as recognition of images, text, and speech, that require a large amount of data to be trained and continuously optimized. Nowadays, all of this is readily available, and big data has played a promoting role in the development of AI, laying a good foundation for its development.

Computational power: AI requires extremely high computational power. In the past, people's research on AI was often limited by the computing power of a single computer. In recent years, with the rapid development of cloud computing technology, its computing power has been greatly improved. Machine learning, especially deep learning, requires high computational resources, while cloud computing has a computing power of up to 1 trillion times per second. In addition, with the continuous progress of graphics processor technology, AI technology has also been greatly developed, and the use of multi core parallel methods can greatly improve the processing speed of AI. With the help of cloud computing technology, image processors can achieve large-scale and low-cost computing power. The key technologies in AI are shown in Fig. 3:

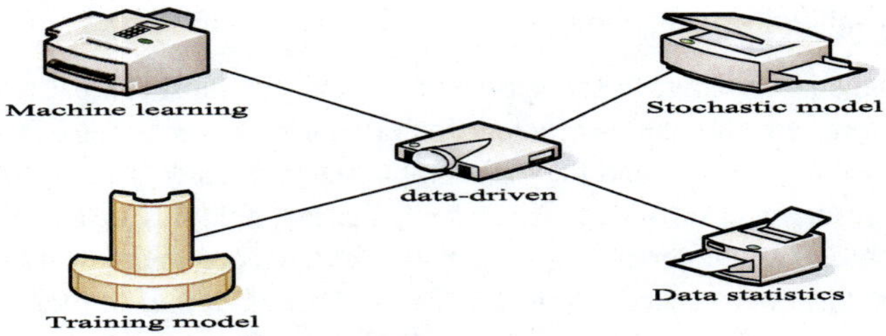

Fig. 3. Key technologies in AI

2.6 Algorithms in AI

Genetic algorithm: Genetic algorithm is an algorithm that imitates the law of survival of the fittest and survival of the fittest in nature and is developed to find the best solution. This algorithm starts from an initial population and restricts the chromosomes above the optimal solution gene. Each chromosome has a corresponding constraint characteristic value, which is manifested as the arrangement and combination of genes and determines the external characteristics of the individual. For example, the characteristics of black hair are determined by alleles on chromosomes, which control hair color. With the emergence of the first generation of groups, according to Darwin's theory of evolution, over time, previously unsuitable groups would continue to be eliminated and gradually produce more adaptable groups to new environments. In each generation, individuals who meet the conditions are selected as fathers, and the crossover and mutation of natural genes are simulated to generate new solutions. Finally, the best individual in the last group is decoded to become the near optimal solution of the problem.

Ant colony algorithm: Ant colony algorithm is a population optimization method that has its advantages and disadvantages compared to other methods. First of all, the ant colony algorithm introduces a positive feedback mechanism, so that it can continue to accelerate, so as to quickly get the optimization results. In this process, each individual can release its own pheromone and achieve communication between individuals. During the search process, multiple individuals can execute in parallel, greatly improving the computational power and execution efficiency of the algorithm. On this basis, a probability based search method is proposed to avoid errors caused by local minima during the search process.

2.7 Text Recognition Under AI

Convolutional based recurrent neural networks are widely used for text recognition in AI. This method first preprocesses the image for text recognition, and then proceeds in two steps: (1) Feature extraction is performed using convolutional neural network method. (2) Neural networks are an AI optical text recognition method that pre measures the position and value of text characters, enabling text recognition. Convolutional layers are widely used in image classification tasks due to their efficient feature extraction ability. Neural networks can be used to detect meaningful edges in images, thereby obtaining (at a higher level) shapes and composite images. For example, compared to fully connected neural networks, convolutional neural networks can repeatedly use different pattern detection filters on images, thereby reducing the complexity of image recognition. Then, text recognition can be achieved by utilizing the relationships between characters. Circular networks have shown good performance in processing data with varying lengths. The most commonly used method is to use long-term short-term memory cells to solve asymptotic disappearance (asymptotic disappearance in a given weighting function, asymptotic disappearance).

2.8 Use of AI Technology

AI technology can grayize color images and process them accordingly. The process of turning an image into an image with only black and white colors is called binarization, which can be obtained through thresholding. Thresholding is the process from input image a to input image b. The calculation method for b is shown in formula 1:

$$b = \begin{cases} 1 \; a \geq T \\ 0 \; a < T \end{cases} \tag{1}$$

In formula 1, T represents the threshold. If the input image a is subjected to segmentation threshold processing, the inter class variance c obtained is shown in formula 2:

$$c = de^2 \tag{2}$$

In formula 2, d represents the proportion of foreground pixel count to the entire image, and e represents the proportion of background pixel count to the entire image. The maximum threshold f of variance c can be obtained by using the traversal method, and the calculation method of f is shown in formula 3:

$$f = \frac{\sqrt{b}}{c^2} \tag{3}$$

3 Simulation Experiment on the Application of AI Technology to Text Recognition Detection Algorithms

The use of AI in text recognition and detection is quite popular. This article conducted an experimental study on the improvement of text recognition and detection algorithms under AI technology, to verify whether the accuracy of text recognition and detection can be improved through AI algorithms. This article conducted an experimental survey on the satisfaction of 100 groups of users who used AI technology for text recognition and detection algorithms with the effectiveness of AI technology (out of 100 points). The satisfaction situation is shown in Fig. 4.

AI technology has better image processing capabilities and is expected to have good results in text recognition. From the experimental results in Fig. 4, it can be seen that users who use text recognition detection algorithms based on AI technology have a

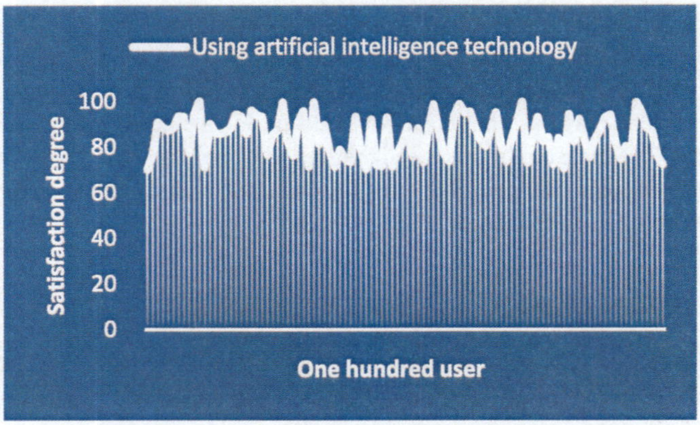

Fig. 4. Employee satisfaction of text recognition detection algorithm using AI technology

satisfaction score of 70 to 100 points with the algorithm. This indicates that users who use text recognition and detection algorithms based on AI technology are quite satisfied with the role of AI algorithms in text recognition and detection algorithms.

This article also tested the improvement of text recognition detection algorithm detection accuracy after using AI technology in five groups. The improvement of accuracy is shown in Fig. 5.

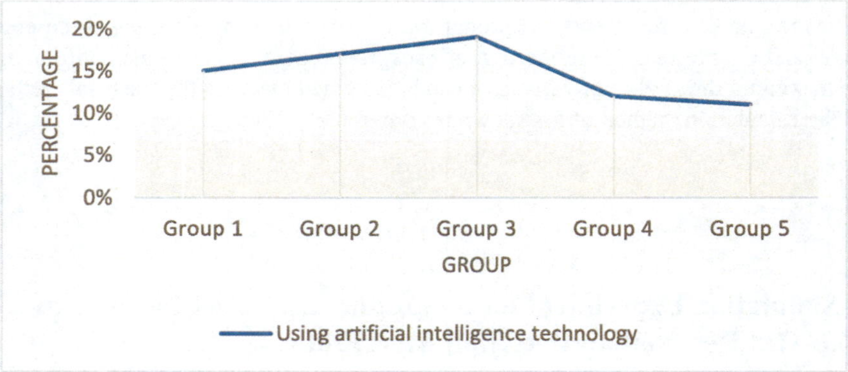

Fig. 5. The improvement of the accuracy of text recognition detection algorithm

From the experimental results in Fig. 5, it can be seen that the accuracy of the text recognition detection algorithm using AI technology has improved by a minimum of 11% and a maximum of 19%. From this experimental data, it can be seen that AI technology can improve the accuracy of text recognition detection algorithms, indicating that AI technology can achieve good results in text recognition detection algorithms.

4 Conclusions

Text recognition is a commonly used technology nowadays. Through text recognition, various types of text can be accurately read out, making it easier for people to obtain the text they need. When people need text that can be obtained by machines without manpower, it greatly reduces the workload of text processing. However, there are still many difficulties in text recognition. Because some texts have uneven or blurry areas, which affect the recognition effect, new technologies are needed to improve the quality of text recognition. This article studied the application of AI technology in text recognition detection algorithms, which aimed to improve the accuracy of text recognition detection algorithms through AI technology. This article tested the improvement of recognition accuracy of text recognition detection algorithms using AI technology through experiments, and found that the effect was good. Therefore, it has been proved that AI technology had a good effect in text recognition detection algorithms. Due to space limitations, there are still many shortcomings in the experiments conducted in this article, and further improvements are needed in the future. Finally, it is wished that the text recognition detection algorithm could be increasingly improved.

Acknowledgements. This work was supported by

Key Research Project of Guangdong Baiyun College, No. 2022BYKYZ02.

Key Research Platform of Guangdong Province, no. 2022GCZX009

Special project in key fields of colleges and universities in Guangdong province, No. 2020ZDZX3009.

References

1. Chen, X., Jin, L., Zhu, Y.: Text recognition in the wild: a survey. ACM Comput. Surv. (CSUR) **54**(2), 1–35 (2021)
2. Lin, H., Yang, P., Zhang, F.: Review of scene text detection and recognition. Arch. Comput. Methods Eng. **27**(2), 433–454 (2020)
3. Petrova, O., Bulatov, K., Arlazarov, V.V.: Weighted combination of per-frame recognition results for text recognition in a video stream. Компьютерная оптика **45**(1), 77–89 (2021)
4. Butt, H., Raza, M.R., Ramzan, M.J.: Attention-based CNN-RNN Arabic text recognition from natural scene images. Forecasting **3**(3), 520–540 (2021)
5. Bulatov, K., Razumnyi, N., Arlazarov, V.V.: On optimal stopping strategies for text recognition in a video stream as an application of a monotone sequential decision model. Int. J. Doc. Anal. Recogn. (IJDAR) **22**(3), 303–314 (2019)
6. Francis, L.M., Sreenath, N.: Robust scene text recognition: using manifold regularized twin-support vector machine. J. King Saud Univ.-Comput. Inf. Sci. **34**(3), 589–604 (2022)
7. Sil, R., Roy, A., Dasmahapatra, M.: An intelligent approach for automated argument based legal text recognition and summarization using machine learning. J. Intell. Fuzzy Syst. **41**(5), 5457–5466 (2021)
8. Liao, M., Shi, B., Bai, X.: Textboxes++: a single-shot oriented scene text detector. IEEE Trans. Image Process. **27**(8), 3676–3690 (2018)
9. Kang, L., Riba, P., Rusinol, M.: Content and style aware generation of text-line images for handwriting recognition. IEEE Trans. Pattern Anal. Mach. Intell. **44**(12), 8846–8860 (2021)

10. Thompson, W.: Using handwritten text recognition (HTR) tools to transcribe historical multilingual lexica. Scripta e-Scripta **21**, 217–231 (2021)
11. Elleuch, M., Kherallah, M.: Boosting of deep convolutional architectures for Arabic handwriting recognition. Int. J. Multimedia Data Eng. Manag. (IJMDEM) **10**(4), 26–45 (2019)
12. Mintz, Y., Brodie, R.: Introduction to artificial intelligence in medicine. Minim. Invasive Ther. Allied Technol. **28**(2), 73–81 (2019)
13. Kaul, V., Enslin, S., Gross, S.A.: History of artificial intelligence in medicine. Gastrointest. Endosc. **92**(4), 807–812 (2020)
14. Gunning, D., Aha, D.: DARPA's explainable artificial intelligence (XAI) program. AI Mag. **40**(2), 44–58 (2019)
15. Davenport, T.H., Ronanki, R.: Artificial intelligence for the real world. Harv. Bus. Rev. **96**(1), 108–116 (2018)
16. He, J., Baxter, S.L., Xu, J.: The practical implementation of artificial intelligence technologies in medicine. Nat. Med. **25**(1), 30–36 (2019)
17. Johnson, K.W., Torres Soto, J., Glicksberg, B.S.: Artificial intelligence in cardiology. J. Am. Coll. Cardiol. **71**(23), 2668–2679 (2018)
18. Haenlein, M., Kaplan, A.: A brief history of artificial intelligence: on the past, present, and future of artificial intelligence. Calif. Manag. Rev. **61**(4), 5–14 (2019)
19. Longoni, C., Bonezzi, A., Morewedge, C.K.: Resistance to medical artificial intelligence. J. Cons. Res. **46**(4), 629–650 (2019)
20. Hosny, A., Parmar, C., Quackenbush, J.: Artificial intelligence in radiology. Nat. Rev. Cancer **18**(8), 500–510 (2018)

Investigation on Data Mining of Intelligent Environmental Protection Big Data Based on Neural Networks

Chao Zhou[1](✉), Junwen Deng[2], and Haoxuan Tang[2]

[1] Powerchina Guiyang Engineering Corporation Limited, Guiyang 550081, Guizhou, China
Zc125@126.com

[2] Guoneng Qinghai Yellow River Maerdang Hydropower Development Co., Ltd., Guoluo 814099, Qinghai, China

Abstract. Intelligent environmental protection refers to the intelligent form of utilizing environmental data information, including comprehensive collection and analysis, continued development and utilization of environmental data, and achieving the goal of supporting environmental management. It is conducive to reducing pollutant emissions, and promoting environmental legitimacy, thus creating an ecological city that is harmonious with the environment and social economy. The development of networked technology has brought many opportunities for production and human life. With the emergence of environmental problems, they have become a global problem, especially severe in Chinese cities. Therefore, in order to achieve rapid socio-economic development, it is necessary to introduce an intelligent environment that is conducive to building green cities. The research on data mining based on neural network-based smart environmental protection big data was of great help in improving the data mining performance of smart environmental protection platforms.

Keywords: Data Mining · Smart Environmental Protection · Neural Network · Big Data

1 Introduction

Considering the actual needs of environmental enterprises at present, the knowledge framework for the development of environmental systems should include the following contents: application systems, environmental information resource systems, infrastructure systems and frameworks, standard regulations systems, information security systems, and operation and maintenance management systems. The application system is the core of this framework. The application system is the core. The development and utilization of information resources are the foundation. Infrastructure is the support, standards, and norms.

A large number of scholars have conducted corresponding research on data mining. Sayah Z summarises the semantics in big data and summarises the various work done on energy savings in smart homes and cities. An efficient architecture is developed

J. C. Hung et al. (Eds.): FC 2023, LNEE 1131, pp. 59–65, 2024.
https://doi.org/10.1007/978-981-99-9299-7_8

based on ontology, big data and the synergy of multiple intelligences. The results of the experiments will provide a useful reference for the development of big data-based architectures. The research highlights that when used jointly, the two will provide a reliable energy-efficient system for smart cities. The results of the research improve energy efficiency and environmental protection while ensuring occupant comfort [1]. Suncare P's extensive research on the Internet of Things and cloud technology made it possible for people to accumulate a large amount of data generated from this heterogeneous environment and transform it into valuable knowledge through the use of data mining techniques. In addition, these generated knowledge would play a crucial role in intelligent decision-making, system performance improvement, and resource and service optimization management. An overview of cloud assisted Internet of Things big data mining systems was also introduced to better understand the importance of data mining in Internet of Things environments [2]. Suma V conducted a detailed analysis of the Indian e-commerce market using data mining technology to predict demand for upgraded e-commerce products. It also analyzed the impact of actual factors on demand and the impact of variables on demand. This analysis studied real datasets from three randomly selected e-commerce websites. Data was collected, processed, and validated using powerful algorithms. Based on the analysis results, it was found that despite the influence of various customer behavior and market factors, the proposed method could predict with high accuracy [3]. Although the above literature studies have proposed technical solutions for data mining problems, the application scope of the technology is relatively narrow.

With the continuous development of science and technology, the district and county environmental protection departments have established a basic computer software, hardware, and internet application environment. The platform construction level is low, and the problem of duplicate construction is very prominent; at the same time, there are also problems of data dispersion, information fragmentation, application fragmentation and service fragmentation among environmental management departments; at present, there is a common phenomenon of "seeing" and "using" "two skins" in China's environmental data, which makes it difficult to support the formulation of environmental management policies. This article would compare and analyze various algorithms of neural networks in big data data mining, so that readers have a broad understanding of data mining technology.

2 Data Mining, Smart Environmental Protection Big Data, and Neural Networks

2.1 Data Mining

Data mining technology:

Data mining technology refers to the step of discovering existing problems (knowledge base), which is an information processing process to automatically discover and extract specific relationships between data from large unstructured data sets. The main task of data mining is to find relationships between objects, cluster analysis, classification analysis, and anomaly analysis [4]. Rough neural network algorithm is a hybrid of rough set algorithm and artificial neural network algorithm. In data mining, when there is uncertainty or fuzziness in the data structure, rough neural network algorithm is often used.

2.2 Smart Environmental Protection

The concept of smart environment is a new method based on digitization and information technology to monitor the urban environment in real-time, while improving the automation, intelligence, and unmanned nature of the urban environment, improving the quality of pollution monitoring data, reducing the use of human and material resources, and fundamentally reducing the prevention of environmental overload caused by various reasons [5].

The challenges faced by implementing smart environmental protection in the construction of ecological cities:

Different Internet of Things products cannot work together well and there is no standardization. However, collecting information at the level of registering an intelligent environment is impossible without Internet of Things technology. Intelligence in a standardized and unified environment provides a unified data collection method and enables effective sharing of environmental data across the entire Internet of Things infrastructure [6, 7].

In order to create an intelligent green city, fast and accurate data processing technology is needed. Modern environmental construction cannot be separated from other building plans, so when establishing a data processing layer, it is necessary to fully consider data support from other fields in order to form a reasonable data processing mode [8]. In this mode, establishing a dynamic data management process is the only way to generate data with high reference value. When constructing a system oriented towards the environmental protection level of smart cities, the first step is to introduce the division of environmental protection areas, and clarify the resource distribution, ecological status, and environmental protection difficulties in this environmental protection task [9].

Overall architecture:

The intelligent environmental protection system would use a reasonable and scientific design concept to comprehensively evaluate the current situation of information technology, business needs, and data requirements. With the help of advanced technologies such as the Internet of Things, cloud computing, SOA (Service-Oriented Architecture), GIS (Geographic Information System), and data storage, it would create a comprehensive platform with intelligent and reasonable architecture, clear hierarchy, and aesthetic practicality. The system consists of four levels: monitoring system, data source, block service, and intelligent application, as shown in Fig. 1.

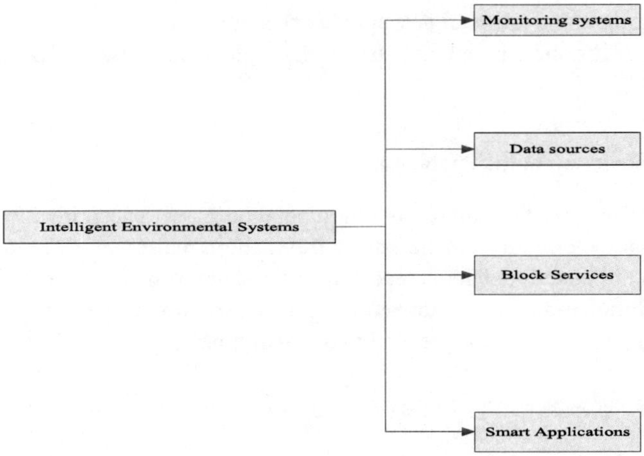

Fig. 1. Architecture of Intelligent Environmental Protection System

2.3 Neural Networks

Artificial Neural Networks (ANNs) have developed rapidly since their inception in the 1930s. This is because they can perform large-scale parallel processing on data and have high fault tolerance and transferability. The BP neural network algorithm is a multi-layer feedforward network trained based on the principle of error backpropagation. BP neural networks are one of the widely used artificial neural networks because they do not require prior knowledge of the mathematical formulas describing the dependency relationships of the problem.

Learning algorithms:

When there is a gap between the network output and the expected result, E represents the output error. The hidden layer is as follows:

$$\mathrm{E} = \frac{1}{2}\sum_{k=1}^{l}\left[d_k - f(net_k)\right]^2 = \frac{1}{2}\sum_{k=1}^{l}\left[d_k - f\left(\sum_{j=0}^{m} w_{jk}y_i\right)\right]^2 \tag{1}$$

The output layer is as follows:

$$E = \frac{1}{2}\sum_{k=1}^{l}\left\{d_k - f\left[\sum_{j=0}^{m} w_{jk}f\left(net_j\right)\right]\right\}^2 \tag{2}$$

According to Formula (1), the expansion of network error to the output layer is only related to d_k and W_{jk}. To reduce the error, it is necessary to modify the weights of neurons y_i [10].

It is assumed that the network has h hidden layers and m neurons. y_1, y_2,….., y_i represent the output of the hidden layer in the network. W_1, W_2,…, W_{h+1} are weights. The weight modification formula is as follows:

The output layer is as follows:

$$\Delta w_{jk}^{h+1} = \alpha \chi_k^{h+1} y_j^h = \alpha(d_k - O_k)O_k(1 - O_k)y_i^h \quad (3)$$

The hth hidden layer is as follows:

$$\Delta W_{ij}^h = \alpha \chi_j^h y_i^{h-1} = \alpha\left(\sum_{k=1}^{l} \chi_k^o W_{jk}^{h+1}\right) y_j^h\left(1 - y_j^h\right) y_j^{h-1} \quad (4)$$

From the recursion, the formula can be obtained as follows:

$$\Delta W_{pq}^1 = \alpha\left(\sum_{r=1}^{m} \chi_r^2 W_{qr}^2\right) y_q^1\left(1 - y_q^1\right) x_p \quad (5)$$

3 Experimental Evaluation of Data Mining Using Neural Networks

In order to effectively compare the advantages of different neural network classification algorithms in data mining, three types of corn seeds were selected from the database, and a subset of SVM (Support Vector Machine) neural network algorithm, BP (Back Propagation) neural network algorithm, and ELM (Extreme Learning Machine) neural network algorithm was used. In order to clean up the samples, 200 records were extracted from the database. Each record contained attributes such as grain density, length, and width, and the corn seed type was represented by binary digits {1, 2, 3}.

(1) SVM neural network algorithm

After using the loading function to import all the raw data into a dedicated storage file, in order to ensure the effective universality of the data, only 80% of the datasets were randomly selected, totaling 160 datasets. They were organized as training sets in the statistical model, while the remaining 20% of the datasets were organized as test sets. After setting the network parameters, the SVMpredict function was directly used to run the experiment. In order to make the experiment more realistic and reliable, a total of five iterations were conducted.

Table 1. Experimental results under SVM neural network algorithm

	First	Second	Third	Fourth	Fifth	Average
Accuracy rate(%)	89.848	94.858	97.2	89.811	94.918	93.327
Modelling time(s)	0.5703	0.122	0.1847	0.1696	0.1952	0.24836

From the analysis of experimental data results in Table 1, it could be shown that the average accuracy and model establishment time of the training set based on SVM neural network algorithm were 93.327% and 0.24836 s. Considering the high reliability of the research equipment used in practice, the data mining technology based on SVM algorithm had high credibility.

(2) BP neural network algorithm

The program for classifying the number of corn seeds using the BP neural network algorithm was basically the same as the program for classifying the number of corn seeds using the SVM neural network algorithm. After normalizing the training data, the number of output and input neurons was 3 and 7, respectively, which corresponded to a total of 3 corn seed types and 7 corn seed features for each type. When designing the BP neural network, the hidden layer node used an S-shaped tangent transfer function, and the output layer node selected a pure linear Newton transfer function. The BP algorithm was used as a learning function. After setting the network parameters separately, the experiment trained the BP neural network using training and simulation functions, and conducted simulation tests on it. The test results after five iterations are shown in Table 2.

Table 2. Experimental results under BP neural network algorithm

	First	Second	Third	Fourth	Fifth	Average
Accuracy rate (%)	82.281	77.831	82.494	87.685	85.546	83.167
Modelling time (s)	1.8007	1.7311	1.7668	1.7864	1.7412	1.7652

After careful statistical analysis of the experimental results, it could be concluded that the average accuracy and average simulation time of using the BP neural network algorithm to classify the dataset were 83.167% and 1.7652, respectively.

(3) ELM neural network algorithm

The method of saving raw data was to import all the raw data into a separate file called a seed, which was created using the load function. In order to ensure the universality of the data, only 80% of the dataset was randomly selected, which meant a total of 160 records. A statistical model was constructed from it as the training set, and the remaining records were used as the test set. The mapminmax function of MATLAB was used to normalize the training dataset, thus resulting in three output and seven input neurons, and representing the number of seed types and attribute features.

In order to create and train ELM neural network, elmtrain function was used to extract attributes and assign them to activation function. After assigning network parameters one by one, the ELM predict function was used to model and test the ELM neural network. In order to make testing more realistic and reliable, it was necessary to repeat testing as much as possible. A total of five iterations were conducted, and the final results are shown in Table 3.

Further statistical analysis of the experimental results shows that the average accuracy of the ELM algorithm was 96.192%, and the average modeling time was 0.119 s.

In this study, several neural network-based classification algorithms were analyzed. The experimental results showed that the ELM algorithm based on neural networks had strong modeling ability and high accuracy. The BP neural network algorithm had the lowest accuracy and requires more time to establish the model. The overall performance

Table 3. Experimental results under ELM neural network algorithm

	First	Second	Third	Fourth	Fifth	Average
Accuracy rate(%)	97.537	95.049	97.495	95.038	95.843	96.192
Modelling time(s)	0.0497	0.4177	0.0476	0.0428	0.0357	0.119

of SVM algorithm and ELM algorithm was vastly different. However, the ELM neural network algorithm used fewer nodes in the hidden layer, resulting in better performance.

4 Conclusions

In the transition from digital environmental protection to intelligent environmental protection, the most important thing is to start from the original digital environmental protection and focus on strengthening the perception layer and building an intelligent layer. Firstly, Internet of Things technology must build sensor systems that can perceive environmental parameters in real-time and adaptively. Secondly, it is necessary to apply intelligent information technologies such as cloud services and fuzzy perception, integrate existing information resources, build high-speed computer systems, storage capabilities, and parallel systems for processing environmental information, and ultimately build support platforms and information services for various applications of "smart environmental protection".

References

1. Sayah, Z., Kazar, O., Lejdel, B., et al.: An intelligent system for energy management in smart cities based on big data and ontology. Smart Sustain. Built Environ. **10**(2), 169–192 (2021)
2. Sunhare, P., Chowdhary, R.R., Chattopadhyay, M.K.: Internet of things and data mining: an application oriented survey. J. King Saud Univ.-Comput. Inf. Sci. **34**(6), 3569–3590 (2022)
3. Suma, V., Hills, S.M.: Data mining based prediction of demand in Indian market for refurbished electronics. J. Soft Comput. Paradigm (JSCP) **2**(02), 101–110 (2020)
4. Arcinas, M.M., Sajja, G.S., Asif, S., et al.: Role of data mining in education for improving students performance for social change. Turk. J. Physiother. Rehabil. **32**(3), 6519–6526 (2021)
5. Shneiderman, B.: Human-centered artificial intelligence: three fresh ideas. AIS Trans. Human-Comput. Interact. **12**(3), 109–124 (2020)
6. Ford, A.T., Ågerstrand, M., Brooks, B.W., et al.: The role of behavioral ecotoxicology in environmental protection. Environ. Sci. Technol. **55**(9), 5620–5628 (2021)
7. Gonzalez, R.A., Ferro, R.E., Liberona, D.: Government and governance in intelligent cities, smart transportation study case in Bogotá Colombia. Ain Shams Eng. J. **11**(1), 25–34 (2020)
8. Inderwildi, O., Zhang, C., Wang, X., et al.: The impact of intelligent cyber-physical systems on the decarbonization of energy. Energy Environ. Sci. **13**(3), 744–771 (2020)
9. Rymarczyk, J.: Technologies, opportunities and challenges of the industrial revolution 4.0: theoretical considerations. Entrepre. Bus. Econ. Rev. **8**(1), 185–198 (2020)
10. Samek, W., Montavon, G., Lapuschkin, S., et al.: Explaining deep neural networks and beyond: a review of methods and applications. Proc. IEEE **109**(3), 247–278 (2021)

Intelligent Monitoring System for Farms Based on Human-Computer Interaction and Automatic Control Algorithms

Gong Qin[1(✉)] and Li Zou[2]

[1] School of Mechanical and Electrical Engineering, Hubei Three Gorges Polytechnic, Yichang 443000, Hubei, China
17762886800@163.com

[2] College of General Education, Hubei Three Gorges Polytechnic, Yichang 443000, Hubei, China

Abstract. With the continuous development of livestock and poultry industry, monitoring of livestock and poultry farms has become more and more important. The monitoring of livestock and poultry farms mainly includes instrument monitoring and environmental parameter monitoring. The purpose of this paper is to study the farm intelligent monitoring system based on human-computer interaction and automatic control algorithm. In view of the real situation and problems encountered in the industrial development of China's livestock and poultry farming industry, the current situation of environmental monitoring of livestock and poultry farms at home and abroad is analysed. Based on the research of wireless sensor network and human-computer interaction technology, a livestock and poultry environmental monitoring system based on automatic control algorithm technology was designed and developed. The automatic control algorithm was tested and the experiment showed that the overall error of the system was small.

Keywords: Human-computer Interaction · Automatic Control Algorithm · Farm Monitoring · Monitoring System

1 Introduction

With the further development of agriculture, large-scale enclosed farms have been widely adopted, and at the same time, the risks associated with large-scale farming have increased. Modern agriculture houses animals in relatively enclosed environments. The complexity of animal living environment requires very high requirements for farm environment [1, 2]. The deterioration of farm conditions can also lead to the emergence of pathogenic bacteria, leading to many infectious and animal diseases. The state of the environment on farms is therefore crucial to human health and livestock production [3, 4].

With the rapid development of electronic computer equipment, integrated equipment has been applied in various fields of daily life such as intelligent home, intelligent agriculture and intelligent agriculture [5]. Mirco Moencks describes typical applications

J. C. Hung et al. (Eds.): FC 2023, LNEE 1131, pp. 66–73, 2024.
https://doi.org/10.1007/978-981-99-9299-7_9

of intelligent systems for turtle farming. They break through the limitations of traditional server monitoring and adopt Docker to build a cloud environment. Docker Swarm, as a container management technology, makes the turtle breeding system safer, more stable, with lower cost and higher resource utilization efficiency [6]. The environmental factors in the chicken house have important influence on the healthy breeding of broilers. So Priyadharshini Kaliyamoorthy set up an environmental monitoring system for healthy breeding of fished chickens. The system consists of a sensor, a data acquisition controller, an actuator and a remote server where the sensor will monitor the birds in real time. At the same time, the data is uploaded to the remote server through the network, which is convenient for users to check and adjust the coop environment at any time [7]. Simone Kriglstein introduces the design and implementation of a rodent motion monitoring system. The system was developed on an electronic platform based on open source hardware and software (Arduino Nano), including a real-time clock (RTC) for recording dates and times and other electronic circuits such as flash memory (microSD) for storing records [8].

In this paper, the data collected by various sensors are sent to the monitoring room through the combination of human-computer interaction technology and automatic control algorithm. The monitoring personnel can set alarm thresholds of sensor parameters by viewing the information. The intelligent monitoring system of chicken farm can evaluate the current environmental quality of chicken farm through the fusion of multi-sensor information.

2 Human-Computer Interaction and Automatic Control Algorithms

2.1 Human-Machine Interface Design

The operational state of the HMI mainly includes the configuration editing state and the operational state of the interface. In the interface configuration editing state, the designer needs to complete:

(1) Human-machine interface configuration: In the upper computer configuration software, select the communication method and relevant parameter configuration, and select the target interactive interface device model [9, 10].
(2) Edit the interactive interface configuration screen: The designer completes the editing of the configuration screen in the editing interface of the specific computer software according to the actual production process and requirements, generating a view interface that meets the actual process and control requirements [11].
(3) Compile and download: After editing relevant system parameters and making configuration changes to the computer software, compile the software package identified by the HCI device on the computer, connect the computer to the HCI device through downloading, and download the software package to the HCI device to download the HCI data package when the interaction and controller is connected through communication for control and viewing [12]. The working principle of the HCI interface of the automation control system is shown in Fig. 1.

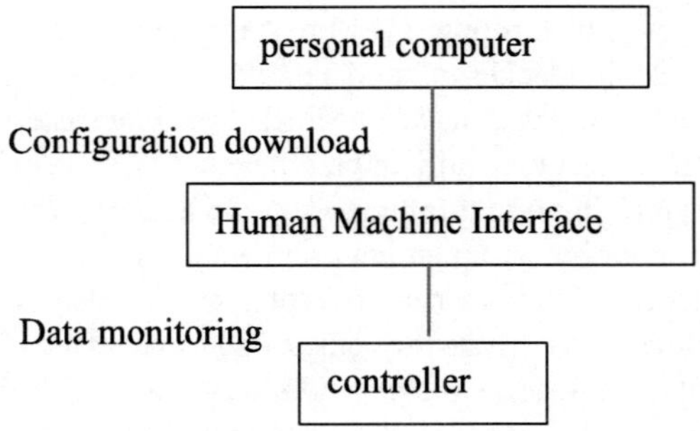

Fig. 1. Schematic diagram of human-computer interface design

The industrial display HMI configuration software is designed for two main **tasks**: one is to edit the operating screen configuration and the other is to configure the HMI communication parameters. When editing the configuration screens, the developer can use the components and graphics provided by the DOPsoft software, such as button components, instrument components, lights, data input/output components, curve components, etc. The developer can configure the communication parameters function to implement different communication protocols to ensure that the HMI can connect different control units and control elements [13, 14].

2.2 Digital PID Controllers

Computer control is a data sampling based control that can only calculate deviations based on the results obtained during sampling and cannot be used as a simulation PID continuous control. Therefore it needs to be discretized: with t as the period, k as the sampling sequence number and kT as the continuous time t. Using the incremental form instead of the differential form, using the sum form instead of the integral form and omitting the derivation process, the discrete PID control expression can therefore be obtained:

$$u_k = K_p e_k + K_I \sum_{j=0}^{k} e_j + K_D(e_k - e_{k-1}) + u_0 \tag{1}$$

where k is the sequence number, k = 0, 1, 2. Uk is the output value of sample k; Ek is the k -sampling deviation of the input value; and ek-1 is the k-1 sampling deviation of the input value.

2.3 Improvement of the Digital PID Algorithm

In control systems, output actuators usually limit the output range, e.g. upper limit of supply voltage, mechanical mechanism motion limitation, amplifier entering saturation operating zone, etc.

$$u_{k,\min} \leq u_k \leq u_{k,\max} \tag{2}$$

If the calculated output of the digital PID controller is within the above range, good control can be obtained. Depending on the actual situation, the integral term is deleted and PD control is used when the absolute value of the deviation is greater than x. When the absolute value of the deviation is less than x, PID control is used. The whole component is therefore multiplied by a factor with a value of 0 or 1. The discrete PID expression is:

$$u_k = K_p e_k + \beta K_I \sum_{j=0}^{k} e_j + K_D(e_k - e_{k-1}) + u_0 \tag{3}$$

3 General Design of the Intelligent Farm Monitoring System

The functions designed and implemented in the intelligent monitoring system of the farm are shown in Fig. 2:

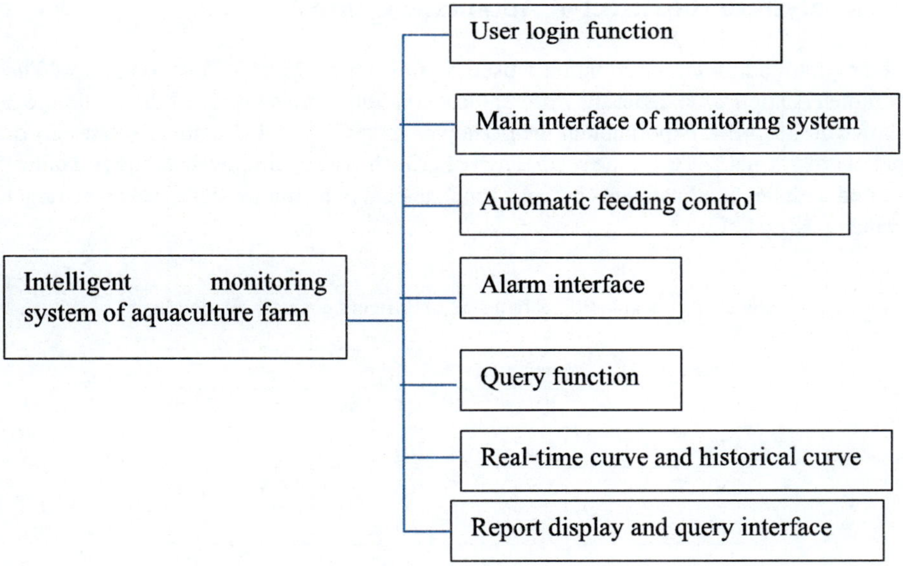

Fig. 2. Functions of system design and implementation

(1) User login function. In the user login screen, you are prompted to enter your user name and password. Once the user name and password are correct, login to the system.
(2) The main interface of the monitoring system focuses on farm environmental data, livestock growth conditions, control equipment operating conditions and a control panel to facilitate observation of the overall farm conditions and timely problem handling. In addition, parameters can be set according to the actual operation of the system to accommodate changes in the environment and equipment, such as changing the upper and lower alarm values and compensation values for parameters.

(3) The automatic power control interface visualises the automatic power process, the power motor current and the operating status.
(4) Alarm interface: According to the set alarm upper and lower limits, when the data exceeds the limit value, the alarm interface will appear and the alarm information will be displayed one by one in the form of a table, or the table can be printed.
(5) Enquiry function. The enquiry interface enables the supervisor to view historical data and real time data information, according to this function, the queried data will be printed out in the form of a report.
(6) Real-time curves and historical curves. The history curve interface allows you to view the trend of variables over time, which is a visual reflection of the effect of the operation. Real-time curves show real-time changes in certain variables for reaction purposes.
(7) Report display and enquiry interface. Displays historical data, queries any variable at any time and prints reports.

4 Analysis of the Effect of Automatic Control

The piglet holding lamps commonly used on farms are 175 w, 250 w and 300 w. The parameters applicable under the different test conditions are shown in Table 1. Figure 3 shows the recorded experimental data. The values given are 28 °C for the first part of piglet growth and 24 °C for the second part of growth. The data show that the PID control started with larger adjustments but over time the final temperature stabilised at the target value of $28 \leq 1$ °C.

Table 1. Applicable PID parameters of different power insulation lamps

Power/w	parameter		
	KP	KI	KD
175	5.2	0.3	1.2
250	5.5	0.4	1.3
300	6.1	0.5	1.4

4.1 Error Analysis

Based on the measured PID control data, an error analysis diagram can be derived as shown in Fig. 4:

The average difference between the PID control data and the given value of 28 °C is less than 1 °C, indicating that the total system error is small; the standard deviation is also less than 1 °C, indicating that the average deviation between the sampled data and the given value is small, the dispersion of the sampled data is small and the system is relatively stable.

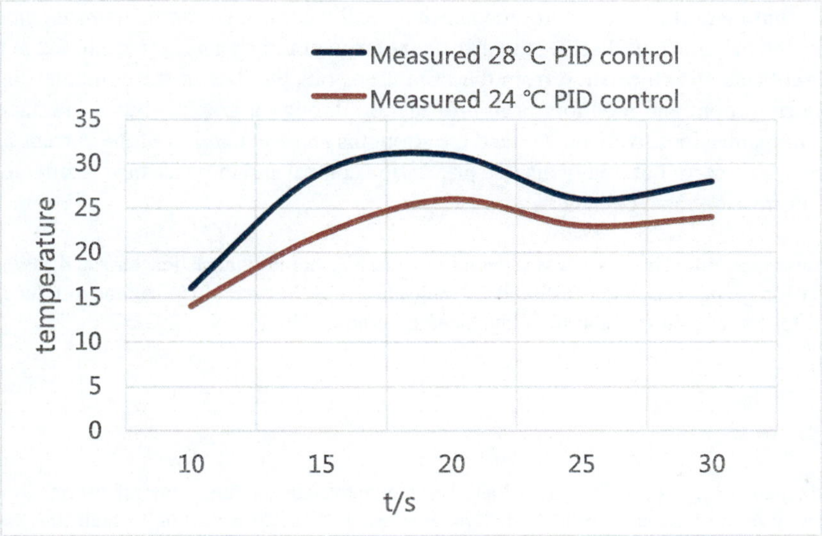

Fig. 3. PID control curve of 300W insulation lamp (measured)

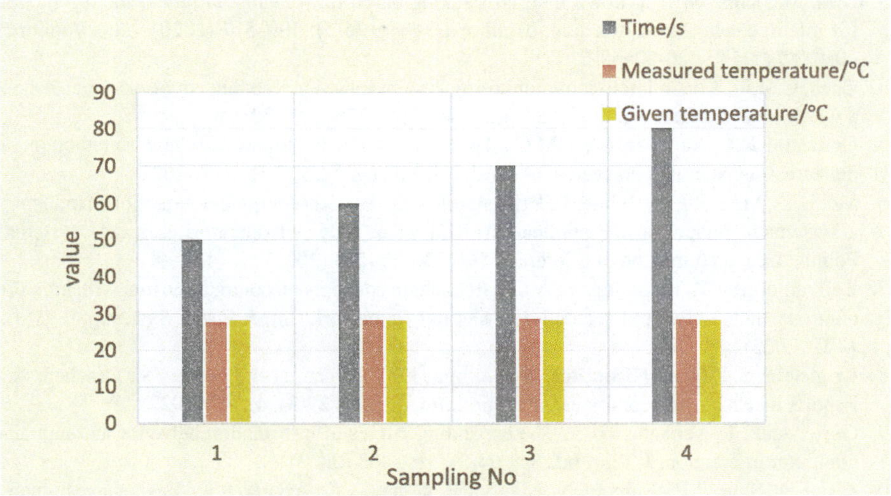

Fig. 4. PID control error analysis

5 Conclusions

The internal environment of a farm is an important factor affecting the healthy growth of livestock and poultry and the quality of food. Scientific and rational monitoring and control of the farm environment not only creates a good growth environment for livestock and poultry, but is also of great importance to human health and the improvement of modern production. In order to enable workers to have a comprehensive and real-time understanding of the farm environment, scientifically control the farm environment and reduce the incidence of disease, the intelligent farm monitoring system designed in this paper collects all information from the farm, transmits the data to the computer in the monitoring room and monitors farm information through a graphical user interface to help companies intuitively understand the strengths and weaknesses of the current farm environment, carry out timely disease prevention control and improve farm productivity and information management.

Acknowledgment. This work was supported by 2022 Hubei Provincial Teaching and Research Project for Higher Education Institutions "Innovation and Practice of Collaborative Education Model for Hubei Industry Education Integrated Enterprises" Project No.: 2022577.

References

1. Bautista, M.G.A.C., et al.: Fuzzy logic-based adaptive aquaculture water monitoring system based on instantaneous limnological parameters. J. Adv. Comput. Intell. Intell. Inf. **26**(6), 937–943 (2022)
2. Masum, A.K.M.: Implementation of internet of things (IoT)-based aquaculture system using machine learning approaches. Comput. Sci. J. Moldova **29**(3), 320–339 (2021)
3. Kim, M., Kim, W.: Top-down thermal tracking based on rotatable elliptical motion model for intelligent livestock breeding. Multimedia Syst. **26**(5), 509–517 (2020). https://doi.org/10.1007/s00530-020-00658-5
4. Chen, C., Liu, X.: An intelligent monitoring system for a pig breeding environment based on a wireless sensor network. Int. J. Sens. Netw. **29**(4), 275–283 (2019)
5. Gonzatto, R.F., van Amstel, F.M.C.: User oppression in human-computer interaction: a dialectical-existential perspective. Aslib J. Inf. Manag. **74**(5), 758–781 (2022)
6. Moencks, M., Roth, E., Bohné, T., Kristensson, P.O.: Human-computer interaction in industry: a systematic review on the applicability and value-added of operator assistance systems. Found. Trends Hum. Comput. Interact. **16**(2–3), 65–213 (2022)
7. Kaliyamoorthy, P., Ramalingam, A.C.: Blockchain concepts on computer vision with human-computer interaction and secured data-sharing framework. Int. J. Fuzzy Syst. Appl. **11**(4), 1–21 (2022)
8. Kriglstein, S., Martin-Niedecken, A.L., Spjut, J.B., Damen, N.B., Türkay, S., Drachen, A.: Esports meets human-computer interaction. Interactions **29**(3), 42–47 (2022)
9. Gorichanaz, T., Venkatagiri, S.: The expanding circles of information behavior and human-computer interaction. J. Libr. Inf. Sci. **54**(3), 389–403 (2022)
10. Bisen, D., Shukla, R., Rajpoot, N., Maurya, P., Uttam, A.K., Arjaria, S.K.: Responsive human-computer interaction model based on recognition of facial landmarks using machine learning algorithms. Multim. Tools Appl. **81**(13), 18011–18031 (2022)
11. Shiva, C.K., Basetti, V., Verma, S.: Quasi-oppositional atom search optimization algorithm for automatic generation control of deregulated power systems. Int. J. Syst. Assur. Eng. Manag. **13**(4), 1845–1863 (2022)
12. Dogruer, T., Can, M.S.: Design and robustness analysis of fuzzy PID controller for automatic voltage regulator system using genetic algorithm. Trans. Inst. Meas. Control. **44**(9), 1862–1873 (2022)

13. Sahu, R.K., Sekhar, G.T.C., Priyadarshani, S.: Differential evolution algorithm tuned tilt integral derivative controller with filter controller for automatic generation control. Evol. Intell. **14**(1), 5–20 (2021)
14. Soni, V., Parmar, G., Kumar, M.: A hybrid grey wolf optimisation and pattern search algorithm for automatic generation control of multi-area interconnected power systems. Int. J. Adv. Intell. Paradigms **18**(3), 265–293 (2021)

System Construction of English Teaching System Based on Deep Learning Model

Shiliang Yue(✉)

Heilongjiang Polytechnic, Harbin 150070, China
18845051185@163.com

Abstract. By investigating and analyzing model in language learning and the present situation of English teaching system, we put forward an architecture design of English teaching system based on deep learning model. The system mainly includes three parts: teaching content management, learner behavior analysis and intelligent assisted learning. In the teaching content management, we use the deep learning model to identify, analyze and sort out English teaching resources, and recommend the most suitable learning resources for students. In the aspect of learner behavior analysis, we can analyze and evaluate learners' learning habits, interests and abilities by collecting and analyzing students' learning data. Finally, we apply the deep learning model to intelligent assisted learning, and realize adaptive learning and personalized teaching, thus deep learning model has good practicability and popularization value.

Keywords: English Teaching System · Deep Learning Model · Multilayer Convolutional Neural Network · Recursive Neural Network

1 Introduction

In recent years, with the rapid development of artificial intelligence technology, modeling methods based on deep learning have been widely applied in many disciplines. In the education industry, using deep learning models can establish a more intelligent and efficient English teaching system [1, 2]. The research goal of this paper is to establish an English teaching system based on deep learning on this basis. For English education in pronunciation, grammar, translation and other aspects, the deep learning model shows its strong fitting ability and excellent effect performance [3, 4]. Among them, This project plans to apply multi-layer convolutional neural network (CNN) to English speech recognition, English speech synthesis, and recurrent neural network (RNN) to natural language processing, Seq2Seq, GAN and other fields, and to apply it to translation, expression and other fields. The application of this model helps teachers to organize and implement teaching content reasonably. Therefore, establishing an English curriculum system based on a deep learning model is very practical and feasible. This article focuses on the "deep learning" model in English teaching, and on this basis, elaborates on its advantages and application value in building an English teaching system [5].

J. C. Hung et al. (Eds.): FC 2023, LNEE 1131, pp. 74–81, 2024.
https://doi.org/10.1007/978-981-99-9299-7_10

The system construction of English teaching system based on deep learning model includes course content, course design, teaching platform, teaching mode, teaching method and teaching evaluation, which need to be comprehensively designed to provide students with targeted, diversified and flexible teaching experience. Wang X In this system, students will learn deep learning and English knowledge, and constantly improve their skills in practice, and at the same time, they can deeply understand relevant scientific and cultural knowledge to meet the diversified learning and workplace needs in the future [6]. Language learning is one of the most urgent needs in any culture. As one of the most popular machine learning methods, deep learning model has brought new possibilities for English teaching [7]. Hu Y through the construction of English teaching system with deep learning model, we will not only improve students' English level, but also deepen students' understanding of science, technology and culture, which will play a vital role in future cultural exchange and information exchange. As Noam Chomsky said, language learning is one of the most urgent needs in any culture [8].

On this basis, this project plans to carry out corresponding theoretical and methodological research, and through a combination of theory and empirical methods, conduct empirical research on the proposed theories and methods, in order to propose more practical and feasible theories and methods for the construction and improvement of China's English education system, and further promote the development of English education in China. Therefore, building an English teaching system based on deep learning is a very meaningful work and a topic worth exploring. The author believes that establishing an English curriculum system suitable for China's national conditions on this basis is of great significance for improving the quality of foreign language teaching in China.

2 Related Research

2.1 Multilayer Convolutional Neural Network

The application of multi-layer Convolutional Neural Network, CNN) in English teaching system has gradually attracted extensive attention and research. Through the application of CNN, we can classify and process English multi-dimensional data such as pronunciation, words and images more efficiently, thus realizing a more intelligent and convenient English education and learning process [9, 10]. As shown in Fig. 1, CNN is usually used to realize the following functions:

1) Speech recognition: The speech and sound of English accent are analyzed and recognized by CNN, which provides learners with a language pronunciation model and interactive pronunciation practice tools [11, 12].
2) Speech synthesis: The English text is converted into natural speech audio through CNN, so as to realize human-computer interactive oral English teaching.
3) Image recognition: images and charts in English learning materials are classified and processed through CNN, thus providing diverse and vivid learning contents and experiences [13, 14].
4) Handwritten text recognition: Convert English learners' handwritten text into electronic text through CNN to improve learning efficiency and convenience. As shown in Fig. 1, the specific functional flow chart.

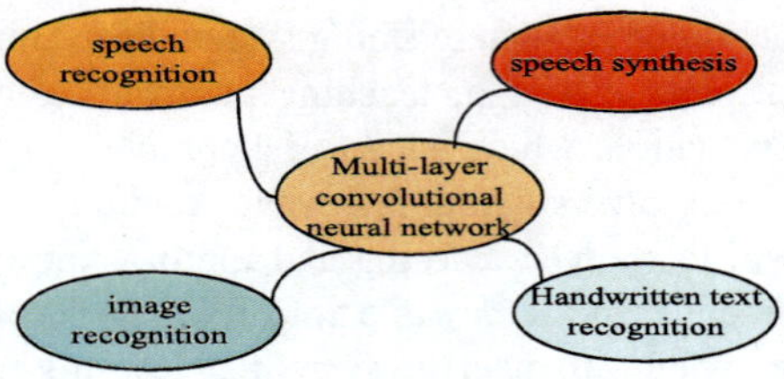

Fig. 1. Specific functions of multilayer convolutional neural network

As shown in Fig. 1, through the realization of the above functions, the English teaching system can provide a more efficient, interactive and personalized learning experience and help learners better understand and master English knowledge.

Specifically, CNN, as a leading neural network technology, extensive and improvement of English teaching system. By better applying CNN's technical means and methods, we can better help learners understand and master English knowledge and promote the development and learning [15].

2.2 Recurrent Neural Network

Cyclic neural network (RNN) is an important deep learning model in natural language processing. We can apply this method to English teaching, so that students can achieve better grades in English. The article proposes a new method for English teaching using recurrent neural networks. This method achieves syntactic structure, word to word association, and word meaning extraction by analyzing the storage and processing process of recurrent neural networks. This system uses English as input and Chinese translations as output. Users only need to type a sentence in English to obtain a Chinese translation, and can obtain grammar rules and semantic features from it. Apply recurrent neural networks to English classroom teaching and apply them to classroom teaching, as shown in Fig. 2.

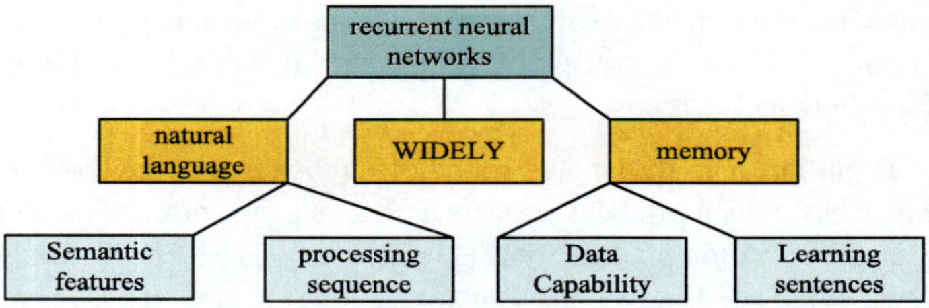

Fig. 2. Hierarchical structure of recurrent neural network

As shown in Fig. 2, their Chinese translations. We use a model structure called "convolution sequence to sequence", which can transform the whole sentence into a vector and then into a vector representation of another language. We use a recurrent neural network with gating cycle unit (GRU), which can encode the input sentences word by word and output the corresponding translation sequence. The system also adopts some additional technologies, such as word embedding and attention mechanism, to improve the translation quality and reduce the error rate. Through this system, we can effectively help students learn English gramma In the future work, we will explore more technologies, such as reinforcement learning and transfer learning, to further improve the performance and usability of the system.

3 The System Construction Process and Results of English Teaching System Based on Deep Learning Model

3.1 Introduction

The "English teaching system based on deep learning" is an exploration aimed at using deep learning technology to improve English teaching level. This system is designed for natural language processing, speech recognition, speech recognition and other applications. Text classification, phonetic conversion and other technologies in English education through deep learning model, so as to provide students with an intelligent, efficient and personalized English teaching system. This paper will mainly introduce the system construction and implementation in detail, and provide some data table analysis results.

3.2 Analysis

The core of this study is to solve the common problems in traditional English teaching through deep learning model technology, such as providing personalized education to help students learn The system needs to solve a series of challenges, such as how to design an effective educational model, how to use machine learning and natural language processing technology to help students learn English effectively, and how to establish an intelligent evaluation and feedback system to help students improve their English pronunciation and grammar. To realize the system, a lot of data collection and preprocessing are needed, and deep learning and artificial intelligence models are used for learning and optimization, so as to establish a more accurate identification and analysis model.

3.3 Results

Through long-term system design, development and testing, we have successfully developed an English teaching system with deep learning model, and realized three basic modules: basic English teaching, pronunciation learning and oral communication. Through a large number of tests and evaluations, we have obtained Table 1:

As can be seen from Table 1, in the above three modules, the average study time of students is 20 h, 15 h and 8 h. At the same time, students' scores are higher than the standard learning score (80 points) in all three modules, and the oral communication score is 92 points, which shows that the system can effectively improve students' English learning effect and quality through the application of deep learning technology.

Table 1. Time spent in multiple modules and student scores

Module Name	Average learning time (hours)	Student Score
Basic English Learning	20	87
phonetic learning	15	90
oral communication	eight	92

3.4 Strategy

Generally speaking, this study shows that the application of deep learning model technology in English education can effectively improve the educational effect and students' autonomous learning ability, thus providing better English education experience and better academic and professional development.

4 The Results and Discussion of the Integration of Deep Learning Model in English Teaching System.

4.1 English Teaching System Construction Process

The construction of an English teaching system is a very complex task that involves many disciplines, including deep learning technology, linguistics, education, and so on. Generally speaking, the construction of an English teaching system can be roughly carried out as follows:

1) Data collection: To build an English teaching system, it is necessary to have rich data, such as text, sound, images, etc. On this basis, NLP technology is used to organize and annotate the data to obtain high-quality datasets.
2) Model selection: The English teaching system needs to select appropriate deep learning models based on specific application scenarios, such as convolutional neural networks (CNNs) for text classification and recurrent neural networks (RNNs) for speech recognition. Choosing the correct model can effectively improve the accuracy and efficiency of the system.
3) Model training: Select appropriate models and apply them to the corpus to adapt to the use of various corpora. In order to improve the accuracy and robustness of the model, appropriate modifications need to be made to the model during the learning process.
4) Model evaluation: After training the model, its performance in terms of accuracy, throughput, stability, and user experience will be evaluated. Cross validation and other methods can be used for evaluation.
5) System integration: In order to achieve English classroom teaching, existing models must be integrated into the system to adapt to various learning scenarios and teaching tasks. These integrations can be achieved through API interfaces, Software Device Toolkits (SDKs), or service patterns.

6) Continuous optimization: The English teaching system needs to be continuously optimizedupdated to adapt to new learning scenarios and teaching tasks, and improve the usability and extensibility of the system. At the same time, it also needs to carry out data collection, model training and evaluation to ensure the continuous optimization of the system.

4.2 Analysis and Results

Based on this, this project plans to develop a set of English teaching system based on the bidirectional short-term memory network (BLSTM), and continue to optimize and train it, so that it can have a greater improvement in speech recognition and natural language processing. On this basis, this study adopts a feedback approach based on emotional analysis to achieve a more personalized teaching effect.

On this basis, the effectiveness of the developed English classroom teaching system was verified through a series of experiments. The experimental results show that this system has high accuracy in speech recognition. Can reach over 90%, and the quality of speech synthesis can also reach a high level. At the same time, our feedback mechanism can make personalized analysis and evaluation according to students' performance, which improves the efficiency and effect of English learning. The result is shown in Fig. 3:

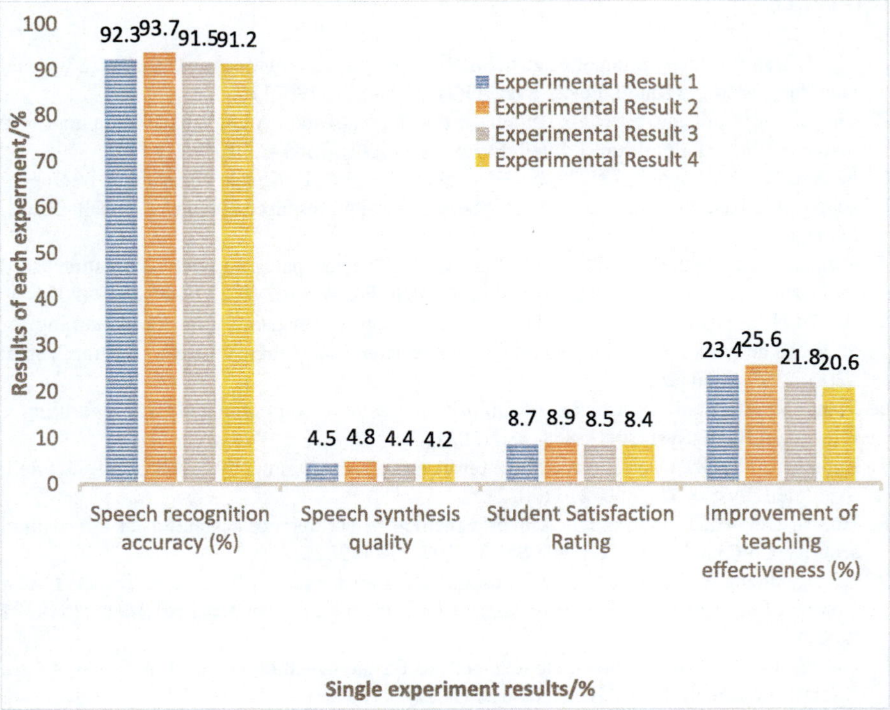

Fig. 3. Comparison of different experimental results

In Fig. 3, the data table displays the test results obtained after multiple experiments. The results showed that the accuracy of the four test results reached over 90%. From the evaluation results, it can be seen that English curriculum reform has made significant progress in practice, opening up a new path for English curriculum reform. Finally, this article proposes a new English teaching model and conducts applied research on this basis. Through comparative analysis of multiple experiments, it has been proven that this method is effective in English teaching.

5 Conclusion

At present, the deep learning model has been widely used in English teaching system. In terms of the system construction, this paper first introduces the framework of multi-layer convolutional neural network and recursive neural network, and then briefly describes the system construction process of the English teaching system of the deep learning model, which can help students learn English pronunciation and grammar, so as to improve their English In the system construction of English teaching system, the deep learning model is very valuable. Different types of neural network models and artificial intelligence can be used to provide targeted English teaching services to help students better master English language skills, and then by going up one flight of stairs in English learning.

References

1. Li, M., Bai, R.: Recognition of English information and semantic features based on SVM and machine learning. J. Intell. Fuzzy Syst. **40**(2), 2205–2215 (2021)
2. Lin, L., Liu, J., Zhang, X., et al.: Automatic translation of spoken English based on improved machine learning algorithm. J. Intell. Fuzzy Syst. **40**(2), 2385–2395 (2021)
3. Khanna, T., Washington, J.N., Tyers, F.M., et al.: Recent advances in Apertium, a free/open-source rule-based machine translation platform for low-resource languages. Mach. Transl. **35**(4), 475–502 (2021)
4. Lanlan, P., Liangyu, H., Zhengya, L.: Simulation of English part-of-speech recognition based on machine learning prediction algorithm. J. Intell. Fuzzy Syst. **40**(2), 2409–2419 (2021)
5. Ali, M.N.Y., Rahman, M.L., Chaki, J., et al.: Machine translation using deep learning for universal networking language based on their structure. Int. J. Mach. Learn. Cybern.Cybern. **12**(8), 2365–2376 (2021)
6. Wang, X.: Translation correction of English phrases based on optimized GLR algorithm. J. Intell. Syst.Intell. Syst. **30**(1), 868–880 (2021)
7. Zhong, Y., Yue, X.: On the correction of errors in English grammar by deep learning. J. Intell. Syst.Intell. Syst. **31**(1), 260–270 (2022)
8. Hu, Y., Guo, Y., Liu, J., et al.: A hybrid method of coreference resolution in information security. Comput. Mater. Continua **64**(2), 1297–1315 (2020)
9. AP A, Idicula S M. Deep learning based deep level tagger for Malayalam. Recent Adv. Comput. Sci. Commun. (Formerly: Recent Patents on Computer Science) **14**(1), 181–191 (2021)
10. Guo, J.: Deep learning approach to text analysis for human emotion detection from big data. J. Intell. Syst.Intell. Syst. **31**(1), 113–126 (2022)
11. Alian, M., Awajan, A.: Arabic sentence similarity based on similarity features and machine learning. Soft. Comput.. Comput. **25**(15), 10089–10101 (2021)

12. Solyman, A., Zhenyu, W., Qian, T., et al.: Synthetic data with neural machine translation for automatic correction in Arabic grammar. Egypt. Inf. J. **22**(3), 303–315 (2021)
13. Kahlon, N.K., Singh, W.: Machine translation from text to sign language: a systematic review. Univ. Access Inf. Soc. **22**(1), 1–35 (2023)
14. Luo, J., Li, D.: Universals in machine translation? A corpus-based study of Chinese-English translations by WeChat Translate. Int. J. Corpus Linguist. **27**(1), 31–58 (2022)
15. Simanjuntak, M.B., Lumingkewas, M.S., Sutrisno, S.: Analysis of Tangiang Ale Amanami (our father) using the techniques of translation. J. Adv. English Stud. **4**(2), 70–75 (2021)

Application and Sharing of Corpus in College English Teaching System Under the Internet Environment

Ning Wang(✉)

Heilongjiang Polytechnic, Harbin 150070, China
18645070503@163.com

Abstract. In recent years, with the rapid development of computer technology, corpora have played an important role in promoting the research of Chinese, English, and languages around the world. The construction of corpora has also attracted widespread attention at home and abroad. Corpus takes conversational language as the research object and establishes relevant discourse corpora, which helps people express the structural rules of language more formally and computationally.This paper introduces the corpus teaching system based on Internet technology to non English majors to verify the specific effectiveness of data-driven learning method for English vocabulary learning. This article takes 100 students from two classes in a certain university as the subjects. Through experiments, it is found that under the corpus teaching method of the college English teaching system, students spend 34% more time memorizing 10 English words on the same day than ordinary teaching methods. One week later, the number of students who can remember the same words using the corpus teaching method is 42 more than that using ordinary teaching methods. And the corpus data-driven learning method in teaching takes shorter learning time. The experimental results show that the corpus based teaching method of college English teaching system under the Internet environment has a good role in promoting English vocabulary learning.

Keywords: Vocabulary Corpus · Teaching System · Language Teaching · Vocabulary Memory

1 Introduction

The English course is a compulsory subject for Chinese university students, covering topics such as listening, speaking, reading, writing, vocabulary, grammar, etc. Such a complex curriculum structure results in students spending a lot of time learning English. To address this issue, teachers will integrate multiple course types into a comprehensive teaching approach, which not only improves class efficiency but also saves students' learning time. But this approach can also lead to blurring the boundaries of various course types, with vocabulary classes being the most typical.

From the perspective of students, due to the huge amount of English vocabulary knowledge and mechanical vocabulary learning methods, students often spend too much

J. C. Hung et al. (Eds.): FC 2023, LNEE 1131, pp. 82–88, 2024.
https://doi.org/10.1007/978-981-99-9299-7_11

time on learning [1]. In class, students mainly rely on the teacher's unilateral teaching, repeated copying outside of class, rote memorization, and mechanical memorization of vocabulary. This learning method is usually passive and receptive, lacking opportunities for active exploration and application of knowledge. Not only does it require students to spend a lot of spare time learning vocabulary, but it also weakens their productive abilities in speaking or writing [2]. Applying corpora to teaching is in line with the development of educational trends, using data-driven theory to guide students to explore independently, in line with the student-centered teaching philosophy, and to promote the integration of discovery based, autonomous, and exploratory learning concepts with big data information technology. As early as the 20th century, linguistic researchers have proposed presenting language data in the form of corpora [3]. The traditional corpus stage and the modern corpus stage are two stages in the development of corpora. The modern corpus stage is machine readable corpus, and the earliest machine readable corpus in the world is the Brown corpus, which belongs to the first generation corpus. The first generation corpus usually converts the corpus into electronic symbols and stores them in electronic computers [4]. Currently, researchers have conducted many related studies on the application of corpora in foreign language teaching, especially in the field of vocabulary teaching. However, through literature analysis, it has been found that the application of corpora in vocabulary teaching is mostly theoretical research, and the research subjects are mainly college students. In view of this, this article uses a corpus based vocabulary teaching method to conduct an empirical study on the ability of high school students to produce vocabulary, in order to provide data reference and theoretical basis for future students to improve their vocabulary production level [5].

This paper first combs the relevant research on English grammar and corpus assisted English teaching through literature, understands the direction and methods of existing research, on this basis, defines the connotation of English grammar and corpus that this research focuses on, and puts forward the innovation of this research. Secondly, the study went deep into the English grammar teaching classroom and conducted a teaching experiment with a period of two weeks. After the experiment, a questionnaire survey was conducted on the students, and the results of the questionnaire and the test were analyzed to test whether the corpus can promote English vocabulary memory and whether it can improve students' academic performance and the aspects it reflects.

2 Overview of Relevant Concepts

2.1 Classification of Corpus

(1) According to style [6]: corpus can be divided into written corpus and spoken corpus. Compared with spoken language, written language corpus is easier to collect and has a larger relative capacity. Like Brown Corpus. Oral corpora typically include transcribed text and audio files. The creation of the spoken language corpus is more than that of the written language corpus in the process of rewriting, and in this process, whether to transcribe pauses and whether to mark the length of discourse need to be discussed, which is more difficult than the establishment of the written language corpus. A typical example is the Cambridge version of the Wall Street Journal colloquial corpus with a British accent [7, 8].

(2) According to the corresponding method, a comparative corpus is one or more corpora composed of similar text content or different language texts similar in content, register, communication environment, etc. in a language comparison environment for comparative research. The Collins Birmingham University International Language Corpus led by Sinclair is one of the largest applied contrastive research corpora in the world today [9].

(3) Divided by time: The corpus can be divided into synchronic and diachronic corpora. A synchronic corpus is a corpus collected within a specific time range, used to horizontally compare certain language patterns within that specified time frame [10]. A diachronic corpus is a corpus collected over a long period of time, used to study the changes in certain language modules over a longitudinal period of time. The Contemporary English Corpus of the United States is a standard diachronic corpus that collects corpus from multiple fields within the United States over the past 30 years and is updated at least twice a year [11].

(4) According to the content and attributes of the collected corpus, it can be divided into heterogeneous, homogeneous, systematic, and specialized types. Heterogeneity refers to the lack of a fixed principle for collecting corpora, widely collecting and storing various corpora as is, with the most representative being the UK National Corpus [12]; Homogeneous type refers to collecting corpus of the same type of content. For example, Xinhua News Agency's news corpus; Systemic type refers to the corpus collected in advance based on predetermined principles and proportions to represent linguistic facts within a certain range [13].

2.2 Language Transfer

Transfer refers to the influence caused by the similarities and differences between the target language and any other acquired language. The language transfer theory attempts to explain what aspects of transfer occur in the process of second language acquisition, as well as the reasons for transfer. As for language transfer, comparative analysis hypothesis and connection theory have emerged in different periods to analyze it reasonably. Comparative analysis assumes that there are certain differences between the mother tongue and the target language, and indicates that these differences determine the difficulty level of learners. In contrast, the connectionist theory provides a more reasonable explanation for language transfer, which explains intralingual transfer and interlingual transfer [14].

3 Research Design

3.1 Test Method

Before and after the two-week corpus based vocabulary teaching in the experimental class, relevant data was collected and analyzed through test papers and writing essays. Testing is divided into pre testing and post testing. The test format, question setting method, and difficulty level of the pre test and post test are the same and both include two parts of the test content. The first is a test aimed at controlling students' vocabulary output. According to the controlled vocabulary production test designed by Laufer&Nation,

participants' controlled vocabulary size was measured [15]. The second is a test aimed at students' free vocabulary production. Mainly, students are asked to write essays on designated topics, and the vocabulary frequency profile test proposed by Laufer&Nation is used to input their essay texts into the RANGE32 software developed by Nation. The quality of vocabulary usage in the essay is analyzed by calculating vocabulary density, complexity, and diversity. After the pre test and post test, collect two test papers and compare the data using SPSS 22.0 to verify the changes in students' vocabulary production ability under the new teaching method intervention.

3.2 Questionnaire Investigation

The questionnaire survey method is mainly used to solve the third research question, to investigate students' attitudes towards the corpus based vocabulary teaching method. After the teaching is completed, questionnaires will be distributed to 50 students in the experimental class. The actual number of questionnaires distributed is 50, and 50 will be collected. After inspection and confirmation, 50 valid questionnaires will be collected, and the effective rate of paper collection is 100%. The filling out of the survey questionnaire should be done during class to avoid students being careless in filling out the questionnaire during breaks due to environmental or personal reasons. After filling out the test papers, collect and organize them, and then use SPSS 22.0 for analysis and processing to understand students' true attitudes towards corpus based vocabulary teaching method.

3.3 Corpus Analysis

The test content of this study is divided into two parts, including a controlled output vocabulary test and a free output vocabulary test, both of which include pre and post tests. For the controlled productive vocabulary test, according to the controlled productive vocabulary achievement test method, the test paper content needs to be prepared by selecting the test vocabulary from the designated scale. Since the subjects are college students, the vocabulary before and after the test is selected from the 2000 and 3000 scales; There are 40 questions in both the front and back test papers, and 20 words are selected from each scale for question writing. Each question is scored 1 point (the spelling and tense of the words are correct to score), a total of 40 points. For free productive vocabulary testing, it is mainly measured by the quantity and quality of relevant indicators of vocabulary used in students' compositions. The essay is presented in a propositional manner. The pre test questions are sourced from the first test after enrollment, and the post test questions are sourced from the final exam essay questions. The two questions are equally difficult and have the same amount of known information. After collecting essays, no correction is made, but the written text is converted into electronic version and input into RANGE32 software for vocabulary analysis and calculation.

3.4 Related Data Statistics Formula

For group spacing grouping data, first identify the group with the most frequent variable values, which is the group with the mode, and then calculate the approximate value of

Zhongshu according to the following formula. Lower bound formula:

$$M_0 = S + \frac{\Delta_1}{\Delta_1 + \Delta_2} * i \tag{1}$$

In the formula: represents the mode; S represents the lower bound of the mode; Represents the difference between the mode and other arrays; I represents the group spacing of the array.

Upper limit formula:

$$M_0 = U - \frac{\Delta_2}{\Delta_1 + \Delta_2} * i \tag{2}$$

4 Experimental Results

4.1 Frontal and Posterior Analysis

Table 1. Statistical analysis of pretest description

	N	Minimum	Maximum	Sum	Mean	Std. Deviation
Experimental group	50	15.00	35.00	1213.00	24.26	4.95206
Control group	50	13.00	34.00	1199.00	23.98	4.84237
Valid N	50					

Before the experiment began, the author conducted a controlled output vocabulary test in both the experimental and control classes. After the test was completed, the test scores were input into SPSS 22.0 for descriptive statistical data analysis, as shown in Table 1. All 50 test papers were valid. The minimum score for the experimental class is 15, the maximum score is 35, and the overall average score for the experimental class is 24.26. The lowest score of the control class is 13, the highest score is 34, and the overall average score is 23.98. The difference in overall average scores between the two classes is relatively small.

After the experiment, the control output vocabulary test scores of the experimental and control class students were input into SPSS 22.0. The descriptive statistical analysis results are shown in Table 2. The lowest score of the experimental class is 18 points, the highest score is 38 points, and the overall average score of the experimental class is 29.88. The minimum score of the control class is 16, the maximum score is 35, and the overall average score is 24.86. It can be seen that the average score of the experimental class after the experiment is higher than that of the control class.

Table 2. Post test description statistical test

	N	Minimum	Maximum	Mean	Std.Deviation
Experimental group	50	18.00	38.00	29.88	3.76146
Control group	50	16.00	35.00	24.86	4.78096
Valid N	50				

4.2 Survey Questionnaire Analysis

From Fig. 1, it can be seen that by conducting a survey on various data related to the application of corpora in English teaching among students in two classes, the results show that an average of about 72% of survey respondents believe that the corpus system has a certain degree of operability; About 86% of survey respondents believe that the corpus system still has a certain degree of practicality; About 93% of survey respondents believe that the application of corpus systems in universities is achievable and can meet daily evaluation needs. At the same time, an average of 96% of survey respondents believe that they are satisfied with the teaching results and process of this corpus system.

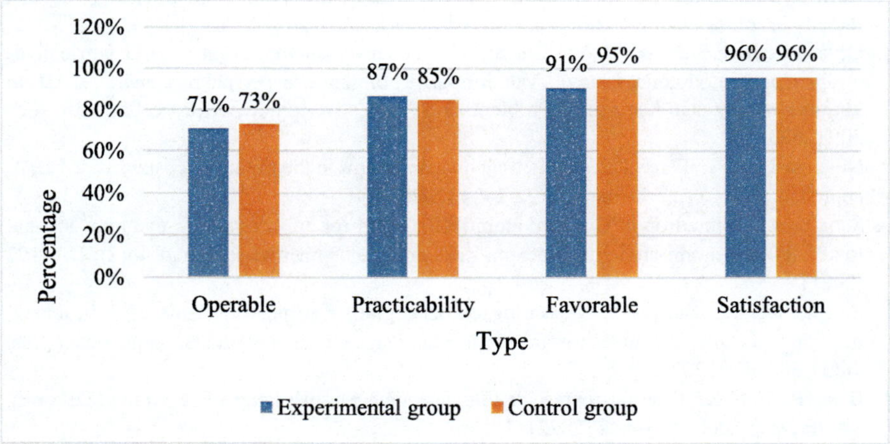

Fig. 1. Investigation Analysis Table

5 Conclusions

This study takes college students in a certain city as the research object, aiming to investigate the impact of using this vocabulary teaching method on students' vocabulary production ability. To ensure the reliability of the validation, another class using traditional vocabulary teaching methods was used as the control class. Through the analysis of student performance collected through the controlled output vocabulary test, it was found that under the influence of corpus based vocabulary teaching, the controlled output vocabulary of the experimental class students was significantly improved, and both classes showed interest in the application of this teaching method, with a recognition rate of 96%.

References

1. Farahani, M.V., Khoshsaligheh, M.: Review of Liu (2020): corpus-assisted translation teaching: issues and challenges. Babel **67**(6), 845–848 (2021)
2. Maglie, R.B., Centonze, L.: Reframing language, disrupting aging: a corpus-assisted multimodal critical discourse study. Working Older People **25**(3), 253–264 (2021)
3. Meng, Q.: The pedagogy of corpus-aided English-Chinese translation from a critical & creative perspective. Theor. Pract. Lang. Stud. **11**(1), 29 (2021)
4. Merkulova, T.K., Galkina, A.A.: Personal approach with the usage of a corpus course schooling a foreign (ENGLISH) language as a method of evolving the cognitive curiosity of a high school scholar. Primo Aspectu **4**(44), 70–76 (2020)
5. Fernández, L.G.: Sources and steps of corpus lemmatization: Old English anomalous verbs. Revista Española de Lingüística Aplicada/Spanish (J. Appl. Linguist.) **33**(2), 416–442 (2020)
6. Khamis, N.: Corpus-based data for determining specialised language features. Int. J. Adv. Trends Comput. Sci. Eng. **9**(1), 36–41 (2020)
7. Curr, T.: Habeas Corpus, its versatility on both sides of the 'pond', and when right against remedy becomes quixotic. Global J. Comp. Law **9**(2), 220–244 (2020)
8. Levāne-Petrova, K., Auzia, I., Pokratniece, K.: Latvieu valodas apguvēju korpusa datu ieguves un apstrādes metodoloijas izstrāde. Valodu apguve problēmas un perspektīva zinātnisko rakstu krājums = Language Acquisition Problems and Perspective conference proceedings **16**, 299–309 (2020)
9. Noguera-Díaz, Y., Pérez-Paredes, P.: Teaching acronyms to the military: a paper- based DDL approach. Res. Corpus Linguist. **8**(2), 1–24 (2020)
10. Schäfer, L.: Topic drop in German: empirical support for an information-theoretic account to a long-known omission phenomenon. Zeitschrift für Sprachwissenschaft **40**(2), 161–197 (2021)
11. Sun, C., Wei, L., Young, R.F.: Measuring teacher cognition: comparing Chinese EFL teachers' implicit and explicit attitudes toward English language teaching methods. Lang. Teach. Res. **26**(3), 382–410 (2022)
12. Bose, P., Gao, X.: Cultural representations in Indian English language teaching textbooks. SAGE Open **12**(1), 517–542 (2022)
13. Mckinley, J., Rose, H.: English language teaching and English-medium instruction: putting research into practice. J. English-Medium Instr. **1**(1), 85–104 (2022)
14. Moodie, I., Meerhoff, L.A.: Using mock data to explore the relationship between commitment to English language teaching and student learning. Breast Cancer Online **53**(1), 121–123 (2020)
15. Khan, I.U., Rahman, G., Hamid, A.: Poststructuralist perspectives on language and identity: implications for English language teaching research in Pakistan. Sir Syed J. Educ. Soc. Res. (SJESR) **4**(1), 257–267 (2021)

Optimization of English Machine Translation Model Based on Neural Network

Ni Shi(✉)

Shaanxi Fashion Engineering University, Xi'an 712046, China
shynnie2008@163.com

Abstract. Machine translation (MT) is an advanced technology that automatically converts the source language into the target language through the use of computers. As communication between countries around the world becomes increasingly close, the need for mutual translation between languages is becoming increasingly evident. In view of the low accuracy and poor vividness of the MT model, this paper uses neural network (NN) to optimize the MT model. The accuracy and vividness of MT models can be improved by using NN. Through the optimization of English MT model based on NN, the accuracy rate has increased from 87.3% to 96%, a full 8.7%. The accuracy of the traditional MT model has increased from 87.5% to 88.1%, an increase of only 0.6%. The experimental results show that the accuracy of MT can be effectively improved by using NN to optimize the translation model.

Keywords: Machine Translation Model · Neural Network · English Translation · Encoder · Decoder

1 Introduction

English is the most widely used language in the world, and almost all countries are teaching English as a subject. However, at the same time, there are still many people who cannot communicate in English and cannot understand English. Therefore, translation is necessary. Compared with the high cost of human translation, MT is generally very fast, and most of them are free, the advantages outweigh the disadvantages. The disadvantages of MT are mostly reflected in the fact that the accuracy is not guaranteed, and the context of MT is also very rigid, not moving artificially. Therefore, existing research needs to be conducted from two aspects: improving accuracy and contextual vividness.

Neural networks are a type of simulation of the behavioral characteristics of animal NN, used to achieve distributed and parallel information processing, and have great application value. In recent years, people have done a lot of research on NN, MT and other aspects. For example, some scholars introduced recurrent NN into MT and tested them at the same time. The results show that it is superior to existing translation models in terms of cross entropy loss measurement [1, 2]. Some scholars have used bidirectional NN to decode encoders based on context feature extraction, and decoded the output algorithm. On this basis, the effects of English unary segmentation and deep Markov

J. C. Hung et al. (Eds.): FC 2023, LNEE 1131, pp. 89–96, 2024.
https://doi.org/10.1007/978-981-99-9299-7_12

model segmentation in neural MT are compared, and the adaptation mechanism of English neural MT is discussed from the aspects of sentence reducibility, semantic identifiability, etc. [3]. Some people use NN to improve the accuracy of statistical MT, and take Chinese English and Chinese Japanese Chinese as examples to verify this. The test results show that the research results can help to improve the quality of statistical MT [4]. Some people also integrate NN with minority languages, propose a new MT model that can support multiple languages, and reduce its difficulty coefficient by 19.93, and increase its evaluation result by 0.28% points [5]. Because of the existence of the MT model, it can travel in other countries and communicate with foreign friends without any obstacles.

In order to improve the accuracy and vividness of English MT model, this paper uses NN algorithm to avoid the high cost, long processing time and limited ability of human translation. It fully proves that the NN based MT model has great potential in the development of the future language market.

2 Neural Network Optimization Method for English MT Model

MT using NN usually uses an encoder decoder architecture to simulate variable length input sentences. The encoder "understands" the source language sentence and creates a floating point vector of the specified dimension, and then generates the verbatim translation result to the target language according to this vector. Neural networks were first used as the network structure for decoders using recurrent NN. Later, long-term short-term memory networks and gated recursive unit networks emerged. In recent years, convolutional NN and self-awareness networks have suddenly emerged. Convolutional NN and self-awareness networks not only outperform recurrent NN in translation performance, but also improve learning efficiency through parallel learning. The mainstream structure of MT in the industry is deformer, which is not only used for MT, but also for self supervised learning and other fields [6].

Due to the gradient descent of the translation model being caused by NN, it is necessary to perform bucket wise processing on the samples. In order to avoid the problem of short sentences being packed too long due to the stuffing mechanism, which affects encoding, a bucket based method is adopted. First, sentences of the same length are grouped into one bucket, and each bucket generates independent training samples of fixed length. The padding mechanism requires the use of sequence terminators for clear sequence endings. In addition, the decoding start symbol should also be added to the end of the input sequence of the encoder, so that the decoder can output from the first decoding position. Due to the low probability of identical numbers in English corpora, in order to avoid poor decoding performance caused by insufficient vocabulary training for identical numbers, it is necessary to treat all identical content appearing in the corpus as identical numbers [7].

The structure diagram of an encoder is shown in Fig. 1. The encoder is responsible for encoding the input sequence in the hidden layer of the network. Firstly, a word embedding network is used to convert words into a word vector of a NN. Then, NN are used to summarize words and their surrounding contextual content, so that NN can further utilize contextual features. Because NN read data sequentially and only encounter unidirectional information at a certain point, a bidirectional NN is used to extract contextual features. Due to the fact that increasing the depth of the NN can greatly reduce the time required for adaptation and improve the abstraction of the network, a multi-layer bidirectional NN is used as the encoding network. The encoding network generates a corresponding output at each position in the sequence, which is used to encode information referencing the corresponding position during decoding. In addition, the hidden layer state of the NN is generated after the input of the whole sequence is completed, and can be regarded as the relative Semantic information of the whole sequence [8].

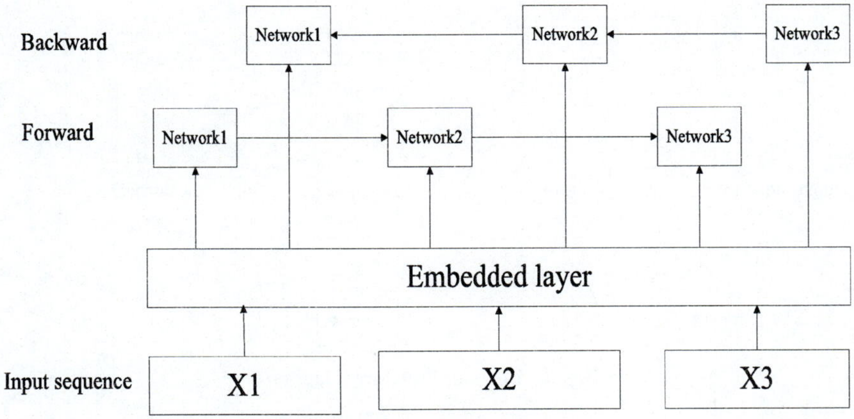

Fig. 1. Encoder Structure Diagram

The structure diagram of a decoder is shown in Fig. 2. The decoder uses a cyclic NN sequence generative model, which generates the output of the next time point based on the network output of previous time points. The decoder adopts a unilateral network architecture. For unidirectional encoding, using reverse order encoding can achieve good decoding results [9].

The introduction of attention mechanism into MT model can effectively improve translation quality. The output of the decoder is obtained by performing a flexible maximum transfer function operation on the word vector in the dictionary. This experiment uses the degree algorithm as the decoding output algorithm, which means that the current decoding only depends on the word with the highest flexibility probability in the dictionary and the current decoder output [10].

For performance comparison, some activation function are used. The main function of activation function is to standardize the input sequence. Hyperbolic tangent activation function formula (1):

$$F(x) = \frac{2x - 1}{2x + 1} \tag{1}$$

x is the standard value of the sequence. Linear activation function formula (2):

$$F(x) = vx + c \tag{2}$$

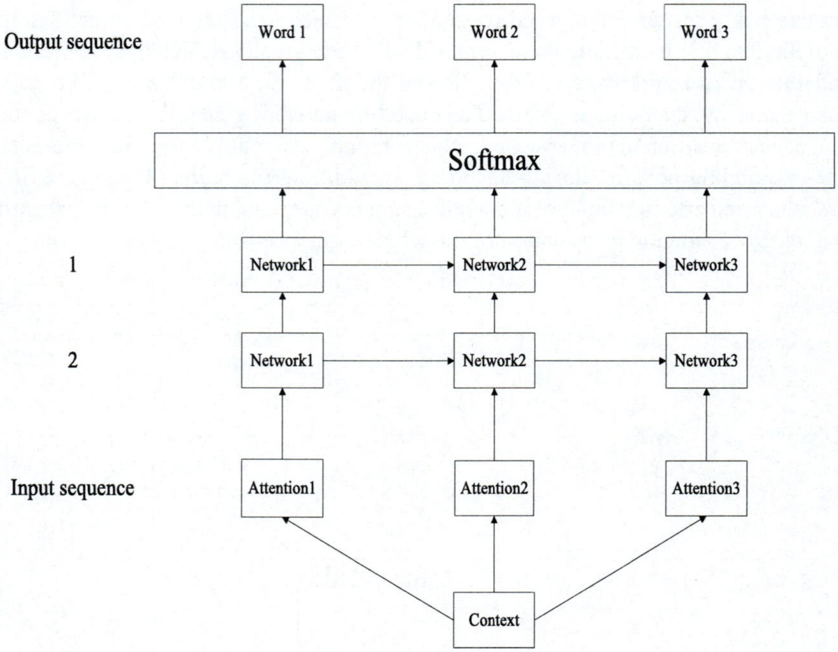

Fig. 2. Decoder Structure Diagram

Among them, v and c represent intercept and slope, respectively. The S-shaped growth curve function and the flexible maximum transfer function are shown in formulas (3) and (4).

$$F(x_i) = \frac{x_i}{\sum_{i=1}^{\delta} x_i} \tag{3}$$

$$F(x) = \frac{1}{1 + (-x)} \tag{4}$$

In the activation function, i = 1, ..., δ; $x = (x_1, x_2, \ldots, x_\delta) \in R^\delta$ represents the tag sequence.

In order to train the model, the error is calculated, and the model is improved through the back propagation loss function. The classification cross entropy function library is used to calculate the error. The classification cross entropy function is shown in Formula (5)

$$Q(v) = -\frac{1}{A}\sum_{i=1}^{A}\left[Z_i \log Z_i + \left(1 + \hat{Z}_i\right)\log\left(1 - \hat{Z}_i\right)\right] \tag{5}$$

Among them, z is the real label, z_i is the predicted label, and the goal is to minimize losses as much as possible to obtain better translations.

3 Optimization Experiment of NN on English MT Model

The experiment would be conducted by implanting a NN into a foreign language translation software, and comparing it in terms of translation accuracy and vividness. The object of comparison with the translation model using NN is another new foreign language translation software from the same software development company that uses traditional translation models. The comparison would be conducted on the performance of the two in the past two months, as shown in Figs. 3 and 4.

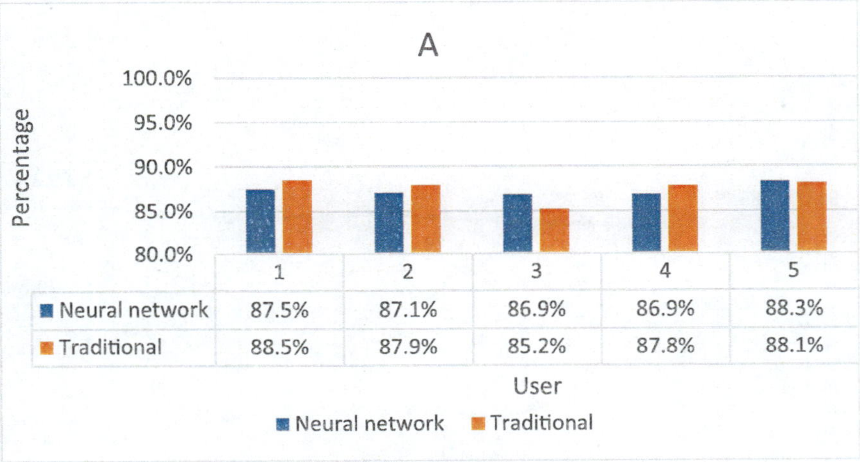

A Results of the experiment after one month

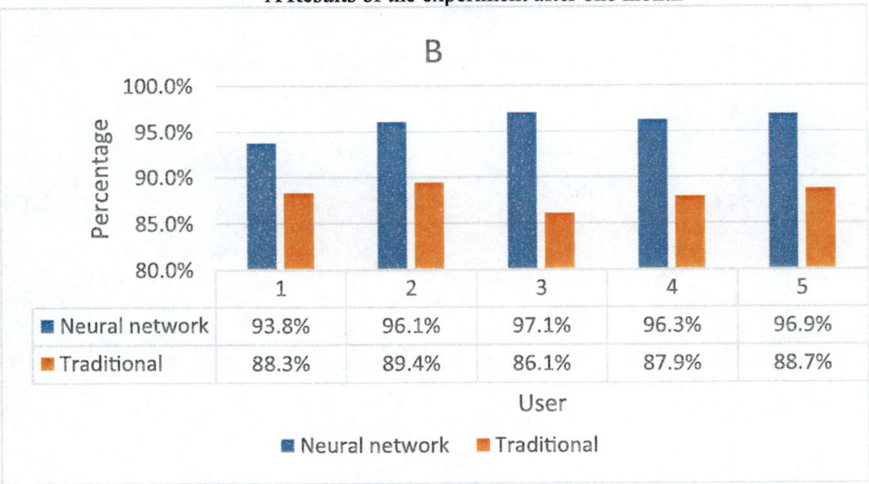

B Results after Two Months of Experiment

Fig. 3. Accuracy Comparison Results

From Fig. 3A, it can be seen that after one month of the experiment, the accuracy of the translation model using NN was the highest for user 5 at 88.3%, while users 3 and 4 had the lowest accuracy at 86.9%. The average accuracy of 5 users is 87.3%. In traditional translation models, the highest accuracy of user 1 is 88.5%, while the lowest accuracy of user 3 is 85.2%, with an average accuracy of 87.5%. From Fig. 3B, it can be seen that after two months of experiment, the translation model using NN has significantly improved its accuracy, with user 3 having the highest accuracy of 97.1% and user 1 having the lowest accuracy of 93.8%. The average accuracy is 96%, which is 8.7% higher than 87.3% a month ago, and the improvement is not small. After two months of experimentation, the average accuracy of traditional translation models is

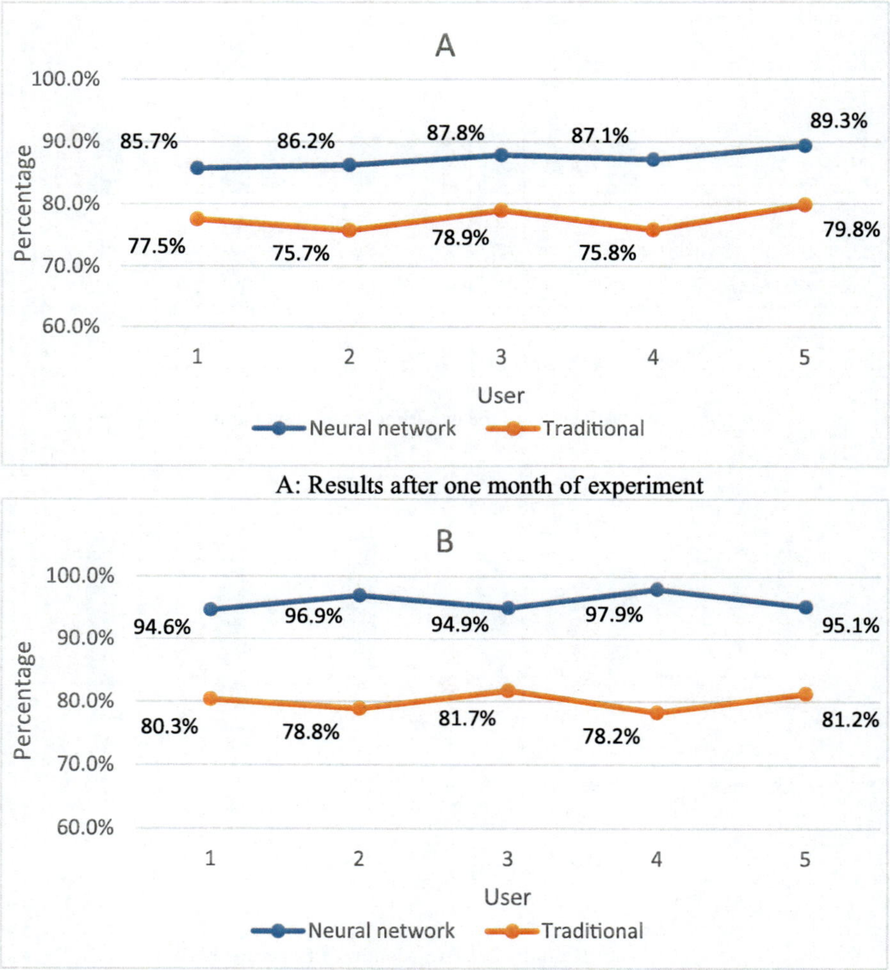

A: Results after one month of experiment

B Results of the experiment after two months

Fig. 4. Comparison Results of Vividness

88.1. Although it has also improved, it still hasn't reached 90%. Therefore, it can be concluded that although the optimization effect is not significant in the short term after the use of NN, the effect would be very significant over time.

From Fig. 4A, it can be seen that after one month of experimentation, among the translation models using NN, User 5 had the highest vividness at 89.3%, while User 1 had the lowest vividness at 85.7%, with an average vividness of 87.2%. In traditional translation models, User 5 had the highest vividness at 79.8%, while User 2 had the lowest vividness at only 75.7%, with an average vividness of 77.5%; From Fig. 4B, it can be seen that after two months of experiment, the vividness of the translation model using NN has significantly improved. The highest value of 97.9% appears in the translation model of User 4, and the lowest value is 94.6% of User 1, with an average vividness of 95.9%. At the same time, traditional translation models have also been counted, but the effect is not very ideal, with User 4 having a lowest value of 78.2%. The average vividness is only 80%, after all, the highest value is only 81.7%, in the translation model of user 3. Therefore, it can be concluded that using NN to optimize translation models can effectively improve the contextual vividness of MT.

4 Conclusions

This paper uses NN to optimize English MT model. In order to improve the accuracy of MT, the encoder and decoder are improved. Neural networks are used to replace the traditional methods to build a new translation model for MT, which reduces the cost and improves the accuracy; The introduction of NN into the translation model would improve the contextual vividness of MT and bring users a very comfortable experience. This method can effectively improve the accuracy and vividness of MT models, but there are also some shortcomings. The experimental data in this article only comes from 5 users, and the sample is still insufficient. Further work would consider using more samples.

References

1. Yang, L., Tongke, F.: Design and implementation of English to Chinese MT model based on recurrent neural network. Comput. Meas. Control **29**(11), 142–147 (2021)
2. Qiao, W., Lei, Y.: English translation method based on recurrent neural network. Autom. Technol. Appl. **39**(11), 37–40+70 (2020)
3. Chen, Z.: Research on English word segmentation based on neural network MT model. Comput. Digit. Eng. **48**(1), 13–18+50 (2020)
4. Li, Z.: Application analysis of statistical MT based on neural network language model. Inf. Comput. **34**(22), 109–111 (2022)
5. Li, Y., Yang, Y., Li, X., Mi, C., Dong, R.: Neural network language model for Hanwei MT. J. Xiamen Univ. Nat. Sci. Edn. **58**(2), 189–194 (2019)
6. Rivera-Trigueros, I.: Machine translation systems and quality assessment: a systematic review. Lang. Resour. Eval. **56**(2), 593–619 (2022)
7. Ranathunga, S.: Neural MT for low-resource languages: a survey. ACM Comput. Surv. **55**(11), 1–37 (2023)

8. Vieira, L.N., O'Hagan, M., O'Sullivan, C.: Understanding the societal impacts of MT: a critical review of the literature on medical and legal use cases. Inf. Commun. Soc. **24**(11), 1515–1532 (2021)
9. Fan, A.: Beyond English-centric multilingual MT. J. Mach. Learn. Res. **22**(1), 4839–4886 (2021)
10. Grechishnikova, D.: Transformer neural network for protein-specific de novo drug generation as a MT problem. Sci. Rep. **11**(1), 1–13 (2021)

Exploration on Text Detection Optimization Algorithm Based on Neural Network Technology

Junxia Liang[1] and Yongjun Qi[1,2](✉)

[1] Faculty of Megadata and Computing, Guangdong Baiyun University, Guangzhou 510450, Guangdong, China
qyj200702022@baiyunu.edu.cn

[2] School of Information and Communication Technology, Mongolian University of Science and Technology, Bayanzurkh District, 13341 Ulaanbaatar, Mongolia

Abstract. In this rapidly changing era of information technology, images have always been the main tool for human communication of information. In daily life, a large number of natural scene images, such as road signs, advertising slogans, etc., often contain important textual information, which is of great help to people's understanding of image content. Text is a type of linguistic information that exists in various aspects of people's lives. The textual information in it can help people understand, describe, and analyze natural scenes, and has important social value. Neural network technology is of great help for optimizing text detection. This paper discussed the optimization performance of neural network algorithm for text detection, and verified the progressiveness of the algorithm through experiments.

Keywords: Text Detection · Algorithm Optimization · Neural Network Technology · Experimental Study

1 Introduction

Due to the complexity of natural scenes and the diversity of words in the scenes, it has a significant impact on the accuracy of text recognition. This is a huge challenge. The traditional machine learning method of step-by-step processing can lead to error accumulation and performance degradation. In recent years, with the continuous development of deep learning technology, text detection technology has ushered in new breakthroughs.

A large number of scholars have proposed related research on text detection optimisation. Wang Y proposed a text detection method for scenes using a pyramidal attention network and a positional attention module. First, a feature extraction method based on cone attention was proposed and semantic segmentation was implemented using this method. Test results show that the accuracy of the algorithm is 2 percentage points higher than that of the traditional backbone network when using a deep backbone network [1]. Kohli H proposes optical character recognition (OCR), an automatic recognition technology that converts documents and images into parsable and editable textual information, which has been widely used in several fields. Based on this, a J&M-based text detection method is proposed. The research work is conducted in MNIST's handwritten digital

J. C. Hung et al. (Eds.): FC 2023, LNEE 1131, pp. 97–103, 2024.
https://doi.org/10.1007/978-981-99-9299-7_13

database using the Python language. The system was experimentally validated to achieve 99.5% training accuracy and 99% experimental accuracy with a training loss of only 1.5% [2]. Dikubab W presented the first large-scale public dataset for text detection and recognition in natural scenes. A large-scale experiment was conducted to evaluate the performance of state-of-the-art Amharic text detection and recognition methods on the dataset. The evaluation results show the robustness of the dataset in benchmark experiments and its potential to facilitate the development of text detection and recognition algorithms in Amharic [3]. The above research proposes new approaches to text recognition, but its application is limited and has only been performed on a small scale.

Multidimensional text in multidimensional space also faces problems such as complex backgrounds, varying shapes and scales, making text region detection and localization in multidimensional space a challenging research topic. The text detection optimization algorithm based on neural network technology discussed in this article has greatly improved the efficiency of text detection, and the technology is widely used and has research significance.

2 Neural Network Technology and Text Detection

2.1 Neural Network Technology

Artificial neural networks are constantly imitating the direction of human cognition. Nowadays, a large number of artificial intelligence products have emerged on the market [4]. Artificial neural networks have successfully overcome the shortcomings of traditional artificial intelligence in fields such as speech recognition, image processing, and unstructured information processing, and can be widely applied in various fields such as pattern recognition, neural expert systems, intelligent control, data processing, weather forecasting, etc. [5]. Neural network technology is a technique used to simulate the structure and function of the human brain, in order to create a computer model that can recognize patterns. The most typical neural network consists of many nodes in the input layer and intermediate layer. Each node processes input information in a nonlinear manner and transmits the digital output structure to other nodes, thus cycling [6]. Artificial neural networks have strong adaptability and self-learning ability, making them widely used for predicting various types of time series [7, 8].

Characteristics of artificial neural networks.

According to existing neural network technology, neural networks have the following characteristics:

(1) The nonlinear characteristics of the research object: Just like the human brain, neural networks are a non-linear relationship.
(2) In the process of processing information in neural networks, they can integrate the patterns and correlations between relevant data, and optimize their topological relationships through continuous iteration, increasing their understanding and knowledge, and improving their ability to analyze and process data [9, 10].
(3) Parallelism: For information input into neural networks, it is not done in a serial manner, but in a parallel manner, similar to series and parallel circuits. In parallel

circuits, electronic components do not affect each other, while the computer motherboard operates in a serial manner [11, 12]. They collaborate to analyze and process data, and transmit the final data obtained from processing and synthesis to other neural units, greatly improving their processing and analysis capabilities. This greatly improves their processing and analysis capabilities [13, 14].

(4) Strong association and error resilience

2.2 Text Detection

The textual information in images is of great value for daily life and provides important clues for computers to understand life scenes. In practical applications, there are many situations that require the use of text information in complex scene images, such as intelligent transportation assistance, language translation for overseas tourism, etc. The intelligent detection and recognition of text information in these scene images using computer technology has become a popular research field. Due to the complexity of scene images, detecting and recognizing textual information generally requires three steps: first, locate the text in the image, then separate the text from the background, and finally perform recognition [15].

In the text recognition stage, scanning the image generates a set of N regions extracted from the original image. The processing of these N independent regions is completed by a ResNet architecture based CNN (Convolutional Neural Networks). In the recognition stage, CNN predicted the probability distribution y in the label space L_k [16].

$$L_k = LU\{\varepsilon\} \tag{1}$$

$$L = \{0 - 9a - Z\} \tag{2}$$

Among them, Ɛ represents an empty label. Depending on the task, this probability distribution is generated by a fixed number of softmax classifiers in T, each of which is used to predict a character in a given word.

The process of text detection includes text detection and recognition, overlapping text merging, correcting entries with only single word errors, title detection, and querying the template library based on the title, as shown in Fig. 1.

2.3 Text Detection Based on Neural Network Technology

Due to the explosive increase in data volume in modern networks, the rapid increase in data volume makes manual auditing impossible, and these behaviors are becoming increasingly difficult to detect. At the same time, with the rapid development of deep learning technology, Natural Language Processing (NLP) technology based on neural networks has gradually entered people's attention and has been gradually applied in multiple fields, promoting the development of NLP technology [17].

After preliminary extraction of the text area, parts similar to the strokes and characteristics of the text would appear, and they would be misjudged as text areas. On this basis, texture analysis technology is used to remove the background in the image. By using the BP neural network for recognition, the true text is ultimately determined [18].

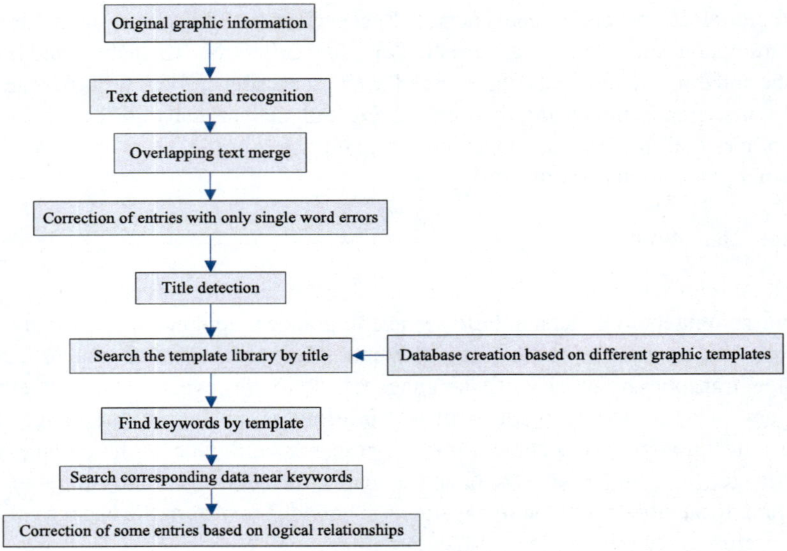

Fig. 1. Path for text detection

3 Experiment on Text Detection Optimization Algorithm Based on CNN Algorithm

For the improved lightweight network structure design scheme, comparative experiments were designed for comparison, and the evaluation indicators used were:

(1) F1ops (floating point operations per second): It refers to the number of floating point calculations performed by the network every second.
(2) Parameters: It refers to the number of parameters in the network during training, that is, the weight parameter values that guide the forward convergence of the network.
(3) FPS (Frames Per Second): refers to the number of images that a network can process per second, often used to measure the computational speed of a model in deep learning tasks.

Further comparison was made between the improved CNN algorithm for lightweight text detection in this article and other text detection models on the ICDAR2015 dataset in terms of Flops, parameter count, and FPS metrics [19, 20]. The experimental comparison results are shown in Table 1.

From Table 1, it can be seen that the number of network weight parameters in the CNN algorithm has been significantly reduced compared to other text detection algorithms, and the network structure has been streamlined. The network computing speed has been improved, with 129 images processed per second, an increase of 8 FPS values compared to EAST. Therefore, the network model lightweight method used in this article is effective and can simplify and compress the network structure, size, and volume. In the process of lightweight improvement in the CNN algorithm, deep separable convolutions were used to replace the original multi-scale feature fusion network structure, and the convolutional

Table 1. Experimental results of improved network structure comparison

Network structure	Flops	Number of parameters	FPS
DRRG	89.20	83.2MB	54
CRPN	79.48	23.1 MB	103
EAST	80.32	10.1 MB	121
Yolo V 3	83.21	34.23 MB	182
CNN	76.12	20 MB	129

and pooling layers at the bottom of the network were removed. Although the computing speed of the network has been improved, the accuracy and performance of text detection have not been verified to be reduced. The CNN algorithm is tested on the MSRA-TD500 dataset to verify that the detection ability of the lightweight model has not significantly decreased. The experimental results are shown in Fig. 2.

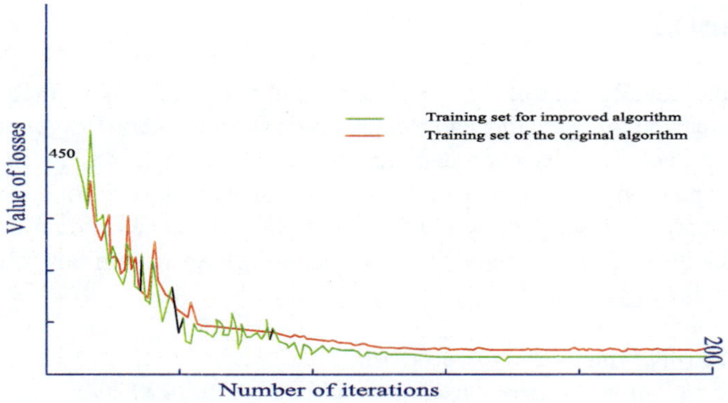

Fig. 2. Curve of loss value variation during network training

The comparison results between the improved CNN algorithm and the original algorithm in terms of parameter quantity are shown in Fig. 3.

The recall rate (percentage of all samples with true positive labels) represents the predicted proportion among all samples with true positive labels. FPN structure can be introduced to combine semantic and positional information to improve the representation of the network, and GAM attention mechanism can be added to improve the robustness of the network to disturbances and feature extraction in complex environments. This is combined with the Context Enhancement Module (CAM) and Feature Refinement Module (FRM), which combine multi-dimensional features and traditional features to extract rich contextual information for feature refinement. However, Fig. 3 shows that after the introduction of CNN, the search recall, accuracy, and processing speed of the images are relatively improved compared to FPN, GAM, and CAM.

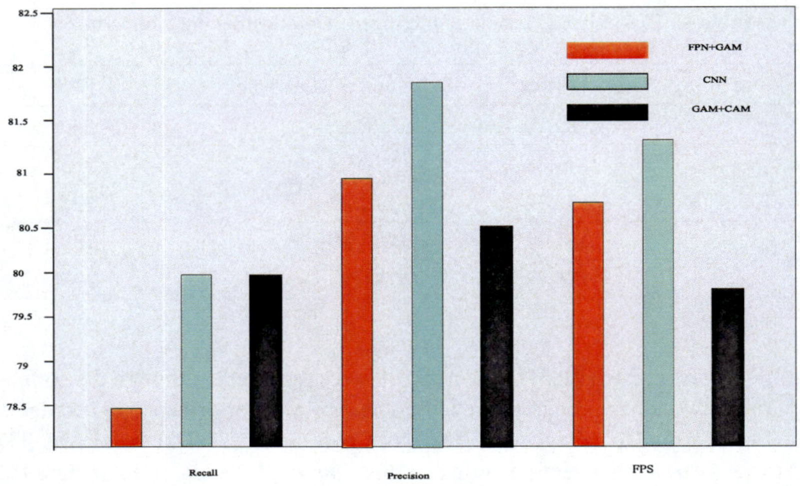

Fig. 3. Analysis of improved network structure model

4 Conclusions

With the rapid development of the big data era and the popularity of mobile storage devices, the demand for extracting and recognizing text from natural images has become increasingly urgent. Text detection and recognition in natural scenes, as an important aspect of computer vision, has also attracted widespread attention from researchers. In this paper, the principle and process of neural network and text detection were analyzed in detail. Finally, experiments proved that the neural network algorithm optimized for text detection is progressiveness.

Acknowledgements. This work was supported by

Key Research Project of Guangdong Baiyun College, No. 2022BYKYZ02.

Key Research Platform of Guangdong Province, no. 2022GCZX009.

Special project in key fields of colleges and universities in Guangdong province, No. 2020ZDZX3009.

References

1. Wang, Y., Gu, X.: Using of attention for scene text detection. J. Comput.-Aided Des. Comput. Graph. **33**(12), 1908–1915 (2021)
2. Kohli, H., Agarwal, J., Kumar, M.: An improved method for text detection using Adam optimization algorithm. Glob. Transit. Proc. **3**(1), 230–234 (2022)
3. Dikubab, W., Liang, D., Liao, M., et al.: Comprehensive benchmark datasets for Amharic scene text detection and recognition. Science China Inf. Sci. **65**(6), 1–2 (2022)
4. Lin, Y., Tu, Y., Dou, Z.: An improved neural network pruning technology for automatic modulation classification in edge devices. IEEE Trans. Veh. Technol. **69**(5), 5703–5706 (2020)
5. Liu, J., Wu, Q., Sui, X., et al.: Research progress in optical neural networks: theory, applications and developments. PhotoniX **2**(1), 1–39 (2021)

6. Zhao, Z., Lu, X.: Research progress of chemical process control and optimization based on neural network. J. Eng. Res. Rep. **21**(12), 10–17 (2021)
7. Ghazal, T.M.: Convolutional neural network based intelligent handwritten document recognition. Comput. Mater. Continua **70**(3), 4563–4581 (2022)
8. Li, D., Xu, S., Wang, D., et al.: Large-scale piston error detection technology for segmented optical mirrors via convolutional neural networks. Opt. Lett. **44**(5), 1170–1173 (2019)
9. Hasibuan, M.S., Nugroho, L.E., Santosa, P.I.: Model detecting learning styles with artificial neural network. J. Technol. Sci. Educ. **9**(1), 85–95 (2019)
10. Kliestik, T., Zvarikova, K., Lazaroiu, G.: Data-driven machine learning and neural network algorithms in the retailing environment: consumer engagement, experience, and purchase behaviors. Econ. Manag. Financ. Markets **17**(1), 57–69 (2022)
11. Chen, Y., Xie, Y., Song, L., et al.: A survey of accelerator architectures for deep neural networks. Engineering **6**(3), 264–274 (2020)
12. Tan, X., Xu, W., Sun, K., et al.: Improving massive MIMO message passing detectors with deep neural network. IEEE Trans. Veh. Technol. **69**(2), 1267–1280 (2019)
13. Wu, H., Li, Y., Zhou, L., et al.: Convolutional neural network and multi-feature fusion for automatic modulation classification. Electron. Lett. **55**(16), 895–897 (2019)
14. Lu, Z., Lu, M., Liang, Y.: A distributed neural network training method based on hybrid gradient computing. Scalable Comput. **21**(2), 323–336 (2020)
15. Pham, D.A., Lai, B.C.: Dataflow and microarchitecture co-optimisation for sparse CNN on distributed processing element accelerator. IET Circuits Devices Syst. **14**(8), 1185–1194 (2020)
16. Wang, Y., Wu, D., Hu, T., et al.: Blind identification of space-time block code based on BP neural networks. J. Signal Process. **38**(8), 1656–1666 (2022)
17. Zhao, G., Zhang, Y., Shi, Y., et al.: The application of BP neural networks to analysis the national vulnerability. Comput. Mater. Continua **58**(2), 421–436 (2019)
18. Li, X., Wang, J., Yang, C.: Risk prediction in financial management of listed companies based on optimized BP neural network under digital economy. Neural Comput. Appl. **35**(3), 2045–2058 (2023)
19. Duan, L., Zhou, J., Wu, Y., et al.: A novel and highly efficient botnet detection algorithm based on network traffic analysis of smart systems. Int. J. Distrib. Sens. Netw. **18**(3), 182459–182476 (2022)
20. Santos, D.C., Silva, F., Pereira, D.R.: Real-time traffic sign detection and recognition using CNN. IEEE Lat. Am. Trans. **18**(3), 522–529 (2020)

CO2 Emission Prediction of Vehicle Fuel Consumption Based on EMD-LSTM

Shuang Lu(✉), Ying Qiao, Jiaming Liu, Xinyao Feng, Yuxi Du, and Mingyu Liu

North China University of Science and Technology, Tangshan 063210, China
shixiaolu0702@163.com

Abstract. In this paper, a prediction model based on Empirical Mode Decomposition (EMD) combined with Long Short-Term Memory (LSTM) is proposed, which can further improve the accuracy of model learning prediction based on only using LSTM. Before model processing, it needs to find indicators that can evaluate vehicle fuel consumption, and then select the indicators that can be learned by the neural network as the initial input. The initial input is substituted into the EMD model, and then the IMF component obtained through EMD decomposition is substituted into the LSTM model. Taking the data of cars of different brands as practical examples, the test results show that the prediction error of this method is only 12.2%. But the prediction error of only LSTM network model is 15.4%, the average absolute error of the former is 7% higher than that of the latter, and the prediction accuracy is higher.

Keywords: Carbon Dioxide Emissions · Automobile Fuel Consumption Indicators · EMD-LSTM

1 Introduction

With the continuous development of China's economy, cars are no longer untouchable commodities for people. More and more surveys show that the rate of car ownership in China continues to increase. China's auto industry has gradually become one of the main supports of the national economy [1, 2]. It has also made remarkable achievements in the process of promoting the rapid development of national economy. However, while we benefit from it, we also bring pressure to the society from environmental pollution, energy consumption [3], traffic congestion and other aspects [4]. Ninety percent of carbon emissions from cars come from fossil fuel burning, which consumes fuel at the age of use. Therefore, it is urgent to reduce vehicle fuel consumption emissions. Based on the current situation in China, a prediction model of CO2 emissions from vehicle fuel consumption is proposed in this paper, which can provide a more direct view of the impact of vehicle fuel consumption on CO2 emissions. It can be provided to automobile manufacturers, which can be improved under the influence index and effectively reduce vehicle CO2 emissions.

Marcelino et al. proposed the issue of excessive CO2 emissions caused by incorrect use of automotive oil [5], Indra Chandra Setiawan et al. proposed a model that can

J. C. Hung et al. (Eds.): FC 2023, LNEE 1131, pp. 104–110, 2024.
https://doi.org/10.1007/978-981-99-9299-7_14

quantitatively predict oil demand, carbon dioxide emissions, and evaluate the impact of proposed policy measures within the framework of activity, mode share, intensity, and fuel selection (ASIF) [6].

2 Construction of the Initial Input Characteristic Set

CO2 emission is closely related to our current life, and the input characteristic set is built with full consideration of vehicle fuel consumption index to maximize the ability to reflect vehicle carbon emissions. In this paper, when collecting vehicle fuel consumption index, the initial data obtained from the external network is sorted into labels, which can be simplified as:

1) Automobile brand: Automobile brand corresponding to different data in data collection.
2) Vehicle model: specific models issued by different brands of vehicles in data collection.
3) Engine size: Different vehicles have adapted engine sizes during manufacturing, and different engine sizes have different relationships to fuel utilization, so engine size is also closely related to its exhaust emissions.
4) The number of cylinders: the more cylinders a car has, the greater its exhaust emissions will be.
5) Fuel type
6) Vehicle transmission device
7) Fuel consumption: In order to better describe the fuel consumption capacity of vehicles, we need comprehensive evaluation data sources, so we collected the following four situations when obtaining data: 1.Fuel consumption of 100 km in city driving 2. Fuel consumption of 100 km in highway driving 3.Combined fuel consumption of 100 km in the first two cases is 4. Combined fuel consumption in MPG units.
8) CO2 emission rating: Greenhouse gases from a vehicle's exhaust pipe and emissions associated with the production of the fuel used to power the vehicle Carbon dioxide emissions from a vehicle's exhaust pipe provides a greenhouse gas rating that rates the vehicle's fuel economy and environmental label on a scale of 1 (worst) to 10 (best).But the rating does not reflect any greenhouse gas emissions associated with the production of the fuel.
9) SOMG emission rating: Based on U.S. vehicle emission standards, we list information about these pollutants as "smog" on the label, using a slider scale of 1 (worst) to 10 (best).

3 EMD-LSTM Model

3.1 Empirical Mode Decomposition

Based on the concept of instantaneous frequency and intrinsic mode function, Empirical Mode Decomposition is a signal can be decomposed into several IMF components with different frequencies and trends, so that each obtained intrinsic mode function (IMF) characterizes the local characteristics of the signal. Signal decomposition is carried out by analyzing the time-scale characteristics of the data itself, so there is no need to set any primary function in advance, so it has self-adaptability. Therefore, this method is widely used in the field of data processing and data mining (especially in the study of non-linear non-stationary time series).IMF should meet the following conditions: 1. The maximum difference between the number of extreme points and the number of zero crossings is less than or equal to 1;2. When in use, the mean value of upper and lower envelope formed by the fitting of local maximum and local minimum of signal is zero [7]. The specific algorithm flow chart is shown as follows in Fig. 1:

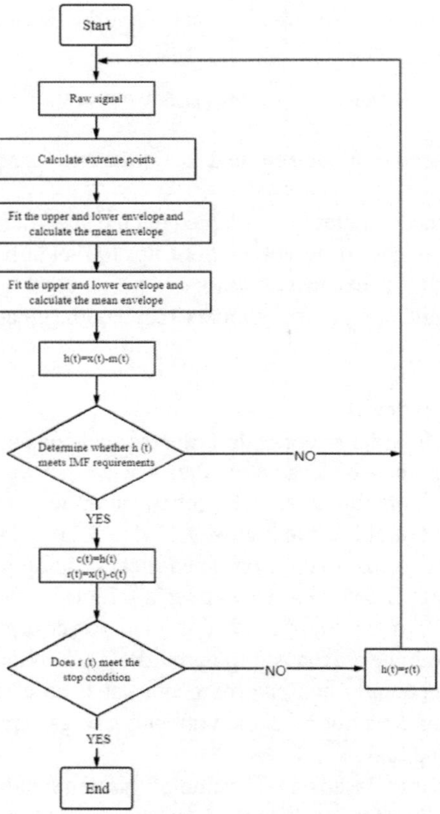

Fig. 1. Flow chart of EMD

3.2 Long Short-Term Memory

The specific working principle of Long Short-Term Memory: At each moment, the input of the LSTM should consist of three parts: the new input of the current moment x_t, the memory of the previous moment (also called short term memory) h_{t-1} and the long term memory line (also called cell state) C_{t-1}, Through the calculation of three layers of the door (sigma as the Sigmoid activation function tanh for hyperbolic tangent activation functions), get a new summary memory $\widetilde{C}_t$, And new memory states C_{t-1} and update h_t. The operation process inside Long Short-Term Memory (LSTM) is as follows [9]:

(1) Through the effect of the forgetting door, I can get f_t as follows:

$$f_t = \sigma * (w_f * [h_{t-1}, x_t] + a_f) \tag{1}$$

(2) In the next step, enter the input layer and get it through the action of the input gate i_t:

$$i_t = \sigma * (w_i * [h_{t-1}, x_t] + a_i) \tag{2}$$

(3) Through the function of output gate, we can get O_t:

$$O_t = \sigma * (w_o * [h_{t-1}, x_t] + a_o) \tag{3}$$

(4) the next state calculation summary $\widetilde{C}_t$:

$$\widetilde{C}_t = \tanh * (w_c * [h_{t-1}, x_t] + a_c) \tag{4}$$

(5) Computed new memory cell state C_t:

$$C_t = f_t * C_{t-1} + i_t * \widetilde{C}_t \tag{5}$$

(6) Update short term Memory h_t:

$$h_t = O_t * \tanh(C_t) \tag{6}$$

In formula, w_f, w_i, w_o, w_c represent weight matrix; a_f, a_i, a_o, a_c indicate different bias for each layer.

3.3 Build the EMD-LSTM Model

Before model building, it is necessary to eliminate the data that the neural network cannot learn. Fuel consumption per 100 km is divided into the following four small labels: fuel consumption per 100 km in cities; Fuel consumption per hundred kilometers of expressway; Fuel consumption per 100 km of mixture; The number of miles a car can travel per mpg of fuel consumption (mixed road conditions).Thus the initial data is selected [8].

(1) Initial data are decomposed by EMD to obtain multiple IMF components. Due to different inherent characteristics of data, IMF components obtained by decomposition are also different in EMD.
(2) Substitute all IMF sequences into the LSTM model as inputs.
(3) Determine the proportion of neural network training set and test set and training model.

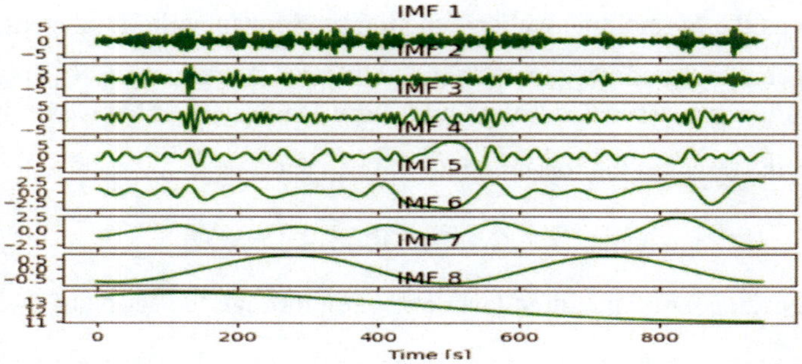

Fig. 2. 8 IMF components were obtained by EMD decomposition of urban fuel consumption per 100 km

4 Algorithm Example

4.1 IMF Component After Input Data Decomposition

The input data is decomposed by EMD, and different IMF components are obtained based on the different characteristics of different data. The following Fig. 2 shows the instance of EMD decomposition.

4.2 EMD-LSTM Model Parameter Settings

Due to the nonlinearity and volatility of data, more neurons are needed to better reflect the law of data change when setting LSTM parameters. Since 62 different IMF components are obtained from the input data after EMD decomposition, the characteristic dimension of the input layer of LSTM is 62, including two hidden layers. And the number of hidden layer neurons is 64 [10]. The LSTM error training effect is shown in Fig. 3.

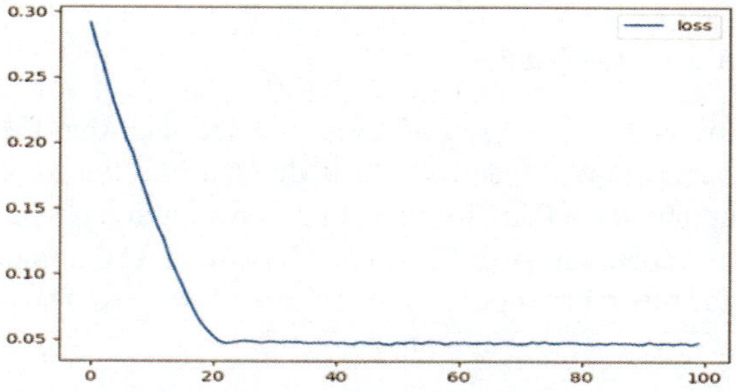

Fig. 3. LSTM error training effect

4.3 Display of Model Results

The comparison results of the two models' predictions are shown in Fig. 4.

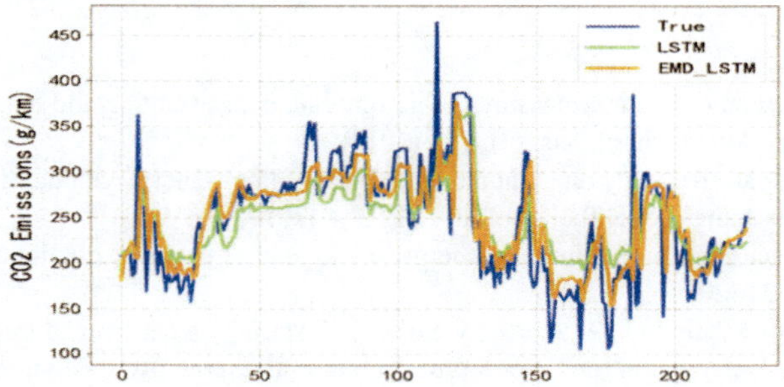

Fig. 4. Model prediction results

In this paper, the EMD-LSTM model is used to compare with the LSTM model. In the LSTM model, parameters are set as follows: the characteristic dimension of the input layer is 8, the hidden layer is 2, and the number of neurons is 64. We set the initial learning rate as 0.0001, the number of iterations as 1000, and the maximum number of input data for a single training as 64.The final evaluation indicators of the fitting degree of the two models are as follows in Table 1:

Table 1. Model accuracy table

	X_{MAE}(g/km)	X_{SMAPE}(%)	R^2
EMD-LSTM	26.83	12.27	0.6013
LSTM	33.57	15.41	0.4936

It can be seen that after the application of EMD for data preprocessing, the degree of fitting of the model is improved to a certain extent, and the accuracy is also improved.

5 Conclusions

In this paper, a CO2 emission prediction method based on EMD-LSTM is proposed. The original input is decomposed by EMD, and the data is sorted from high frequency to low frequency to fully reflect the local characteristics of IMF component of each data, which can effectively improve the prediction accuracy of the model. The IMF component after EMD decomposition is brought into the LSTM model for calculation, and the CO2 emission of automobiles is predicted by the trained model. Compared with the direct use of LSTM, this method can obtain better prediction accuracy and better fitting effect.

References

1. Cui, Y.: Research on countermeasures of automobile energy saving and emission reduction in our country. Sci. Technol. Vis. (4), 190 (2014)
2. Lan, C.: Comparative study on carbon credits Policy of new energy vehicles in China and the United States. Jiangsu Sci. Technol. Inf. (24), 3–4 (2017)
3. Fang, Z.: Discussion on automobile energy saving and emission reduction. Energy Saving **38**(9), 102–103 (2019)
4. Ma, D., Zeng, S., Zhu, Q., et al.: "Low-carbon +" strategy research and implementation of "low-carbon + transportation" development countermeasures. Rev. Econ. Res. (62), 28–33 (2016)
5. Carrera-Rodríguez, M., et al.: Monitoring of oil lubrication limits, fuel consumption, and excess CO2 production on civilian vehicles in Mexico. Energy **257**, 124765 (2022). ISSN 0360-5442
6. Setiawan, I.C., Indarto, Deendarlianto: Quantitative analysis of automobile sector in Indonesian automotive roadmap for achieving national oil and CO2 emission reduction targets by 2030. Energy Policy **150**, 112135 (2021). ISSN 0301-4215
7. Sarangi, S., Biswal, C., Sahu, B.K., Samanta, I.S., Rout, P.K.: Fault detection technique using time-varying filter-EMD and differential-CUSUM for LVDC microgrid system. Electric Power Syst. Res. **219**, 109254 (2023). ISSN 0378-7796
8. Mounir, N., Ouadi, H., Jrhilifa, I.: Short-term electric load forecasting using an EMD-BI-LSTM approach for smart grid energy management system. Energy Build. **288**, 113022 (2023). ISSN 0378-7788
9. Li, P., He, S., Han, P., et al.: Short-term load forecasting of smart grid based on long and short term memory real-time electricity price. Power Grid Technol. **42**(12), 4045–4052 (2018). Li, P., He, S., Han, P., et al.: Short-term load forecasting of smart grid based on long-short-term memory. Power Syst. Technol. **10**(1), 1–8 (2018)
10. Johny, K., Pai, M.L., Adarsh, S.: A multivariate EMD-LSTM model aided with time dependent intrinsic cross-correlation for monthly rainfall prediction. Appl. Soft Comput. **123**, 108941 (2022). ISSN 1568-4946

Application of Data Fusion Algorithms in the Data Processing of Intelligent Greenhouses

Jinguang Li[1] and Dan Li[2](✉)

[1] College of Information and Control Engineering, Jilin Institute of Chemical Technology, Jilin City, Jilin, China

[2] Electrical and Information Engineering College, Jilin Agricultural Science and Technology University, Jilin City, Jilin, China

lidan800401@126.com

Abstract. Greenhouses can ensure the normal growth of crops in extreme environments. With the popularization of social intelligence, how to build an efficient and accurate environmental monitoring system to ensure that crops have suitable growth conditions is a top priority. In order to effectively solve the problem of data missing and large data errors caused by sensor collection in intelligent greenhouses under extreme environments, this paper uses data preprocessing and multi-sensor fusion algorithms to ensure the accuracy of data. The effectiveness of the data fusion algorithm is verified through data transmission experiments.

Keywords: Intelligent Greenhouse · Zigbee · Data Fusion

1 Introduction

In agriculture, crops exhibit different growth states under the influence of various factors including temperature, humidity, light intensity, CO2 concentration, and water. However, smart agriculture is a product of the integration of knowledge from various disciplines and fields such as communication, computer science, and agriculture. It can monitor the factors that affect crop growth and regulate them based on their levels. People can perceive the growth status of crops more intuitively through displays and control their growth status accordingly. Through the Internet of Things, smart agriculture can achieve a more harmonious communication between humans and crops [1, 2].

With the explosive development of computer, communication, and big data technologies, it is a trend to apply cutting-edge computing technologies to real-time monitoring systems in agriculture. The collected data is processed, transmitted, and used for refined management and growth prediction in different stages of agricultural production, achieving optimal management, consuming fewer resources, and gaining more output. This transformation enables agriculture production to shift from uncontrollable to controllable.

In this paper, the mature Zigbee wireless sensing technology is adopted, which features low cost, high efficiency, low energy consumption, and strong reliability. It has been widely applied in various fields [3]. Compared with traditional wired networks,

J. C. Hung et al. (Eds.): FC 2023, LNEE 1131, pp. 111–117, 2024.
https://doi.org/10.1007/978-981-99-9299-7_15

ZigBee technology does not require wiring and is less affected by the external environment. Therefore, this paper designs and experiments the intelligent greenhouse system based on ZigBee technology.

2 Materials and Methods

2.1 Overall Design

The main framework of the greenhouse, as shown in Fig. 1, is divided into four layers: data acquisition layer, network communication layer, data processing layer, and system application layer [4].

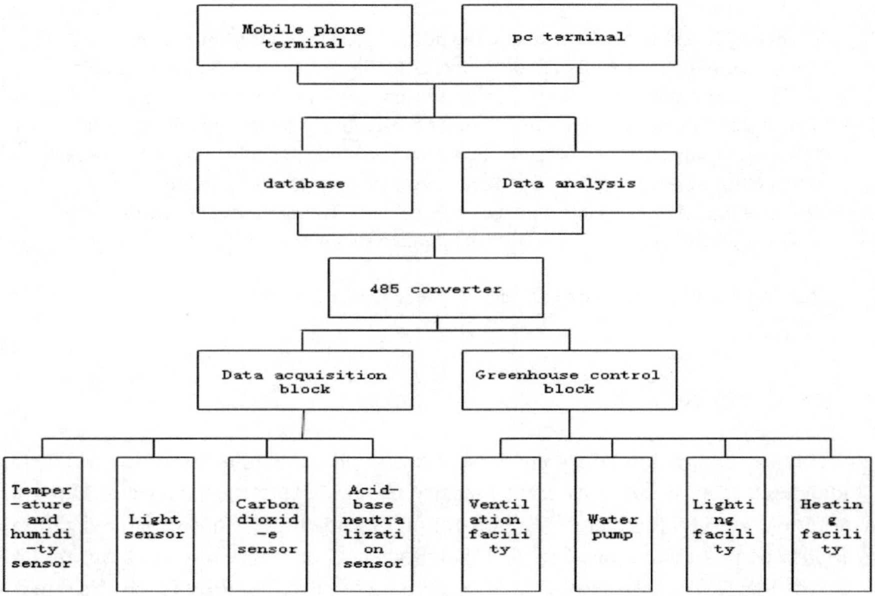

Fig. 1. Main frame of greenhouse

2.2 Data Acquisition Layer

The data acquisition module in intelligent greenhouse adopts Zigbee module for data acquisition and preprocessing, and finally completes wireless transmission of the preprocessed data through the Zigbee module. The data acquisition module uses the wireless node of the Zigbee module CC2530 for data collection of light intensity, pH value, CO_2 concentration, and temperature and humidity. The data is then aggregated to the Zigbee terminal node and transmitted to the PLC module through R485 (MODBUS protocol) for data storage on the touch screen, and ultimately transmitted to the server to reach the mobile and PC terminals.

2.3 Network Communication Layer

The network communication layer mainly consists of an R485 converter. In the intelligent greenhouse, data collected through Zigbee wireless transmission nodes is transmitted to the touch screen for data storage through the PLC using an R485 converter. This method not only avoids the limitations brought by wired nodes but also ensures stability during data transmission. Uploading the data to server can also enable communication with mobile and PC terminals, achieving real-time monitoring.

2.4 System Application Layer

The system application layer mainly uses an app to convey information to mobile and PC terminals. The PC terminal uses JAVA as the software development environment to display system transmission information in real-time. Meanwhile, the system sets alarm values. When the monitored environmental information exceeds the alarm value, the mobile or PC terminal sends an alert message to the customer [5].

3 Sensor Data Preprocessing

In practical measurements, errors are inevitably present. In the indoor monitoring of the smart greenhouse, the omission error elimination algorithm is used for preprocessing to promptly eliminate measurement results with excessively large errors or errors, thereby making the data more reliable. This article proposes to use the Grubbs' test criterion to check the data consistency [6, 7].

Assuming that a certain collection node i is independent in measuring specific environmental data, the measurement results collected during a certain period are $X_i = \{X_{i1}, X_{i2}, X_{i3}, \cdots\cdots X_{ik}\}$, and the measurement results follow a normal distribution, where K is the number of times the sensor is measured during a certain period.

- Firstly, the measurement data within a period are arranged in ascending order according to $\tilde{X} = 76.79$.
- Then, the arithmetic mean and standard deviation of the measurement data x_{ik} are calculated as shown in the equation:

$$\varepsilon_i' = \sqrt{\frac{\sum_{j=1}^{k} \varepsilon_{ij}^2}{k-1}} \tag{1}$$

According to the Grubbs criterion statistics, the exact distribution of the Grubbs of statistic can be obtained.

$$g_{ij} = \frac{x_{ij} - \overline{x_i}}{\varepsilon_i'} \tag{2}$$

After a significant level α is given, usually taking values such as $\alpha = 0.05, \alpha = 0.25$ or $\alpha = 0.01$, the critical value $g_i(k, \alpha)$ of the Grubbs statistic can be determined by table lookup. Since the measurement sequence X_i follows a normal distribution, it can be obtained.

$$P\Big[g_{ij} \geq g_i\big(k, \alpha\big) \Big] = \alpha \tag{3}$$

It can be seen that the probability of this event occurring is extremely small and unreasonable in the current situation, and needs to be screened out. It can also be understood that for any $j \in \big[1, k\big]$, if there exists $g_{ij} \geq g_i(k, \alpha)$, the Grubbs criterion can be used to preprocess the measurement data of a single node sensor during a certain period of time and screen out abnormal values. Using this criterion to eliminate sparse errors effectively solves the interference of abnormal data on monitoring results and increases the robustness of the data.

4 Multi-sensor Data Fusion Algorithm

After preprocessing to eliminate missing errors, a set of reliable measurement sequences has been obtained. To improve measurement accuracy, it is necessary to conduct intra-group fusion based on the batch estimation theory on k measurements of a single sensor within a certain time period. This can result in the optimal estimated value of the data in that time period, which can more reliably reflect the situation of the data segment [8–10].

In this paper, the following steps were taken: the data x collected in a certain time period after screening out missing errors are divided into 3 groups, and the i-th group can be represented as $X_{i1}, X_{i2}, X_{i3}, \cdots\cdots, X_{ij}$ $\left(i = 1, 2, 3, \sum_{i}^{3} j = x\right)$. The mean and variance of each group were calculated.

The mean and variance of each group of data can be obtained through the above formula. Then, the optimal estimation value of data fusion for a single sensor during this time period can be obtained through the theory of batch estimation, which is:

$$\varepsilon_i' = \sqrt{\frac{\sum_{j=1}^{k} \varepsilon_{ij}^2}{k-1}} = \sqrt{\frac{\sum_{j=1}^{6} \varepsilon_{ij}^2}{5}} = 1.3 \tag{4}$$

The total variance σ^2 of the sensor data after fusion can be calculated as:

$$\sigma^2 = \frac{\sigma_1^2 \sigma_2^2 \sigma_3^2}{\sigma_1^2 + \sigma_2^2 + \sigma_3^2} \tag{5}$$

In the above, this paper provides a solution to improve the accuracy and reliability of individual sensor node data measurements, but in a greenhouse, multiple sensors of the same type need to be deployed locally for data collection. Since the sensors are arranged in different orientations, they are affected by different environmental factors, resulting in different measurement results. In order to improve the accuracy of data acquisition, after calculating the optimal estimate value of a single sensor, according to the principle of minimizing the total mean square error, the weights of each sensor are reasonably allocated, and the weights of each node's sensors of the same type are calculated. Finally, adaptive weighted fusion is performed to obtain the best fusion result.

Assuming there are n sensors of the same type in the greenhouse, and the optimal estimate X and total variance σ^2 of a single sensor are calculated using the above formula. Let the weighted values of each sensor be $\omega_1, \omega_2, \omega_3 \cdots\cdots \omega_n$, and the fused result be $\hat{X}$. Then we have:

$$\begin{cases} \hat{X} = \sum_{i=1}^{n} \omega_i X_i \\ \sum_{i=1}^{n} \omega_i = 1 \\ \sigma^2 = \sum_{i=1}^{n} \omega_i^2 \sigma_i^2 \end{cases} \tag{6}$$

The function form of σ^2 in the above equation is the total mean square error.

$$f(\omega_1, \omega_2, \omega_3, \cdots\cdots \omega_n) = \sigma^2 = \sum_{i=1}^{n} \omega_i^2 \sigma_i^2 \tag{7}$$

The above equation shows that the total mean square error is a quadratic function of multiple variables. To minimize the total mean square error σ_{min}^2, we need to find the minimum value of function f, and the obtained ω_i values will be the optimal weights. The formula for weight allocation is:

$$\omega_i = \frac{1}{\sigma_i^2 \sum_{i=1}^{n} \frac{1}{\sigma_i^2}} \tag{8}$$

The above demonstration shows that data fusion of collection sensors in greenhouse is feasible. It can be used for data denoising, reducing redundant communication, and improving data robustness through environment data fusion.

5 Results and Analysis

In order to verify the effectiveness of the data fusion algorithm for collecting data in the intelligent greenhouse, experimental validation is required. In this paper, the collected data from the sensors, pre-processed data from the Zigbee sensor terminal, and data after fusion algorithm processing on the server will be compared to verify its feasibility.

Table 1 records humidity data from three sensors over an hour.

After preprocessing, Sensor 3, 19:00:00 data is screened.

After calculation of the data in Table 1, the mean square deviation is obtained: 1.3.

After verification, it was found that by consulting the critical values table of Grubbs, $g_1(6, 0.05) = 1.822$. Therefore, $g_1(6, 0.05) \cdot \varepsilon_i' = 1.822 \times 1.3 = 2.3686$ for the data at 19:00:00 in Sensor3, $|\varepsilon_6| = 2.6 > 2.3686$.

Table 1. The Humidity Sensor Collects Data

Time	Sensor1(%)	Sensor2(%)	Sensor3(%)
19:50:00	77.70	75.70	77.50
19:40:00	77.00	74.60	77.80
19:30:00	77.40	75.40	77.70
19:20:00	77.20	75.00	78.20
19:10:00	76.90	75.90	78.20
19:00:00	77.20	76.00	81.00

Therefore, $x_6 = 81.00$ is a suspicious value, judged as an omission error data and removed. After removing this data, there are 5 remaining data points in Sensor3. It has been verified that for Sensor1 and Sensor2, the data all satisfy $g_{1j} < g_1(6, 0.05)$, and for Sensor3, the data satisfy $g_{1j} < g_1(5, 0.05)$. It can be proven that the remaining data sequence meets the requirements.

The final result of the data after the multi-sensor data fusion algorithm is shown in Table 2.

Table 2. Humidity Data After Multi-Sensor Data Fusion Algorithm

Azimuth sensor	Estimated value	Measurement variance	weight	Fusion result $\hat{X}$
Sensor1	77.30	0.00071	0.31	76.02
Sensor2	75.16	0.00035	0.62	
Sensor3	77.90	0.0029	0.07	

It can be seen from Table 2 that the monitoring estimated value of the multi-sensor data fusion algorithm is $\hat{X} = 76.02$, while the result of simple arithmetic averaging of the estimated values is $\tilde{X} = 76.79$. Comparing the two, it can be found that the data obtained by the weighted fusion algorithm is closer to the true value.

6 Conclusion

This article argues that future greenhouse monitoring and control systems will become more intelligent, automated, and efficient. By incorporating artificial intelligence and big data technologies, the system can automatically learn and adapt to environmental changes, achieving automatic adjustment and optimization of the production environment. At the same time, through the Internet of Things (IoT) technology, real-time monitoring and remote control of various parameters inside the greenhouse can be realized, improving production efficiency and quality. The significance of this research is to provide a more stable and reliable production environment for agricultural production and promote sustainable development in agriculture. The future greenhouse monitoring and control system will become an important part of agricultural production, providing more reliable and secure food supply for humanity.

Acknowledgment. This work was supported by the science and technology research project of Jilin Province Education Department, "Research on Key Technologies of Disease Visualization in Plant Factory driven by Computer Vision", No. : JJKH20230432KJ.

References

1. Anuradha, M.R., Raghunandan, A.: Intelligent technology for agricultural parameter monitoring. J. Agric. Hortic. **2**(2), 7–16 (2020)
2. Sahour, A., Boumehrez, F., Benouaret, M., Mokhneche, A.: Greenhouse climate controller by using of internet of things technology and fuzzy logic. I2M **20**(1), 29–38 (2021)
3. Qu, D., Yang, J., Chen, W.: Design of intelligent mobile monitor system based on ZigBee and android. Microcontrollers Embed. Syst. (5), 10–12 (2012)
4. Wu, Y., et al.: Remote-control system for greenhouse based on open source hardware. IFAC Papers OnLine **52**(30), 178–183 (2019)
5. Jie, C., Wei, L., Junhui, W., et al.: Research and construction of intelligent production control platform for plastic film greenhouse based on internet plus. Dalian Maritime University, Hong Kong Society of Mechanical Engineers (HKSME).Proceedings of 5th International Conference on Automation, Control and Robotics Engineering (CACRE 2020). IEEE Express Conference Publishing (2020)
6. Adilkaram, K.K.L.B., Hussein, M.A., Effenberger, M., et al.: Data transformation technique to improve the outlier detection power of grubs' test for data expected to follow linear relation. J. Appl. Math. **5**, 511–515 (2015)
7. Keykhosravi, K., Hamednia, A., Rastegarfar, H., Agrell, E.: Data preprocessing for machine-learning-based adaptive data center transmission. ICT Exp. **8**(1), 37–43 (2022)
8. Sung, W.T., Lin, J.S.: Design and implementation of a smart LED lighting system using a self adaptive weighted data fusion algorithm. Sensors **13**(12), 16915–16939 (2013)
9. Tong, Y., Bai, J., Chen, X.: Research on multi-sensor data fusion technology. Advanced Science and Industry Research Center. In: Proceedings of 2020 2nd International Conference on Computer Modeling, Simulation and Algorithm (CMSA 2020), pp. 846–851 (2020)
10. Ting, Y.-Y., Wang, H.-S.: Develop a multi-detection security system using multi-sensor fusion algorithms. Artif. Life Robot. **18**(1–2), 83–88 (2013)

Intelligent Film and Television Communication Optical Technology Based on Network New Media

Zhaoqi Wang(✉)

Department of Media Arts, Wuhan Qingchuan University, Wuhan, Hubei, China
zhaoqiwang2023@163.com

Abstract. New media, as an emerging means of communication, has become an indispensable part of people's lives. It plays an important role in the film and television industry, advertising industry, and cultural research. This article analyzed the internal aspects of online new media. Firstly, it elaborated on the current development status and future trends of new online movies. Secondly, this article also focused on the problems and solutions in the process of intelligent film and television dissemination using Internet Protocol (IP) as the carrier. Finally, a solution and specific implementation methods for building a new media platform using IP technology were proposed, including establishing a complete website system, optimizing website structure, and updating webpage interfaces. The performance of the model was tested. The test results showed that the anti-interference ability of the model ranged from 92% to 95% .

Keywords: Network New Media · Intelligent Film and Television · Film and Television Communication · Optical Technology

1 Introduction

The emergence of new media has brought tremendous changes to film and television communication. Traditional media such as movies and television can no longer meet people's information and cultural needs. Especially with the advent of network technology and the digital age, humanity has entered the stage of "digital earth" and has had a significant impact. With its unique and powerful vitality, new media has penetrated into all aspects of life, sparking a new revolution in the field of mass media. Social platforms represented by Weibo have opened up the era of all media and become a new mode of communication.

In the context of new media, the research on intelligent film and television communication technology mainly revolves around digital image imaging and transmission in the network new media environment. Scholars have used models to construct 3D animation scenes based on channel coding methods. They have established a multi view projection system to simulate various complex image phenomena that have appeared in the original movie, and based on these data results, they have analyzed various types of film and television visual effects that may occur in the future [1, 2]. Some scholars

J. C. Hung et al. (Eds.): FC 2023, LNEE 1131, pp. 118–124, 2024.
https://doi.org/10.1007/978-981-99-9299-7_16

have proposed a new method of virtual scene 3D photography. This model is built on the basis of digital optical flow, with 3D laser and computer image processing software as the main components [3, 4]. Therefore, this article conducted research on intelligent film and television communication optical technology based on network new media.

With the development of new media technology, online new media has gradually become an essential and indispensable link in film and television communication. This article took intelligent film and television communication based on online new media as the research object. After organizing and analyzing relevant literature on the digital film industry and IP process based production, it was found that there are currently some urgent problems that need to be solved in this field, and corresponding solutions and countermeasures were proposed.

2 Exploration of Intelligent Film and Television Communication Optical Technology Based on Network New Media

2.1 Intelligent Film and Television Communication

With the increasing application of digital imaging technology and multimedia service industry in film production, film production can also monitor the shooting scene environment in real-time through video live streaming and other methods. At the same time, high-definition image resources are used to obtain relevant film and television segment images and convert them into film and television works. The camera movement track can be simulated with software to form an image of a complete three-dimensional space environment. This technology not only allows the audience to intuitively experience the visual image, but also enables real-time processing and reproduction of the captured content [5, 6]. New media can provide various forms, contents, and rich and colorful information to meet the needs of different people for watching film and television works. Communication through online platforms can also improve work efficiency and reduce costs. Intelligent film and television dissemination can also achieve video resource sharing and interactive dialogue services between users, making it more convenient and efficient for people to browse web pages. Figure 1 shows the process of film and television dissemination.

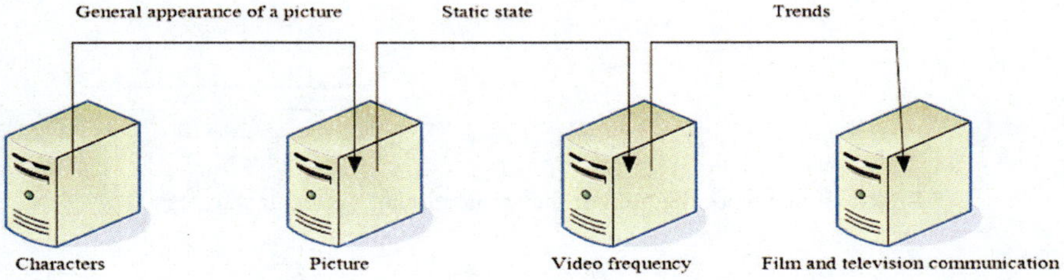

Fig. 1. Film and television communication process

Video websites can provide users with real-time updates and interactive communication functions. For example, uploading images or audio and other materials during filming, and adding text and sound effects to attract audience attention during playback can bring different effects to the audience. It can meet the needs of different audience groups. With the advent of the Internet era, people are increasingly relying on intelligent devices such as computer terminals to watch movies and related entertainment and leisure activities [7, 8]. Audiences who watch movies or variety show on video websites would also receive service information and feedback anytime and anywhere through the network platform. It can also interact and communicate with relevant websites to achieve maximum dissemination effects, thereby promoting the development of new media technology and enhancing its application value and actual social benefits [9, 10].

2.2 Optical Technology

Optical technology combines the principles of optical systems with computer simulation systems to achieve the calculation, storage, and processing of three-dimensional spatial light information. By using laser beams as focal objects, polycrystalline materials with advantages such as high brightness and narrow broadening are manufactured. In the production process, it is necessary to consider the distribution of refractive index under different intensities and frequencies generated by the interaction between light and atoms, as well as the variation of reflected light intensity. Therefore, the required three-dimensional spatial information can be obtained by using lasers of different wavelengths to cause diffraction distortion or extinction. By utilizing the computing power of the computer, the image can be analyzed. Figure 2 is the operational framework diagram of optical technology.

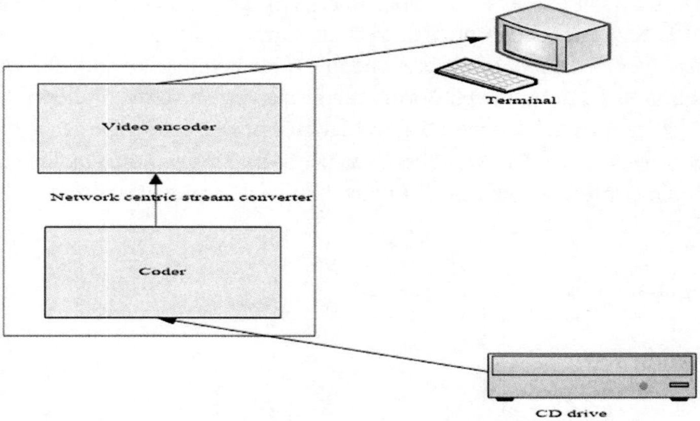

Fig. 2. Framework diagram of the application of optical technology

In film and television dramas, due to the huge amount of data that cameras need to process and the high requirements, in order to improve shooting efficiency and quality, it is necessary to fully utilize the camera lens to complete this task. After obtaining the required video stream information through a high-speed camera, the computer software analyzes and generates image images, and converts them into corresponding formats and stores them in the computer for later analysis and research, thus achieving the evaluation and detection of image quality. If the luminous intensity of an extended light source is B, that is, its brightness is independent of direction, it is called a cosine emitter. The above law of emitting luminous flux according to the cos law is known as Lambert cosine law, and its emission law satisfies Lambert cosine law. The formula for calculating light intensity is:

$$I = I_{\max} \cos\theta (-\pi/2) \leq \delta \leq (\pi/2) \tag{1}$$

Among them, the unit of I is cd. New media has achieved rapid development based on its high-speed data transmission and information processing capabilities. When implementing basic operations such as image compression and projection, digital signal encoding methods and large screen technology are used. By using a computer network communication system, images of different types and performance requirements are converted into images with a certain resolution and visual display, in order to achieve the best visual effect. The higher the surface temperature of a substance, the lower the temperature.

2.3 Network New Media

Network new media is an emerging medium that utilizes digital and information technology for information dissemination. In this new environment, the traditional film and television industry has merged with the internet, forming a new type of multimedia interactive media. Network new media is based on computer technology and transmits traditional dissemination information to the audience through means such as the Internet, mobile communication networks, digital television, and multimedia terminals.With the support of new media technology, online new media, due to its unique advantages, has become an essential tool for people to obtain information, exchange and interact, and engage in cultural dissemination activities. Two independent coded signals X and Y of the network are respectively set as H(X) and H(Y), and the corresponding code rates are Rx and Ry. If the joint entropy of the two signals is set as H(X, Y), the total code rate is:

$$\mathrm{R} = \mathrm{Rx} + \mathrm{Ry}\ \mathrm{H(X)} + \mathrm{H(Y)}\ \ \mathrm{H(X,Y)} \tag{2}$$

If only the reference information is obtained at the decoding end, this proves that the compression limit of X is the same as the encoding efficiency of obtaining the reference information at the encoding end, that is, encoding two signals X and Y independently of each other, and the total bit rate $\mathrm{R} = \mathrm{Rx} + \mathrm{Ry}$ can still reach the joint entropy H(X, Y) of the two signals, namely:

$$\begin{aligned} R_X &\geq H(X|Y) \\ R_Y &\geq H(Y|X) \\ R_X + R_Y &\geq H(X, Y) \end{aligned} \tag{3}$$

The traditional film and television production process usually requires multiple departments to work together to complete the content shooting and post editing work of the entire project. Internet film and television can achieve synchronous transmission and playback across the entire network. By receiving uploads to servers or terminal devices anytime and anywhere through mobile clients, there is no time limit, greatly saving costs for online new media advertising and operation.

3 Experimental Process of Intelligent Film and Television Communication Optical Technology Based on Network New Media

3.1 Model of Intelligent Film and Television Communication Optical Technology Based on Network New Media

The intelligent film and television communication optical technology model based on network new media (as shown in Fig. 3) utilizes new digital communication transmission media such as computers and the internet to exchange and interact data in the traditional film and television production process. In this model, optical technology is mainly applied in the transmission process, including fiber optic transmission technology, optical wireless communication, etc., aiming to improve transmission speed, bandwidth, and quality. At the same time, optical technology can also be applied to film and television production processes, such as using optical imaging technology, spectral imaging technology, etc.

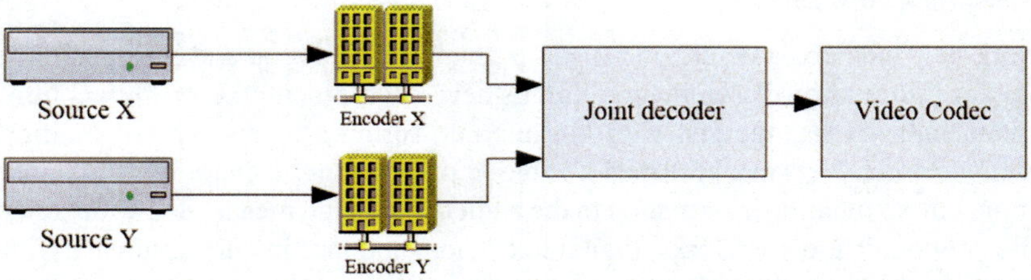

Fig. 3. Model of intelligent film and television communication optical technology based on network new media

3.2 Model Performance Testing of Intelligent Film and Television Communication Optical Technology Based on Network New Media

With the support of new media technology, the development and progress of the film and television industry require the support of advanced equipment and technology. Intelligent film and television communication optics is a digital and multimedia operation mode based on network new media platforms. Therefore, in practical applications, it is necessary to strictly control its performance.

4 Experiment on Intelligent Film and Television Communication Optical Technology Based on New Network Media

Table 1 shows the testing parameters of intelligent film and television communication optical technology. Intelligent film and television communication optical technology based on network new media can achieve 3D spatial scene generation using advanced methods such as modern computers, digital images, video compression and storage. The whole 3D scene image processing and kinematics analysis can be realized by using projector, camera, video and other equipment.

Table 1. Performance test parameters of intelligent film and television propagation optical technology

Test node	Resolution ratio(px)	Sensitivity(ISO)	Correction ability(%)
1	1280*720	200	86
2	1280*720	100	85
3	1280*720	200	89
4	1280*720	200	84
5	1280*720	400	87

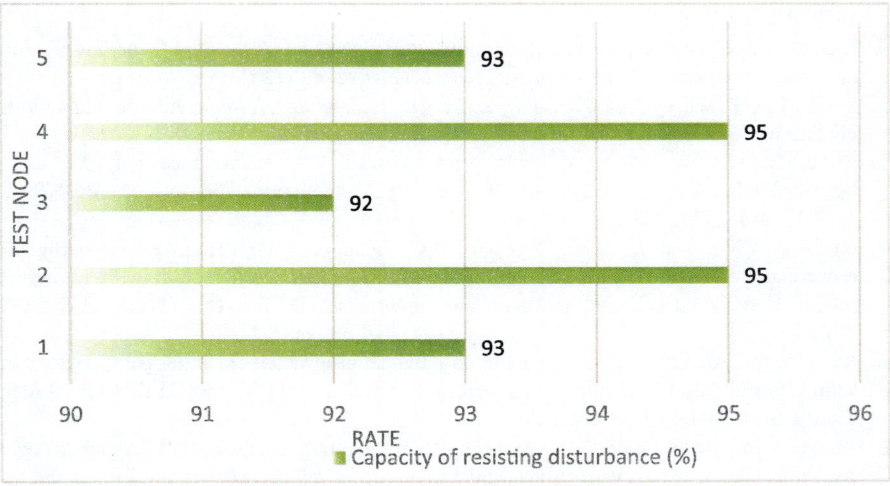

Fig. 4. Capacity of resisting disturbance

The process simulation simulation of film and television production can be realized through network software. Virtual reality was used to verify that the model met the requirements, and the virtual reality system was used to study the effect of image processing and the variation of various performance parameters in different scenarios. In the new media environment, combining existing algorithms with traditional optical methods, the optimization design goal was achieved. Therefore, this article optimized and redesigned the network data collection equipment to meet the above requirements and achieve high-speed real-time transmission. In Fig. 4, the anti-interference ability of this model ranged from 92% to 95%.

5 Conclusions

In the era of new media, as a new type of communication medium, the film and television industry is developing at an astonishing speed, and is widely used in digitization, network technology, and multimedia. This article studied intelligent film and television communication optical materials based on network new media systems, and analyzed the existing research achievements and application status in this field. After that, the basic principles and commonly used methods were elaborated.

References

1. Zhang, H., et al.: Intelligent omni-surfaces for full-dimensional wireless communications: principles, technology, and implementation. IEEE Commun. Mag. **60**(2), 39–45 (2022)
2. Chao, M., Pan, Y., Zeng, C.Y.: Intelligent interaction design research based on block chain communication technology and fuzzy system. J. Intell. Fuzzy Syst. **39**(2), 1685–1691 (2020)
3. Zhu, L., Karim, M.M., Sharif, K., Xu, C., Li, F.: Traffic flow optimization for UAVs in multi-layer information-centric software-defined FANET. IEEE Trans. Veh. Technol. **72**(2), 2453–2467 (2023)
4. Erçetin, Ö.: Computational and communication aspects of digital twin: an information theoretical perspective. IEEE Commun. Lett. **27**(2), 492–496 (2023)
5. Lei, Y., Liu, Y., Wu, Q., Yuan, X., Ning, J., Duo, B.: Enhancing UAV-enabled communications via multiple intelligent omni-surfaces. IEEE Commun. Lett. **27**(2), 655–660 (2023)
6. Badarneh, O.S., Awad, M.K., Muhaidat, S., Almehmadi, F.S.: Performance analysis of intelligent reflecting surface-aided decode-and-forward UAV communication systems. IEEE Syst. J. **17**(1), 246–257 (2023)
7. Nguyen, T.N., Thang, N.N., Nguyen, B.C., Hoang, T.M., Tran, P.T.: Intelligent-reflecting-surface-aided bidirectional full-duplex communication system with imperfect self-interference cancellation and hardware impairments. IEEE Syst. J. **17**(1), 1352–1362 (2023)
8. Kavousifard, A.: Editorial special issue on application of advanced intelligent methods in vehicle to smart grid communications: security, reliability, and resiliency. IEEE Trans. Intell. Transp. Syst. **24**(1), 891–892 (2023)
9. Kumar, P.M., Konstantinou, C., Basheer, S., Manogaran, G., Rawal, B.S., Babu, G.C.: Agreement-induced data verification model for securing vehicular communication in intelligent transportation systems. IEEE Trans. Intell. Transp. Syst. **24**(1), 980–989 (2023)
10. Liu, R., Liu, A., Qu, Z., Xiong, N.N.: An UAV-enabled intelligent connected transportation system with 6G communications for internet of vehicles. IEEE Trans. Intell. Transp. Syst. **24**(2), 2045–2059 (2023)

Design of Heterogeneous Database Encryption and Decryption System Based on Data Mining

Yanhui Bai(✉)

School of Mathematics and Computer Science, Chifeng University, Nei Monggol, Inner Mongolia, Chifeng 024000, China
cf_byh@163.com

Abstract. The main idea behind this method is to build a model to describe the relationship between encrypted data and plaintext data, as well as the relationship between the two models. This information can be used to construct an encryption key. In addition, it will help to find out which parts of plaintext are similar to some parts of encrypted data. When an application has a large amount of sensitive information, you may need to protect this information. In this case, it is necessary to encrypt all or part of the sensitive information so that only authorized users can access them. The second point is how to manage encrypted data.

Keywords: Data mining · database Encryption system

1 Introduction

Security and confidentiality has long been a bottleneck restricting the development of e-commerce and government affairs. So far, the solutions to ensure information security are mainly cryptographic: one is the symmetric encryption mode represented by DESu encryption algorithm; "Symmetrical" party use the same key p. Symmetrical encryption algorithm is simple and fast to use. This encryption technology is widely used at present [1]. Although symmetric encryption algorithm has the advantages of high efficiency and speed, the number of keys increases with the number of users, and the generation, storage and distribution of keys will be a difficult problem to solve. The other is the asymmetric cipher mode represented by RSApy encryption algorithm. Here, "public key" refers to the public key, while "private key" refers to the public key, which can only be known by the holder. However, the password mode has some fatal defects, such as easy to forget, easy to lose and easy to be stolen, and the security is not satisfactory [2]. The uncertainty of communication objects in e-commerce and government affairs determines that asymmetric encryption mode is the main way to realize network encryption and decryption. Asymmetric encryption system can publicly build a secure communication channel by issuing public key and hiding private key, but the public key issued by the publisher does not contain information indicating the topology of the issuer's fingerprint. Based on this, this paper studies.

J. C. Hung et al. (Eds.): FC 2023, LNEE 1131, pp. 125–131, 2024.
https://doi.org/10.1007/978-981-99-9299-7_17

2 Related Work

2.1 Format of Data Transmission Package

RapidIO operations are request and response based. Data packet is the basic unit of communication between terminal devices in the system. After the transaction initiator initiates a request transaction and sends it to the target device, the target device generates a response and returns it to the transaction initiator to complete the handshake operation. In hardware implementation, RapidIO terminals are usually not directly connected to each other, but connected through one or more switching devices. The switching device sends a check symbol to confirm the receipt of the request packet and sends it to the target device. At this point, the request phase of the operation has been completed. When the target device completes the operation corresponding to the request, a response transaction is generated. The packet corresponding to the response transaction is retransmitted to the initiator device through the switching device. Use the check symbol to view each packet in the transmission. When the response packet arrives at the initiator's equipment and is confirmed, the operation is considered to be completed [3].

RapidIO packet format is mainly composed of several fields in the physical, transmission and logical layer protocol specifications. The physical layer is the header of the RapidIO packet, followed by the transport layer and the logical layer. Common message can be divided into request message and response message. Figure 1 shows the RapidlO request packet format.

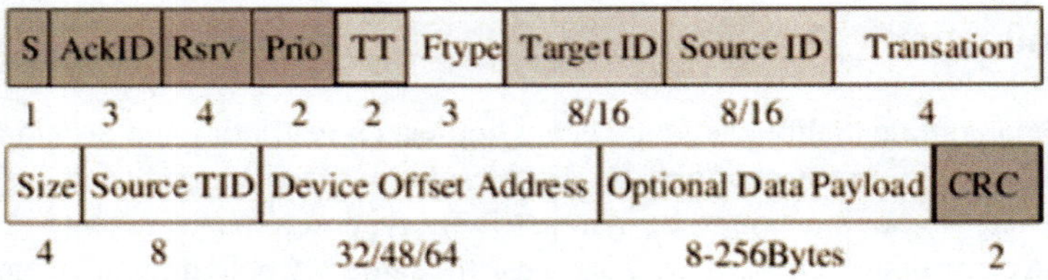

Fig. 1. RapidlO request packet format

The link interface mainly includes two ports: receive port and send port, and is designed to connect with the physical layer or buffered application conforming to RapidIO.

There are mainly two types of maintenance interfaces: maintenance request or response port and configuration record interface, which are used to read and write the logical layer configuration register. The configuration register port also controls reading and writing data from user-defined or connected physical layer configuration registers.

2.2 Decryption Algorithm

It is precisely because of the special structure of data mining that it is possible to define an equivalent decryption algorithm. In the data mining specification, InvSubBytes, InvShiftRows, InvMixCloumns and AddRoundKey are used to represent each stage of the operation, corresponding to each step of the encryption, but it is an inverse and reverse order operation process. The decryption process is different from the encryption process in steps, and the sequence of steps is slightly different [4].

SubBytes is a nonlinear operation. First, you need to define an S-BOX, that is, an S-Box. The S-Box is a matrix of 16 * 16 bytes, including 28 = 256 possible transformations. Each byte in the State is mapped to a new byte by searching the S box. The method is to first take out the high 4 bits of the byte and record them as the value of the row, then take out the low 4 bits of the byte and record them as the value of the column. Through the value of the row and column, you can find and take out the corresponding element of the row and column in the S box as the output. For example, suppose the original word in the State is {0x19}, which corresponds to 1 row and 9 columns of the S box, and then replace it with the value corresponding to the position in the S box, such as {Ox D4}. How to search S boxes quickly and efficiently is a difficult point of data mining algorithm. The encryption and decryption flow chart is shown in Fig. 2 below.

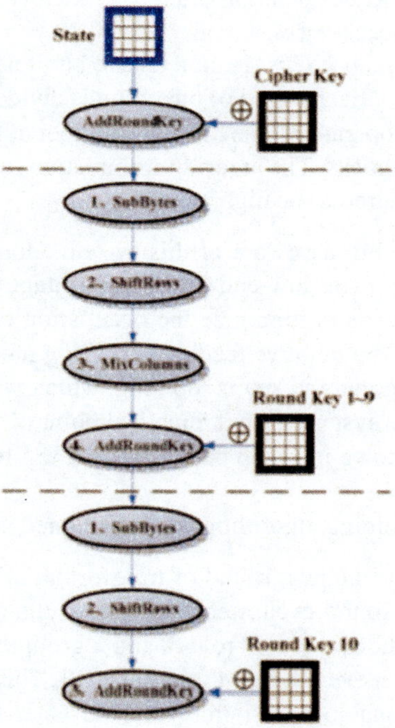

Fig. 2. Encryption and decryption flow chart

There is a round constant in data mining, that is, l Rcon [j], which is a word. The three bytes on the right of Rcon [j] are always "O". In the encryption/decryption algorithm of data mining, there is an operation process of key expansion through the seed key, so that multiple keys can be generated from one original key. Although the complexity of the program is increased, this process greatly increases the spread of bits [5]. For example, if the input key is four words, a 44 word one-dimensional array can be obtained through key expansion, Thus, a 16 byte round key is provided for each round of data mining operation.

3 Design of Heterogeneous Database Encryption and Decryption System Based on Data Mining

The hardware of data mining algorithm can be designed in many ways. r351:

1) Serial mode. In this way, the next round of transformation is implemented by combining logic. In the 10 round cycle iteration, the results of the previous round are directly used as the input of the next round, and a grouping operation is completed in one cycle, so as to achieve the best throughput [6]. This method requires a lot of memory and combinational logic resources, and the general FPGA chip can not meet the requirements, and the clock frequency is low;
2) Basic iterative feedback mode. In all iterations, only one round transformation module is used to complete a packet operation after 10 clock cycles. Corresponding to the first method, less resources are consumed;
3) In wheel assembly line mode. This method is fast, but it needs to insert a register in the round transformation. Each round of operation is divided into multiple operation segments for pipelining operation. The more the number of pipeline stages in a round, the more the number of clocks. The algorithm simulation frequency can be very high, but the algorithm is required to be high.

The data mining algorithm hardware of this system adopts the implementation of EP2C8Q208, which belongs to the low-end chip. Its advantage is cost performance rather than performance [7]. It needs to minimize the occupation of resources. Therefore, it was decided to adopt the basic iterative feedback working mode for design.

In addition, considering that the key is too long, which will lead to too many input and output interfaces of the system, from a practical point of view, the key selected for this system is 128bit. As shown in Fig. 3 below, it is the top-level file of the encryption and decryption module.

The hardware of data mining algorithm can be designed in many ways:

1) Serial mode. In this way, the next round of transformation is implemented by combining logic. In the 10 round cycle iteration, the results of the previous round are directly used as the input of the next round, and a grouping operation is completed in one cycle, so as to achieve the best throughput [8]. This method requires a lot of memory and combinational logic resources, and the general FPGA chip can not meet the requirements, and the clock frequency is low;
2) Basic iterative feedback mode. In all iterations, only one round transformation module is used to complete a packet operation after 10 clock cycles. Corresponding to the first method, less resources are consumed;
3) In wheel assembly line mode. This method is fast, but it needs to insert a register in the round transformation. Each round of operation is divided into multiple operation segments for pipelining operation. The more the number of pipeline stages in a round, the more the number of clocks. The algorithm simulation frequency can be very high, but the algorithm is required to be high [9].

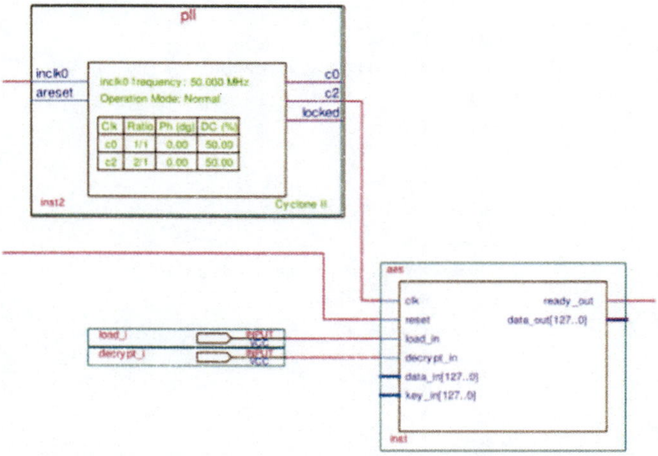

Fig. 3. Encryption and decryption module top-level file

The data mining algorithm hardware of this system adopts the implementation of EP2C8Q208, which belongs to the low-end chip. Its advantage is cost performance rather than performance. It needs to minimize the occupation of resources. Therefore, it was decided to adopt the basic iterative feedback working mode for design.

4 Simulation Analysis

In the decryption module, except for column mixed program design, other program design processes are roughly the same as encryption. The byte transformation in the decryption process changes the value in the replacement table to the inverse of Sbox. The program design of the line transformation module during decryption changes the circular left shift in the rules during encryption into circular right shift.

The biggest difference between decryption module programming and decryption is the multiplication operation when columns are mixed. Since the elements in the fixed matrix in the column mixed multiplication during decryption are 0x09, 0x0B, 0x0D, 0xOE, and the functions mult9, multB, multD, multE are defined in turn, the data interface code of the decryption module is shown in Fig. 4 below:

When the CPU configures the DMA descriptor, the CPU acts as the task initiator and the FPGA acts as the task target, so the data is transmitted through the Treq interface. The internal register of FPGA is 16 bit access, and its address is srio_ Addr [15:0] The address Treq in the Treq bus_ Addr and reg_ low_ Addr is composed of two parts. Treq_ Only [15:3] is valid for addr, which is the upper 13 bits of the internal register address and the lower 3 bits are determined by reg_ low_ Addr padding[10].

After the CPU sends the NWRITE request_ byte_ en_ The judgment of n completes the address mapping from the address on the Treq bus to the internal register of FPGA, and Terq_ The valid bytes in data are stored in the register.

```
   i=0
   while(I<Nk){
   w[i]=word[ key[4*i], key[4*i+1], key[4*i+2], key[4*i+3] ]
   i=i+1}
end while
i=Nk
 while( i<Nb*(Nr+1) ){
     word temp=w[i-1]
     if ( I mod Nk=0 )
        temp=SubWord( RotWord(temp) ) xor Rcon[i/Nk]
     else if ( Nk=8 and I mod Nk=4)
        temp=SubWord( temp )
end if
 w[i]=w[i-Nk] xor temp
 i=i+1}
 end while
 end
 }
```

Fig. 4. Decrypting module data interface code

5 Conclusion

The system aims to protect sensitive data in existing relational databases by using various types of encryption technologies with different security levels. It can be used for symmetric and asymmetric encryption. First, describe the design process in detail by describing the functions of each component (such as user interface, encryption engine, decryption engine, etc.), and then explain how to integrate these components into a single system. Finally, some performance results are given to prove that the design proposed by us works well under actual conditions.

References

1. Xie, J.: Design and implementation of marine massive heterogeneous data integration system based on cloud platform. Ship Sci. Technol. (2016)
2. Hou, S., Ye, Y., Song, Y., et al.: HinDroid: an intelligent android malware detection system based on structured heterogeneous information network. SIGKDD Explor. (2017)
3. Yu, Y., Xin, L., Dan, L., et al.: An innovative library decision-making support service system based on multi-source heterogeneous database integration. Res. Libr. Sci.(2017)
4. Li, X., et al.: Distributed Data Mining for Multiple Sourced Heterogeneous Datasets: A Survey (2018)
5. Li, K., Chen, Y.: Fuzzy clustering-based financial data mining system analysis and design. Int. J. Found. Comput. Sci. **33**(06n07), 603–624 (2022)
6. Mutabaruka, E.: Enhancing Data Security by Using Hybrid Encryption Technique (Advanced Encryption Standard and Rivest Shamir Adleman) (2017)
7. Luo, N., Tao, J., Liu, J.: Research on heterogeneous monitoring between business systems based on big data outlier mining algorithm. Autom. Instrum. (2019)
8. Shi, Y., Yang, Z.: Discussion on electric power user information mining based on big data. Electron. Test (2018)

9. Garcia, A., et al.: Heterogeneous data source integration for smart grid ecosystems based on metadata mining. Expert Syst. Appl. **79**(Aug), 254–268 (2017)
10. Kang, L.: Research and Implementation of the Heterogeneous System Integration in Universities Libraries Based on Network Monitoring. Library Work in Colleges and Universities (2017)

Research and Implementation of Music Recommendation System Based on Particle Swarm Algorithm

Yawen Chen(✉)

Jiangxi Metallurgical Vocational and Technical College, Xinyu 338000, Jiangxi, China
c350906807@163.com

Abstract. The role and significance of recommendation system in music teaching is very important, but there is a problem of low management level. The recommendation system cannot solve the problem of processing multi-note data in music teaching, and the recommendation accuracy is poor. Therefore, this paper proposes particle swarm optimization to optimize the music recommendation system. Firstly, music teaching standards are used to classify music data, and selected according to the degree of compliance to realize the preprocessing of music data. Then, according to the degree of compliance, a systematic review collection is formed, and the evaluation results are analyzed. MATLAB simulation shows that the particle swarm algorithm has a higher degree of optimization for the music recommendation system and improves the compliance rate of music selection, which is better than the single system method.

Keywords: music teaching · Standards · Recommended · particle swarm algorithm · System

1 Introduction

The role of the recommendation system in music selection is very important, and the content of music teaching can be reasonably selected [1]. However, in the process of building the music recommendation system, there is a problem of low system selection accuracy, and the recommendation cannot be played effectively [2]. Some scholars believe that the application of particle swarm algorithm to the recommendation system can effectively select information, achieve stability analysis [3], and provide support for the optimization of the recommendation system [4]. On this basis, particle swarm the recommendation system and better select music.

2 Related Concepts

2.1 Mathematical Description of Particle Swarm Arithmetic

The particle swarm algorithm uses the beat, melody and music conformance to optimize the music data, and finds the recommendation system according to the recommendation indicators and form views, data tables. By integrating the analysis results [5],

J. C. Hung et al. (Eds.): FC 2023, LNEE 1131, pp. 132–139, 2024.
https://doi.org/10.1007/978-981-99-9299-7_18

the optimization results are finally obtained. Particle swarm algorithm combined with music teaching standards to optimize music data can improve the management level of recommendation systems [6].

Hypothesis 1: The music data is $\sum x_i$, the recommendation data set is y_i, the music conformance degree is x_i, and the recommendation system optimization function is $f(x_i)$ as shown in Eq. (1).

$$f(x_i) = \lim_{i \to \infty} \sum x_i \Leftrightarrow y_i \cup \xi^2 \tag{1}$$

2.2 Optimize the Selection of Content

Hypothesis 2: The optimization function of the recommended system is $F(x_i)$, the optimization view in the SQL database is z_i, then the optimization content selection is shown in Eq. (2).

$$F(x_i) = \frac{-x^3 \pm 4x\xi}{2\Delta x} \to z_i \cdot \xi \tag{2}$$

2.3 Handling of Abnormal Recommendation Information

The melody and syllable in the recommendation system are analyzed standardly, and the music data is mapped to the SQL view to determine the abnormal recommendation information. Comprehensively analyze music data and set constraints to support particle swarm optimization [7]. Music data needs to be preprocessed, and the processed results are included in the data collection if they meet the music selection requirements, otherwise After the calculation, the exception recommendation information is handled in Fig. 1.

The music data in Fig. 1 shows that the analysis results are uniform, line with the objective facts. Moreover, the proportion of abnormal recommendation information is small, indicating that the particle swarm algorithm has high rationality and can be studied as a recommendation system [8]. The identification of abnormal recommendation information can reduce the processing of music information and lay the foundation for further analysis.

The Correlation between different music. The particle swarm algorithm comprehensively analyzes the music data and adjusts the relevant parameters to optimize the method of the recommendation system [9]. The particle swarm algorithm classifies the recommendation information and randomly selects different methods to achieve correlation analysis of music. In the correlation analysis process, the nature of the music can be identified and the music selection can be better made.

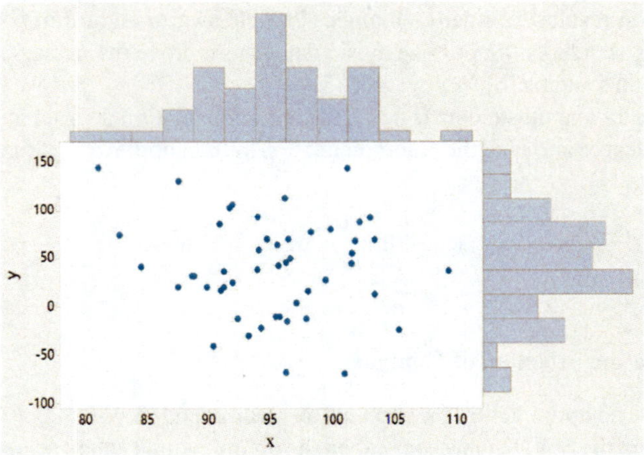

Fig. 1. Analysis results of particle swarm algorithm

3 Actual Examples of Recommendation Systems

3.1 Data on Music Recommendations

In order to facilitate the analysis of information systems, different types of music information are studied in this paper, and the data are shown in Table 1.

Table 1. Recommended parameters for music

Music genre	range	Amount of data	rationality	Constraints
popular	recommend	49.47	42.11	normal
	normal	22.11	46.32	normal
Classical	recommend	24.21	33.68	normal
	normal	31.58	41.05	normal
Rock	recommend	18.95	43.16	normal
	normal	27.37	49.47	normal

The processing of different music genres in Table 1 is shown in Table 2.

Table 2. Beat processing

source	degree of freedom	Adj SS	Adj MS	F-number	P-value
regression	30.53	53.68	18.95	21.05	30.53
3	56.84	25.26	21.05	37.89	56.84
5	37.89	54.74	49.47	32.63	37.89

It from Table 1 that the single system, the analysis results of particle swarm optimization are closer to the actual requirements. In terms of F value and P value in the recommendation system, the results of particle swarm optimization are better than the single system method. From the beat change in Fig. 4, the particle swarm is better and the judgment speed is faster [10]. Therefore, the information processing speed, stability and optimization degree of particle swarm optimization are better.

3.2 Optimized Ratio of Music Recommendations

The optimization of the information includes abnormal recommendation information, beat, speed. After the constraint standard screening of particle swarm algorithm [21], the preliminary management results are obtained, and the correlation of the management results is analyzed. In order to verify the effect more accurately, select different abnormal recommendation information and calculate the overall stability of the system information, as shown in Table 3.

Table 3. Analyzes the overall situation

Optimize the proportion	Degree of systematization	Outlier recognition rate
25%	63.16	40.00
50%	36.84	45.26
70%	51.58	36.84
mean	48.42	62.11
X^2	32.63	18.95
25%	52.63	34.74
50%	38.95	29.47
70%	58.95	36.84
mean	22.11	45.26
25%	31.58	26.32

(continued)

Table 3. (*continued*)

Optimize the proportion	Degree of systematization	Outlier recognition rate
50%	60.00	34.74
70%	56.84	28.42
mean	60.00	43.16

3.3 Stability and Accuracy of the Recommendation System

To verify the accuracy, the stability and accuracy are compared with the recommendation system, and the results in Fig. 2.

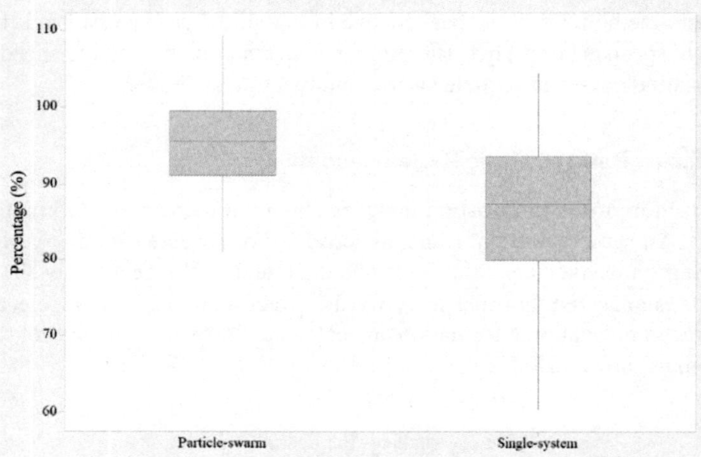

Fig. 2. Optimization stability of different algorithms

It can be seen from Figs. 3 and 4 that the optimization stability of particle swarm optimization is shorter than that of the single system method, but the error rate is lower, indicating that the selection of particle swarm optimization algorithm is relatively stable, while the optimization degree of the single system method is uneven.

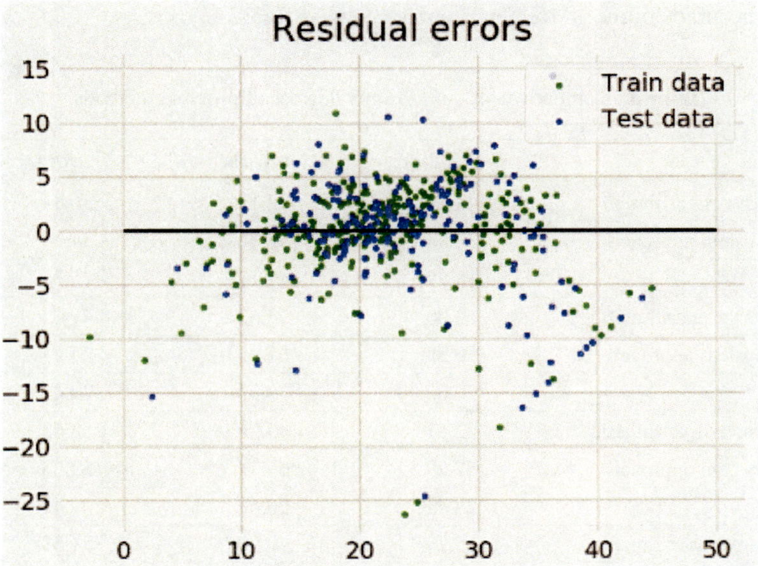

Fig. 3. Optimization stability of different algorithms

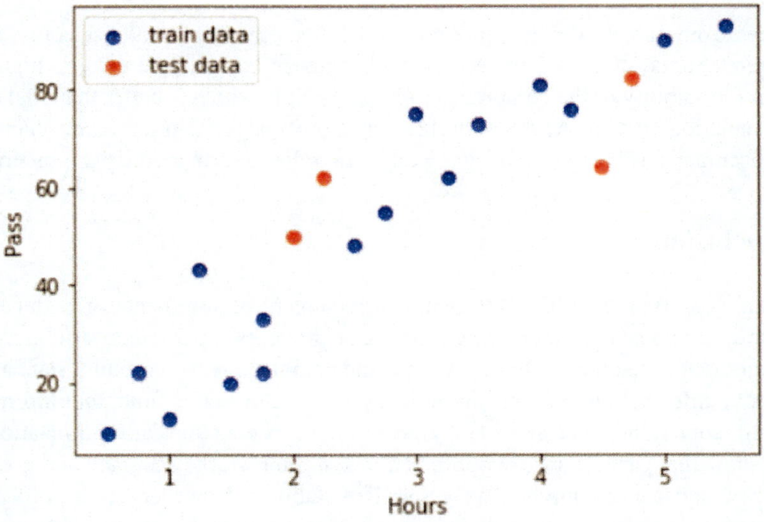

Fig. 4. Optimization stability of different algorithms

The accuracy of the above algorithm is shown in Table 4.

Table 4. Comparison of optimization degrees of different methods

algorithm	stability	accuracy	error
Particle swarm arithmetic	44.21	22.11	36.84
Single system approach	30.53	56.84	34.74
P	15.79	42.11	25.26
Particle swarm arithmetic	60.00	52.63	41.05
Single system approach	50.53	61.05	24.21
P	20.00	30.53	29.47
Particle swarm arithmetic	22.11	45.26	52.63
Single system approach	57.89	54.74	52.63
P	28.42	29.47	29.47
Particle swarm arithmetic	15.79	40.00	51.58
Single system approach	17.89	54.74	51.58
P	52.63	17.89	41.05

It seen from Table 3 that has shortcomings in the optimization the selection of the recommendation system, and the accuracy changes greatly and the error is high. The optimization stability of the comprehensive results high, which is better than that of the recommendation system. At the optimization stability algorithm is greater than 90%, and the accuracy has not changed significantly. In order to further verify the superiority.

4 Conclusion

Under the condition that the information management requirements of colleges and universities are constantly increasing, this paper proposes a particle swarm algorithm for the problem of recommendation system, and combines music teaching standards to improve the information relationship in the system. At the same time, the information processing constraint criteria are analyzed in depth to construct an optimization set. Studies show that particle swarm optimization can improve the accuracy and accuracy of stability, and can comprehensively stabilize stability. However, in the process of particle swarm arithmetic, too much attention is paid to the ability of unilateral index analysis, and the proportion of stability is ignored.

References

1. Hu, J., Xie, C.: Research and implementation of e-commerce intelligent recommendation system based on fuzzy clustering algorithm. J. Intell. Fuzzy Syst. **3**, 1–10 (2021)

2. Tang, J.: Optimization of English learning platform based on a collaborative filtering algorithm. Complexity **2021**, 1–14 (2021)
3. Zhu, Y.: Personalized recommendation of educational resource information based on adaptive genetic algorithm. Int. J. Reliab. Qual. Saf. Eng. **30**(02) (2023)
4. Yu, H.: Apriori algorithm optimization based on spark platform under big data. Microprocess. Microsyst. **80**(11), 103528 (2021)
5. Chen, H., Yu, J., Zhou, H., et al.: SmartStore: a blockchain and clustering based intelligent edge storage system with fairness and resilience. Int. J. Intell. Syst. (2021)
6. Zhang, L.: Optimization of an intelligent music-playing system based on network communication. Complexity (2021)
7. Guerrini, G., Romeo, L., Alessandrini, D., et al.: Analysis, design and implementation of a forecasting system for parking lots occupation (2021)
8. Zhang, S., Tang, M., Zhang, Q., et al.: Given users recommendations based on reviews on Yelp (2021)
9. Minuto, A., Celi, E., Timo, G., et al.: New maximum power point tracking MPPT algorithm based on research of a target voltage range and its implementation in a commercial inverter for photovoltaic systems. In: European Photovoltaic Solar Energy Conference and Exhibition (2021)
10. Malinowski, M.: Implementation of recommendation algorithm based on recommendation sessions in e-Commerce IT system (2021)

Design of High-Speed Multi-channel Data Transmission System Based on Single-Chip Microcomputer

Rong Rong Cui(✉) and Yanyan Ren

College of Mathematics and Computer Science, Chifeng University, Chifeng 024000, China
yan0203140203@163.com

Abstract. The role of speed multiplexing the design of data transmission system is very important, but there is a problem of unstable data transmission. Single data transfer does not solve the problem of multiple types of data transfer and is less accurate. Therefore, this paper proposes a single-chip microcomputer high-speed multiplexing method to construct a data transmission model. First, classify according to the transmission plan and transmission data, and select the scheme according to the transmission standard to realize the transmission of Standardized data treatment. Then, according to the classification of transmitted data, a collection of transmitted data is formed, and an iterative the transmitted data. MATLAB simulation that under the condition that the transmission plan is unchanged, the integrity of single-channel data transmission by the high-speed multiplexing method of a single-microcomputer microcomputer The transfer time is better than the standard transmission method.

Keywords: high-speed multiplexing · transmission system · Single-chip microcomputer high-speed multiplexing method · Optimize

1 Introduction

Transmission optimization is one of the important data of high-speed multiplexing, which the improvement of transmission system [1]. However, in the process of transmission system construction [2], high-speed multiplexing has the problem of unstable data transmission, and the role of high-speed multiplexing cannot be effectively played [3]. Some scholars believe that applying intelligent methods such as decision trees to single data transmission can quickly classify schemes [4], save high-speed multiplexing time analysis, and provide support for transmission optimization [5]. On this basis, this paper proposes a single-chip high-speed multiplexing method to optimize the single-channel data transmission and verify the effectiveness of the model implementation.

J. C. Hung et al. (Eds.): FC 2023, LNEE 1131, pp. 140–145, 2024.
https://doi.org/10.1007/978-981-99-9299-7_19

2 Related Concepts

2.1 Mathematical Description of Single-Chip Microcomputer High-Speed Multiplexing Method

The single-chip high-speed multiplexing method sets the transmission target, transmission scheme and transmission standard, makes decisions on the transmission data, and discovers the transmission according to the high-speed multiplexing index and form a program rectification plan. The correlation between the solutions is analyzed by integrating the transmission optimization results. The single-chip high-speed multiplexing method combines transmission theory to optimize the transmission results and improve the level of high-speed multiplexing.

Hypothesis 1: The scenario is l_i, the set of optimization results is $\sum l_i$, the importance of the scheme is ϕ_i, and the transfer judgment function is $f(l_i)$ as shown in Eq. (1).

$$f(l_i) = \sum l_i \not\subset \phi_i \rightleftarrows \xi \tag{1}$$

ξ Adjust the factor for the scheme to reduce the impact of redundant data.

2.2 Selection of Transmission Optimization Scheme

Hypothesis 2: The selection function of the transfer optimization method is $F(x_i)$ and the scheme weight coefficient is ι_i, then the selection of the transfer optimization method is shown in Eq. (2).

$$F(x_i) = z_i \xrightarrow{\iota_i \cdot \xi} f(x_i|y_i) \tag{2}$$

2.3 Processing of Redundant Data

Before analyzing the single-chip microcomputer high-speed multiplexing method, the transmission time and single transmission data should be cleaned, and the scheme should be mapped to the selection table to judge the data Redundancy. Firstly, the scheme is comprehensively analyzed, and the transmission standard and weight are set to support the accurate analysis of the high-speed multiplexing method of the single-chip microcomputer. The scheme data needs to be standardized and processed if the processed results meet the transmission requirements, indicating that the processing is valid, otherwise, it is carried out again. Scenario revisions. In the single-chip high-speed multiplexing method should be used to select the scheme, and the specific method selection in Fig. 1.

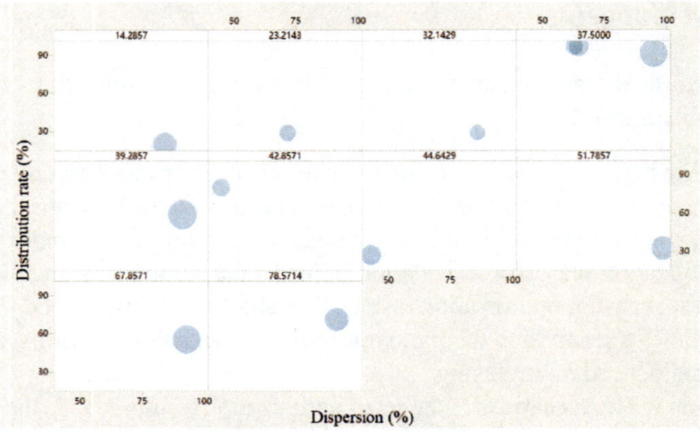

Fig. 1. The process of high-speed multiplexing analysis of single-chip microcomputer

3 Practical Examples of Single-Channel Data Transmission

3.1 Transmission System Situation

In order to facilitate the analysis of transmission systems, the different types of transmission systems in this paper are studied as the research objects, and the number of test orders is 2421, and the time is 450 This is shown in Table 1.

Table 1. Evaluation of single data transmission

parameter	path	Amount of data	Loss rate	Transmission standard
path	single	91	7.14	7.86
	many	89	0.36	5.36
Transfer metrics	single	64	8.93	5.00
	many	57	2.86	0.36
Comprehensive transmission effect	single	94	9.64	9.64
	many	71	4.29	3.21

The transfer processing of the different paths in Table 1 is shown in Fig. 2.

It from Table 1 the standard transmission method, the scheme of the single-chip high-speed multiplexing method is closer to the actual time. In terms of high-speed multiplexing selection scheme node selection rate and accuracy, single-chip microcomputer high-speed multiplexing method single-channel data transmission. The change of scheme node in Fig. 3 shows that the accuracy of the single-chip high-speed multiplexing method is better, and the judgment speed is faster. Therefore, the processing speed, time, and completeness of the single-chip high-speed multiplexing method are better.

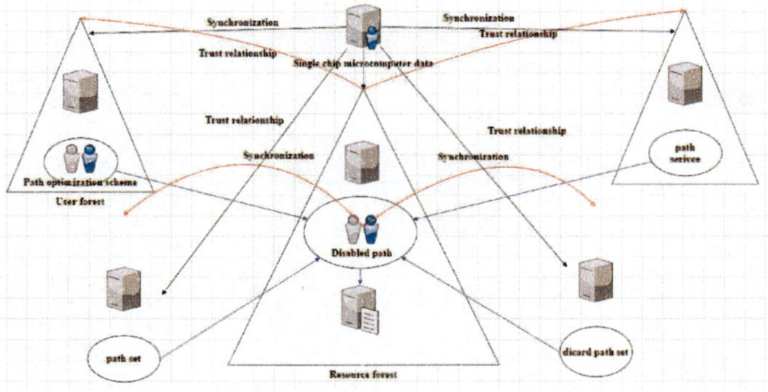

Fig. 2. Multipathing

3.2 Optimization Ratio of the Scheme

The optimization of the scheme includes redundant data, the number of paths, and speed. After the standard screening of the single-chip high-speed multiplexing method, the preliminary optimization results are obtained, and the correlation of scheme data is analyzed. In order to verify the effect more accurately, select different redundant data and calculate the overall time of high-speed multiplexing, as shown in Table 2.

Table 2. Processing results of high-speed multiplexing

Path optimization rate	Scenario match rate	The recognition rate of redundant data
25%	96.43	71.43
50%	62.50	25.00
70%	75.00	87.50
mean	87.50	28.57
X^2	17.86	94.64
P = 0. 031		

3.3 Time and Accuracy of Transmission Optimization

In order to verify the accuracy multiplexing method, the time and accuracy are compared with single-channel data transmission, and the in Fig. 3.

It from Fig. 3 the transmission of the single-chip high-speed multiplexing method is shorter than that of the standard transmission method, but the error rate is lower, indicating that the choice of the single-chip high-speed multiplexing method is relatively stable, while the standard transmission method The completeness of is uneven. The accuracy of the above algorithm is shown in Table 3.

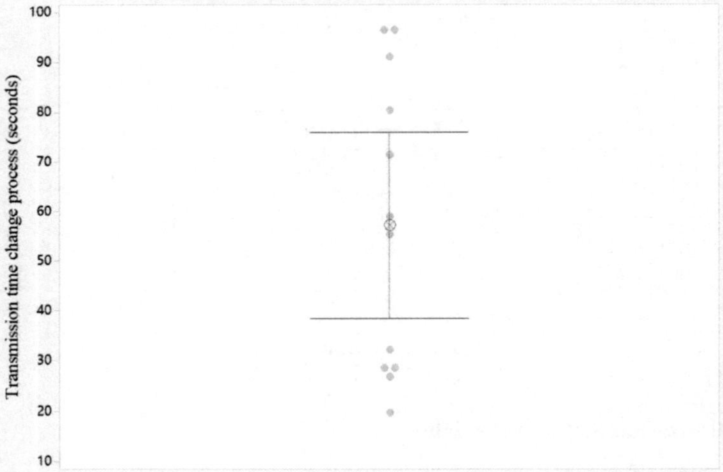

Fig. 3. Transmission time of high-speed multiplexing method of the single-chip microcomputer

Table 3. Time and accurate letter comparison of different methods

algorithm	Transfer time	accuracy	error
The single-chip high-speed multiplexing method	14.29	90.00	2.36
Standard transfer method	66.07	88.57	3.93
P	35.71	51.79	7.86

It Table 3 that the selection data transmission in high-speed multi-channel transmission has shortcomings in transmission time and accuracy, and the accuracy changes greatly, and the error rate is high. The transmission time of high-speed multiplexing of the single-chip high-speed multiplexing method is shorter, which is better than the standard high-speed multiplexing method. At the transmission the single-chip high-speed multiplexing method is greater than 90%, and the accuracy has not changed significantly. In the single-chip microcomputer high-speed multiplexing. In further the sustainability of the single-chip high-speed multiplexing method, the high-speed multiplexing method is comprehensively analyzed by different methods, and the result 4 is shown.

It from Fig. 4 that the single-chip high-speed multiplexing method are significantly better than the standard transmission method, and the reason is that the single-chip high-speed multiplexing method increases the time adjustment coefficient and is set Corresponding standards, proposing results that do not meet the requirements.

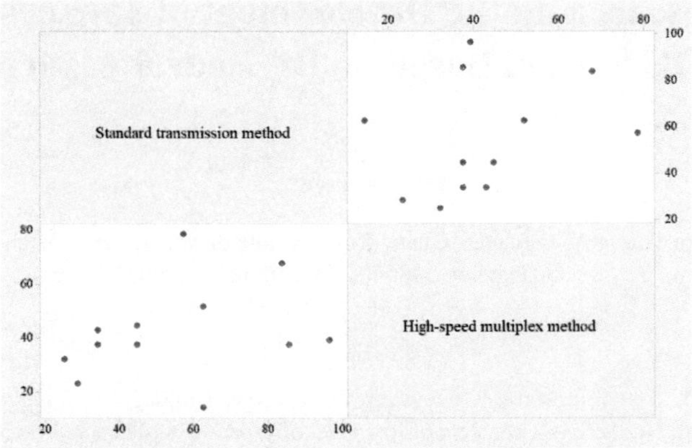

Fig. 4. Synthesis of inter-program evaluations with different methodologies

4 Conclusion

In the case of increasing transmission standards, aiming at the problem of low high-speed multiplexing level, this paper proposes a single-chip high-speed multiplexing method and combines transmission theory to transmit data to the scheme The relationship is analyzed. At the same time, the program processing standards are corrected and the optimization set is constructed. Studies show that the single-chip high-speed multiplexing method can improve the accuracy of time, shorten the comprehensive time, and is better than the standard high-speed multiplexing Method. However, in the process of single-chip microcomputer high-speed multiplexing method, too much attention is paid to the analysis ability of scheme data and the independence of scheme data.

References

1. Vahidi, V.: Uplink data transmission for high speed trains in severe doubly selective channels of 6G communication systems. Phys. Commun. **49**, 101489 (2021)
2. Jiang, L., Yan, L., Yi, A., et al.: Integrated components and solutions for high-speed short-reach data transmission. Photonics **8**(3), 77 (2021)
3. Zbinden, E.: Vertical interconnect system for high-speed data transmission. WO2022081682A1 (2022)
4. Weng, D., Wallace, J., Xie, X., et al.: High-speed data transmission via RJ45-connectored optical cable system. In: Conference on Ultra-High-Definition Imaging Systems (2021)
5. Pei, X.: Design of a multi-function high-speed digital baseband data acquisition system **21**(10), 61–72 (2021)

Research on the Development of Japanese MOOC System Based on BP Neural Algorithm

Weizhou Feng(✉)

School of Foreign Languages, Guangzhou Institute of Science and Technology, Guangzhou 510540, Guangdong, China
jojoconti@126.com

Abstract. The role of MOOC system in Japanese language teaching is very important, but there is a problem of low level of Japanese assistance. The MOOC system cannot solve the problem of translating grammar and words in Japanese language teaching, and it is less logical. Therefore, this paper proposes a BP neural algorithm to construct an optimization model for MOOC system. First of all, the transfer standard is used to divide the Japanese teaching content, and the Japanese teaching content is carried out according to the translation requirements Pre-treatment for Japanese language teaching. Then, the transfer criteria are divided according to the teaching standards, forming an optimized set of Japanese language teaching, and the Japanese content is carried out Dig deeper. MATLAB simulation shows that the optimization degree and stability of BP neural algorithm are better than those of online translation methods under the condition that the teaching standards are consistent.

Keywords: transfer criteria · stability · BP neural algorithm · Optimize the results

1 Introduction

Japanese language teaching optimization is one of the important contents of the MOOC system [1], which plays a very important role in the improvement of the Japanese language teaching optimization system [2]. However, in the process of building the Japanese language teaching content system, the MOOC system has the problem of low level of Japanese language assistance [3], and the MOOC system cannot be effectively functioned. Some scholars believe that the application of BP neural algorithm to Japanese language teaching can effectively carry out redundant Japanese teaching and stability analysis [4], and provide corresponding support for the optimization and verification. On paper proposes a BP neural algorithm to optimize the Japanese teaching content.

2 Related Concepts

2.1 Mathematical Description of the BP Neural Algorithm

The BP neural algorithm optimizes the content of Japanese teaching by using the key points of Japanese language teaching, the relationship between Japanese language teaching, and the Japanese teaching standards, and translates according to Japanese language

J. C. Hung et al. (Eds.): FC 2023, LNEE 1131, pp. 146–152, 2024.
https://doi.org/10.1007/978-981-99-9299-7_20

teaching indicators, find outliers in Japanese language teaching optimization, and form a path table. By integrating the results of Japanese language teaching optimization, the correlation of MOOC system results is finally judged. The BP neural algorithm combines the transfer criteria and optimizes the translation results using the BP neural algorithm, which can improve the level of the MOOC system [5].

Hypothesis 1: The content of Japanese language teaching is $\sum x_i$, the set of Japanese language teaching optimization results is x_i, the Japanese language teaching standard is y_i, the judgment function of the MOOC system results is $f(x_i)$ shown in Eq. (1).

$$f(x_i) = \lim_{i \to \infty} \sum x_i | y_i \Leftrightarrow \sqrt{\xi^2 - 4xy} \tag{1}$$

2.2 Choice of Japanese Language Teaching Program

Hypothesis 2: The translation function for Japanese teaching is $F(x_i)$ and the check coefficient for Japanese teaching content is z_i, then the selection of Japanese language teaching optimization method is shown in Eq. (2).

$$F(x_i) = \frac{-x \pm \sqrt{x^2 - 4x\xi}}{2x} + [z_i \cdot \xi] \tag{2}$$

2.3 Processing of Redundant Japanese Language Teaching

Before the analysis of the BP neural algorithm, the stability and single standard analysis of the MOOC system results are performed, and the Japanese teaching content is mapped to the selection table to determine the content with semantic abnormality. First, the Japanese teaching content is comprehensively analyzed, and the translation logic and content verification of the Japanese teaching content are set as the BP neural algorithm is supported by accurate analysis. The content of Japanese teaching needs to be preprocessed, and if the processed results meet the requirements of Japanese teaching in universities, the processing is effective, otherwise the data structure is deepened. The accuracy of the BP neural and improve the translation level, the BP neural algorithm scheme should be selected, and the specific method selection is shown in Fig. 1.

The Japanese teaching content in Fig. 1 shows that the BP neural algorithm analysis is uniform and consistent with objective facts. The selection method is not directional, indicating that BP neural algorithm analysis has high accuracy, so it is studied as a Japanese teaching content. The selection method meets the mapping requirements, mainly because the selection method is adjusted by the transfer standard, the redundant Japanese teaching is eliminated, and the key points of Japanese language teaching are revised, so that the entire Japanese teaching content is highly selective.

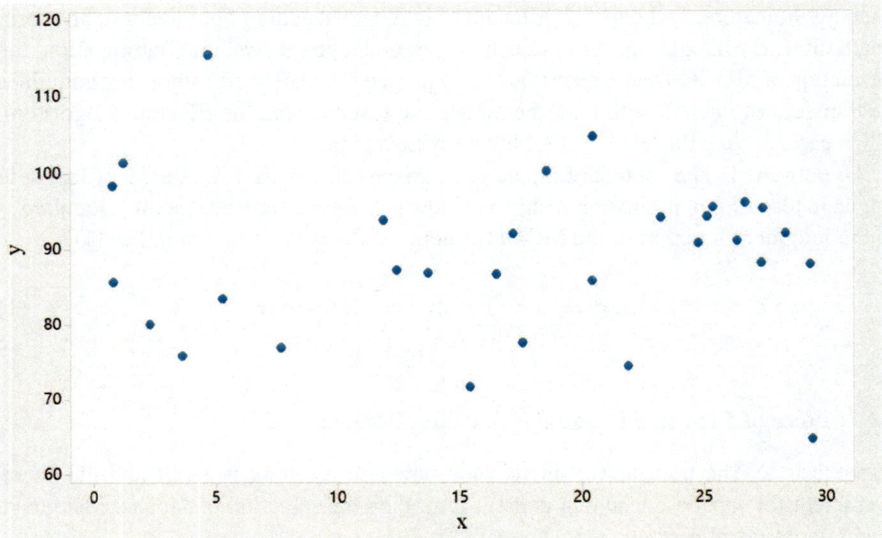

Fig. 1. Analysis results of BP neural algorithm

The correlation between different key points of Japanese language teaching. The BP neural algorithm adopts the accuracy judgment of stability, and adjusts the corresponding redundant Japanese teaching relationship to optimize the Japanese teaching content method. The BP neural algorithm divides the content of Japanese teaching into transfer classes and randomly selects different methods. In the semantic mining process, the translation of the transfer class requires correlation processing with the selection method. After the correlation processing is completed, different methods are compared for Japanese teaching content, and the translation results with the highest accuracy are stored.

3 Actual Examples of Japanese Language Teaching Content Systems

3.1 Japanese Language Teaching System

In order to facilitate the analysis of Japanese language teaching systems, the number of test orders is 2421 and the stability is 450 based on different types of Japanese language teaching systems This is shown in Table 1.

Table 1. Characteristics of different structures

Japanese language teaching system	range	Amount of words	logicality	Translation logic
SQL database	exterior	9.47	17.89	8.42
	interior	11.58	13.68	6.32
School system	exterior	16.84	9.47	9.47
	interior	21.05	20.00	18.95
client	exterior	13.68	15.79	8.42
	interior	21.05	12.63	5.26

Table 1 shows the processing process between the different key points of Japanese language teaching, as shown in Table 2.

Table 2. Handling of key points of Japanese language teaching

source	degree of freedom	Adj SS	Adj MS	F-number	P-value
regression	6.32	18.95	16.84	15.79	14.74
3	15.79	12.63	11.58	17.89	18.95
5	11.58	8.42	21.05	12.63	9.47
error	7.37	12.63	13.68	16.84	16.84

As seen from Table 1, the results of BP neural algorithm are closer to the actual stability compared with a single MOOC system. In terms of Japanese teaching content, selection rate and accuracy of key points of Japanese teaching, BP neural algorithm MOOC system. The changes in key points of Japanese language teaching in Fig. 4 show that the BP neural algorithm has better accuracy and faster judgment speed. Therefore, the Japanese assistance speed and optimization of BP neural algorithm are better.

3.2 Optimal Ratio of Japanese Language Teaching

The optimization of Japanese language teaching includes redundant Japanese language teaching, key points of Japanese language teaching, and speed. After the translation logic standard screening of BP neural algorithm, the preliminary translation results were obtained, and the translations were made The correlation of the results was analyzed. In order to verify the effect more accurately, different redundant Japanese teaching is selected to calculate the overall stability of the MOOC system, as shown in Table 3.

3.3 Stability and Accuracy of Japanese Escaping

To verify the accuracy of the BP neural algorithm, the optimization stability and accuracy comparison with the MOOC system is shown in Fig. 2.

Table 3. Overall situation of Japanese language teaching optimization

Escape ratio	The degree of escape optimization	False escape recognition rate
25%	5.26	15.79
50%	11.58	9.47
70%	14.74	8.42
mean	7.37	21.05
X^2	12.63	11.58
P = 0.042		

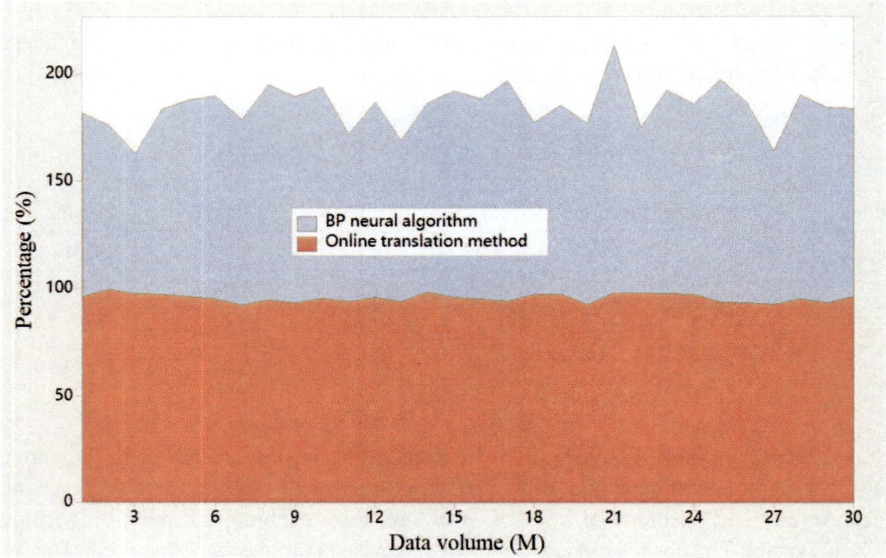

Fig. 2. The degree of optimization of different algorithms

It from Fig. 3 that the optimization stability of BP neural is shorter than that of online translation method, but the error rate is lower, indicating that the choice of BP neural algorithm is relatively stable, while the optimization degree of online translation method is uneven. The accuracy of the above algorithm is shown in Table 4.

Table 4. Comparison of optimization degrees of different methods

algorithm	Optimize stability	Escape key points	error
BP neural algorithm	93.68	92.63	5.26
Online translation method	86.84	84.74	17.89
P	6.32	7.37	7.37

Table 3 shows that the MOOC system has shortcomings in optimizing stability and accuracy in terms of logic and accuracy of Japanese teaching content, and the accuracy of data processing has changed greatly and the error is high. The optimization stability of the comprehensive results of BP neural algorithm is high, which is better than that of MOOC system. At the same time, the optimization stability of the BP neural algorithm is greater than 90%, and the accuracy has not changed significantly. In order to further verify the superiority of the BP neural algorithm. In order to further verify the continuity of the method, the BP neural algorithm was comprehensively analyzed by different methods, and the results are shown in Fig. 3.

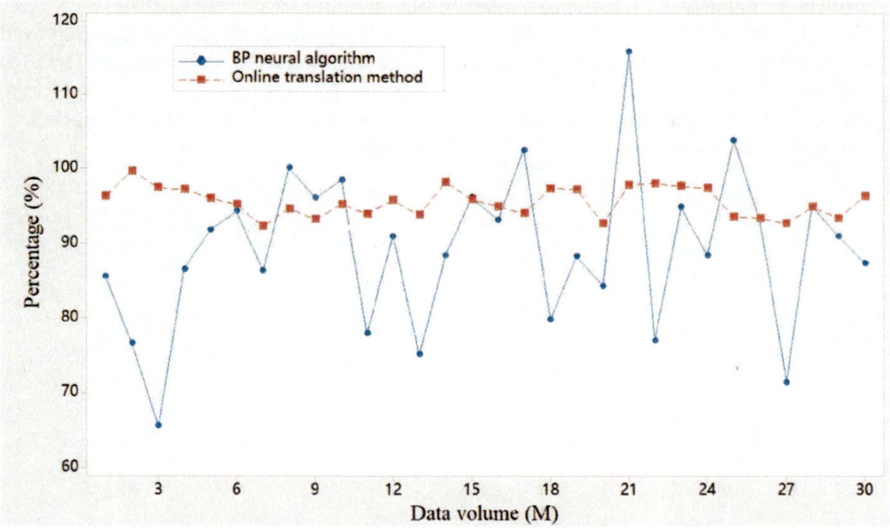

Fig. 3. BP neural algorithm evaluation results

It can be seen from Fig. 3 that the results of the BP neural algorithm are significantly better than the online translation method, and the reason is that the BP neural algorithm increases the stability adjustment coefficient and sets the corresponding translation logic to present a result that does not meet the requirements.

4 Conclusion

In view of the increasing requirements of Japanese teaching in colleges and universities, BP neural algorithm for the problem of Japanese teaching content, and combines the transfer standard to understand the relationship between Japanese teaching in the system Make improvements. At the same time, the logical standards of Japanese assisted translation are analyzed in depth to construct an optimization set. Studies have shown that BP neural algorithm can improve the accuracy and accuracy of stability, and can comprehensively stabilize stability. However, in the process of BP neural algorithm, too much attention is paid to escape indicators and comprehensive escape analysis is neglected.

References

1. Wang, Y.: Research on E-commerce big data classification and mining algorithm based on BP neural network technology (2021)
2. Zhao, N., Tsai, S.B.: Research on prediction model of hotels 鈥development scale based on BP artificial neural network algorithm. Math. Prob. Eng. (2021)
3. Yan, W., Shabaz, M., Rakhra, M.: Research on nonlinear distorted image recognition based on artificial neural network algorithm. J. Interconnection Netw. **22**(Supp06) (2022)
4. He, J., Liu, N., Han, M., et al.: Research on danjiang water quality prediction based on improved artificial bee colony algorithm and optimized BP neural network. Sci. Programm. **2021**(Pt.14) (2021)
5. Zhao, X.: Implementation of English ICAI MOOC system based on BP neural network. J. Ambient Intell. Hum. Comput. **14**(1), 3177–3186 (2023)

Research on Default Risk Prediction of Listed Companies' Green Credit Based on Deep Learning Algorithm

Jing Fu(✉)

Software Engineering Institute of Guangzhou, Guangzhou 501900, Guangdong, China
fj17991008@163.com

Abstract. The prediction of green credit default on deep learning algorithm is to predict the in the green credit portfolio of enterprises. We use recursive feature elimination (RFE) algorithm to train the DL model, and then apply it to predict the default risk of green credit of listed companies. The RFE algorithm iteratively deletes features from the input data until the performance on the verification set cannot be improved. In our research, we found that the combination of RFE algorithm and logical regression as outlier detector can improve the prediction performance by 2%, and the prediction probability as the input of the risk control system can be used to evaluate the effectiveness of the management's decision-making ability.

Keywords: Risk prediction · Credit default · Deep learning

1 Introduction

Risk management is one of the core businesses of financial institutions, which is crucial to their sustainable development. In the past, financial institutions mainly focused on risks directly related to finance, etc. However, with the frequent occurrence of global extreme climate and environmental events, climate and environmental risk has increasingly become one of the important risk sources concerned by most financial institutions and financial regulators in the world. Research on the Network for Greening the Financial System (NGFS) of the Central Bank and regulators shows that the climate and environmental risks faced by Financial institutions mainly come from transformation risks and physical risks [1].

With the proposal of China's dual carbon goals, the impact of transformation risks on financial institutions will also be significantly enhanced.

Physical risk refers to the risk generated by the change related disasters and the vulnerability of human and natural systems, such as the economic losses caused by the increasingly frequent and serious extreme weather events (such as storms, floods and heat waves), and the impacts caused.

J. C. Hung et al. (Eds.): FC 2023, LNEE 1131, pp. 153–159, 2024.
https://doi.org/10.1007/978-981-99-9299-7_21

For financial institutions, physical risks are manifested through direct and indirect ways: the loan default or asset devaluation brought by enterprises, households and countries suffering from climate impact [2]. The indirect. Based on this, this paper studies the prediction of green credit default on deep learning algorithm.

2 Related Work

2.1 Green Credit Mechanism

Green credit was first derived from environmental finance and sustainable development. Among the banks, it was the policy oriented environmental protection bank established in Germany in 1974 to issue loans for environmental improvement. The concept of green credit is put forward less abroad, mainly for the discussion of green finance, of which the more important part is green finance [3].

The Bank attaches great importance to the construction of green finance and puts forward requirements for green finance construction in the following five aspects. In terms of corporate governance, in accordance with various rules and regulations such as the Interim Measures for Green Credit Management, a green credit leading group was established, a green financial management system was established, the whole process management was strictly controlled, and green approval was strengthened. In terms of policy incentives, relevant indicators in the field of green finance will be included in the assessment system to promote development through assessment. In terms of information sharing: improve the information sharing mechanism, and release the distribution of green finance in the province in the form of notice throughout the year; In terms of innovation and R&D: carry out systematic financing planning for key strategic customers, and seek ideas for high-quality development of key clean energy enterprises. Professional training: continue to strengthen special training, deepen the understanding of green policies, and establish the green financial value concept of employees.

At present, the risk disposal of the branch is mainly based on the principle of "responding to litigation and fulfilling litigation", adopting the "dual path" approach, that is, collecting loans according to law and selling them to asset management companies in packages to complete the task of risk resolution, observation list and post audit recourse [4]. However, compared with other businesses, green credit business needs more ways of risk disposal, and the current conventional solutions are difficult to meet the characteristics of green credit and the diversity of project repayment sources. At present, disposal measures should be taken for all green non-performing loans to provide a basis for asset settlement and create conditions for the subsequent good green credit assets.

The conventional green credit mechanism is shown in Fig. 1.

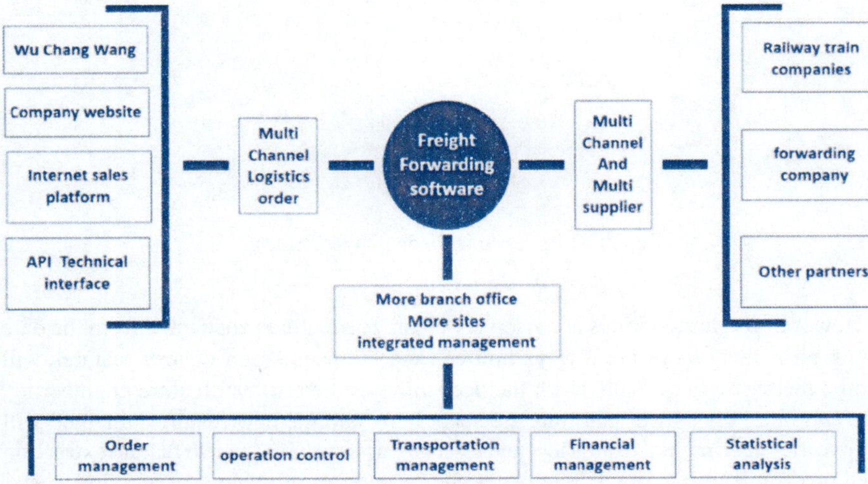

Fig. 1. Green credit mechanism

2.2 Application of Deep Learning in Prediction Tasks

Deep learning is one of the hot research directions in computer science. Based on deep learning, computers have made exciting progress in image, voice, NLP and other fields. And deep learning itself is also in constant exploration and development, and its potential in various fields has not been fully developed. At present, more mature applications of deep learning focus on such topics as machine vision, speech recognition and natural language processing. The common feature of these fields is that their data sets are continuous. For example, in image recognition, each layer is closely related to the part of the layer behind it. There is also a strong correlation before and after pronunciation; Although each word is discrete in natural language processing, the data before and after it is also strongly correlated. People can easily understand these data, but for general machine learning algorithms, it is very difficult to process this data. At present, deep learning has made great breakthroughs in personalized recommendation and prediction. In the future, the field of computer advertising and recommendation will also be an important application field for in-depth learning [5]. The depends on data and features, and the goal of optimization is to infinitely. Existing prediction systems depend on the effect of feature engineering. Mature credit models still rely on feature engineering with a lot of manpower, which is based on continuous in-depth understanding of problems and obtaining additional data sources. Figure 2 below shows the deep learning prediction mechanism.

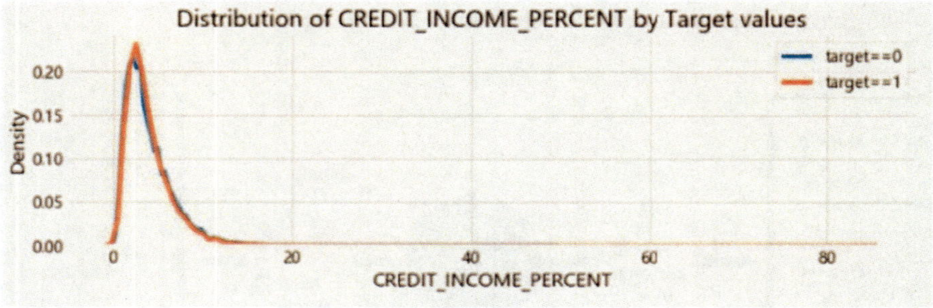

Fig. 2. Deep learning prediction mechanism

However, its shortcomings are also very clear. The features abstracted from the data relying on manual work are always limited, and the acquisition of new features will become increasingly difficult. With the deepening of the artificial feature engineering, more and more manpower and time are spent here, and the new features obtained will improve the experimental results less and less. At this time, if we can do feature extraction based on deep learning, it will be a better choice in terms of manpower and time costs.

3 Research on the Prediction of Green Credit Default Risk of Listed Companies Based on Deep Learning Algorithm

GKQ Bank's green credit risk follows the basic principles of comprehensiveness, independence and effectiveness. First, the principle of comprehensiveness. The risk management system covers all businesses and all risk types. Risk management includes risk identification, measurement, assessment, monitoring, reporting, control or mitigation and runs through all processes of business development, decision-making, implementation, management and supervision. It covers all financial products and businesses in various business fields, including on and off balance sheet, domestic and foreign, domestic and foreign currency. Regularly assess and dynamically adjust according to the regional financial environment, asset and risk characteristics, business development and risk management needs, and environmental changes; The second is the principle of independence. In terms of the functional setting of departments and the division of responsibilities of the management, it is necessary to ensure the separation of the front, middle and back offices, and separate control of loan review and release, so as to achieve effective checks and balances; The third is the principle of effectiveness. Establish a parallel operation mechanism for integrating risk management into business processes, and strengthen the mutual influence between business development and risk prevention and control. Risk management personnel are required to use advanced risk measurement technology and analysis tools to establish a series of risk management mechanisms to ensure the full implementation of risk management objectives. The green credit risk assessment process is shown in Fig. 3 below.

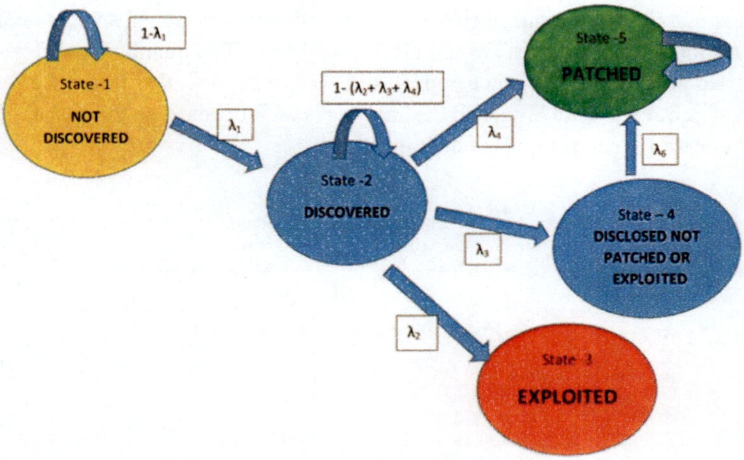

Fig. 3. Green credit risk evaluation process

From the perspective of rating, because there is no special rating standard for green projects at present, the same rating method as other projects is still used for the rating of green projects, and their ability and willingness to repay debts in a timely manner are comprehensively evaluated through the established rating procedures, which cannot highlight the difference between green loans and ordinary loans.

From the perspective of externalities of evaluation indicators, the externalities of some project environmental problems cannot be effectively calculated. At present, the externalities of GKQ Bank's environmental problems are mainly based on the feasibility study report of the project, and there is no external evaluation system. The long-term trend of the government to improve the weight of environmental protection and sustainable development remains unchanged. If GKQ Bank wants to make a long-term layout, the positive externalities of the environment need to be reassessed. For development financial institutions, special support should be given to projects such as garbage incineration power generation with better benchmarking function, better treatment effect and higher utilization rate. From the trend, as China's national policy continues to pay more attention to green industry, it is expected that enterprises with dividend policies will continue to improve their profitability in the future, and there should be differentiated evaluation policies. On the one hand, for projects with positive externalities, the impact on environment and resources cannot be fully evaluated. If the project only relies on its own income, it cannot pass the review. However, the social effect of the actual project itself is far higher than the electricity fee income of the project; On the other hand, projects such as coal power generation cannot fully reflect the negative external effects on the environment under the "3060" carbon peak and carbon neutral policy.

As shown in Fig. 4, the code of risk prediction indicators is shown, and the qualitative indicators adopted by branches are not clearly analyzed. The qualitative indicators are mainly listed in the due diligence report and review report. The reference value of subsequent approvers is not high, and the analysis is mainly based on the past experience of practitioners in green projects, which is unfair.

```
def plot_stats(feature,label_rotation=False,horizontal_layout=True):
    temp = app_train[feature].value_counts()
    df1 = pd.DataFrame({feature:temp.index,'Number of contracts':temp.values})

    cat_perc = app_train[[feature,'TARGET']].groupby([feature],as_index=False).mean()
    cat_perc.sort_values(by='TARGET',ascending=False,inplace=True)

    if(horizontal_layout):
        fig,(ax1,ax2) = plt.subplots(ncols=2,figsize=(12,6))
    else:
        fig,(ax1,ax2) = plt.subplots(nrows=2,figsize=(12,14))
    sns.set_color_codes('pastel')
    s = sns.barplot(ax=ax1,x=feature,y='Number of contracts',data=df1)
    if(label_rotation):
        s.set_xticklabels(s.get_xticklabels(),rotation=90)

    s = sns.barplot(ax=ax2,x=feature,y='TARGET',order=cat_perc[feature],data=cat_perc)
    if(label_rotation):
        s.set_xticklabels(s.get_xticklabels(),rotation=90)
    plt.ylabel('Percent of target with value 1[%]',fontsize=10)
    plt.tick_params(axis='both',which='major',labelsize=10)
    plt.show()
```

Fig. 4. Risk forecast indicator code

Due to the large scale of the branch's early green credit projects and the fact that certain risks did occur, in the subsequent business operations, the approvers were afraid of some projects due to their subjective judgment and previous green non-performing loans, which led to a certain deviation between the approval results and the actual situation of the project. Some projects that could have been passed were artificially denied. In green projects, The relatively important rated indicators will indeed be diluted in the approval process, which cannot fully reflect the uniqueness of green projects, and ultimately rely too much on the evaluation of quantitative indicators.

4 Conclusion

Deep learning is one of the most promising and powerful algorithms in machine learning. It has been applied in many fields, and has been successfully applied in the financial industry. In this paper, we will introduce a deep neural network model, which can predict default risk based on green credit data of listed companies. Green credit = good debt + bad debt (credit risk) The main components of green credit are good debt and bad debt.

Acknowledgements. Scientific research and teaching project of Software Engineering Institute of Guangzhou, "Applied Undergraduate Practical Course Teaching Effect Evaluation Research (KY202055)".

References

1. Youhui, L.I., Huang, Y., Wang, Y.: Research on credit risk of listed companies in paper industry: based on the analysis of green factors. Issues Forest. Econ. (2018)
2. Cao, Y., Shao, Y., Zhang, H., et al.: Study on early warning of E-commerce enterprise financial risk based on deep learning algorithm (2022)
3. Xu, S., Wang, X., Liu, Z.: Innovation of credit risk measurement of listed coal companies based on FSVM-KMV. J. Nonlinear Convex Anal. **9**, 22 (2021)
4. Hu, Y., Miao, Q., Lu, L.: Impact of default distance on financial warning of listed companies (2) (2018)
5. Wang, H., Zhu, J.: Financial risk assessment of listed companies in environmental protection industry based on self-organizing neural network (2018)

Improvement of Cloud Platform Utilization Based on Evaluation and Optimization of Computing Resource Runtime Surplus Capacity

Shuai Gong(✉), Xiaoyu Yin, Min Zhang, and Wanwan Cao

Information and Communication Branch of State Grid Anhui Electric Power Co., Ltd., Hefei 241204, Anhui Province, China
gongs5038@163.com

Abstract. In cloud computing, users may face the problem of resource shortage. In this regard, overcapacity is an important issue. In traditional computer systems, when CPU and memory resources are scarce, it will lead to serious performance degradation or even crash. However, in a cloud computing environment, if the client has no resource constraints (for example, CPU and memory), it can be used to improve the utilization efficiency of other resources such as storage and network bandwidth. This paper proposes a new evaluation method, which is based on the evaluation results of runtime residual capacity obtained from real data and simulation model analysis. The results obtained in each area will indicate whether these computing resources are worth utilizing. In addition, we need to consider other factors, such as cost and safety.

Keywords: Evaluation and optimization · Utilization rate · Resource operation

1 Introduction

The history of data center is not very long, and it belongs to the product of information technology revolution in the 21st century. In the early days of the data center, it was mainly aimed at large enterprises to provide sites, power supply, bandwidth, maintenance services and other projects. Subsequently, the data center market changed dramatically and its business scope was expanded. At this stage, IDC was not limited to rental services, but value-added services emerged, including website hosting, server hosting, application hosting, and network acceleration. Computing resource utilization is an important performance indicator of the cloud computing platform. Deep learning is a computing intensive task. How to efficiently schedule deep learning processors (NPUs) to meet the computing needs of deep learning reasoning tasks from different tenants and applications is crucial. This paper breaks the boundaries of different tenants and different applications in the use of resources, and innovatively proposes and implements a virtualization system based on NPU resource pooling. Special job scheduling and acceleration platform, NPU resource pooling method and NPU fine granularity scheduling

J. C. Hung et al. (Eds.): FC 2023, LNEE 1131, pp. 160–166, 2024.
https://doi.org/10.1007/978-981-99-9299-7_22

method are mainly studied. The experimental data shows that after a single NPU device has realized resource pooling, different neural network models and parameter ratios can achieve different performance improvements in batch processing reasoning tasks, ranging from 493% to 915% [1]. The research results have been applied to SothisAI, a deep learning service platform developed by Sunlight for the field of artificial intelligence. Based on this, this paper studies the improvement of cloud platform utilization based on the evaluation and optimization of computing resource runtime residual capacity.

2 Related Work

2.1 Optimized Allocation of Computing Resources

In the centralized resource provision model represented by cloud computing, all kinds of hosts distributed in the same region are connected through network technology and uniformly regulated and arranged through the centralized data center. At this time, instruments, sensors, electrical appliances and other terminal devices are only responsible for input and output interaction tasks, sending a large number of sensor data and business requests to the data center for processing. In recent years, the field of cloud computing has developed rapidly. There are more and more researches on cloud computing resource management and scheduling algorithms, and they continue to develop iteratively [2].

From the perspective of cloud computing platform, the classic open source cloud computing platform is Hadoop with centralized scheduling architecture, which encapsulates a fixed amount of resources into a fixed number of slots, and takes slots as the basic unit of resource allocation. However, as the computing load of the cloud platform becomes more diversified and presents different resource preferences, this fixed unit coarse granularity resource allocation method cannot meet the load needs of different resource demands, and resource fragments are easily generated during the operation process, resulting in resource waste. Under this background, cloud computing service users who request computing resources can apply for CPU, memory, disk and other resources in units of one thousandth of a core, byte, etc. The granularity of computing resource allocation has been refined and developed to meet the load needs of different resource preferences. However, the amount of resource allocation depends on the amount of resources requested by users, and the amount of resources requested is usually the maximum amount of resources used. However, most business applications are dynamic and will not remain at the peak level all the time, so there is a certain difference between the resource allocation and the actual resource usage, which limits the continuous improvement of resource utilization. In order to further improve resource utilization and adapt to the personalized and differentiated development of resource demand, scholars have adopted collaborative scheduling, sharing status, scalable allocation mechanisms and frameworks to improve the granularity of resource allocation and optimize the matching between demand and resources. It solves the problem of demand and resource matching degree from two aspects of task resource demand quantification and fine-grained resource allocation [3]. On the one hand, it infers the task resource demand in different stages through similar task operation, and on the other hand, it refines the granularity of resource allocation from the perspective of multi-stage and multi type resource allocation. The application of virtualization technology can transform a single physical machine resource of the cloud platform into multiple virtual machines, and each virtual machine operating environment is isolated from each other. Figure 1 below shows the resource optimization framework.

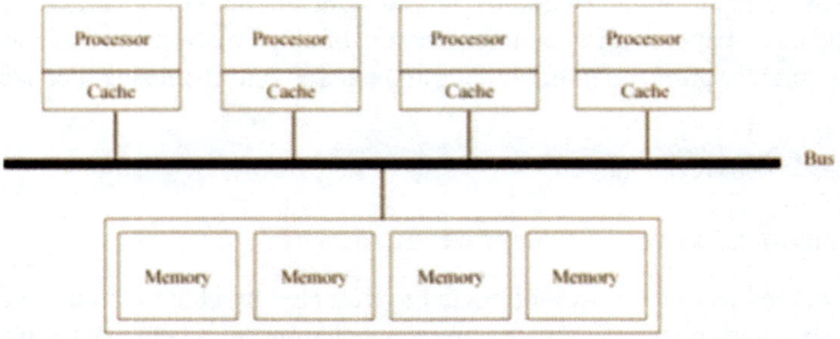

Fig. 1. Resource optimization framework

2.2 Cloud Platform

The power distribution IoT cloud master station is the information gathering and application processing center, and is one of the main distribution areas of computing resources. The cloud master station uses cloud computing technology to virtualize the internal computing resources, network resources and storage resources of the master station system, and forms a huge resource pool, which completes the mapping of the hardware resources of the master station in the software system, thus realizing the decoupling of physical hardware and virtual software. In the resource pool, the cloud master station realizes the full life cycle management, dynamic resource migration, dynamic resource configuration and other functions of virtual machines through centralized management and control, and provides services for business applications subsequently loaded into the cloud master station on demand [4]. The service level of the IoT cloud master station is divided into three levels: Infrastructure as a Service (laaS), Platform as a Service (PaaS) and Software as a Service (SaaS), as shown in Fig. 2. In the actual process of building the cloud master station, the three service levels of the cloud master station are stacked on each other, as shown below:

1) LaaS service layer: It virtualizes computing, network and storage resources, builds a huge cloud master resource pool, and is equipped with a cloud operating system to allocate resources according to business needs.
2) PaaS service layer: monitor and manage the objects and corresponding connections of the access platform, and provide support services from data source acquisition to application for power distribution business applications. The main links of PaaS layer include data acquisition and preprocessing, data management and analysis, and basic services. In the link of data management and analysis, in order to achieve accurate management of data classification, the data management mode is further divided into real-time monitoring center, graphic model data center and application analysis center [5].

3) SaaS service layer: for the vast number of power distribution IoT service users, based on unified standards, it provides common functions and interfaces in the form of microservices to support the development and implementation of different business applications. The SaaS layer can manage the access rights of applications, so that each user and application can only operate the basic services, data and resources with access rights.

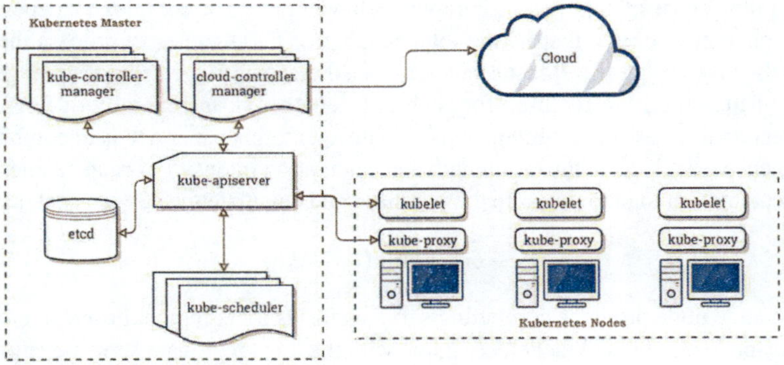

Fig. 2. Cloud master station service hierarchy

3 Improvement of Cloud Platform Utilization Based on Evaluation and Optimization of Computing Resource Runtime Surplus Capacity

The communication network is located at the network layer of the technical architecture, which is divided into local communication network and remote communication network according to the scope of application to realize information transmission and interaction between different side notes of the power distribution Internet of Things. The local communication network is used between the mass sensing terminal, the response terminal and the edge computing terminal, and is the network medium to realize the transmission of information and control instructions of the three [6]. The difficulty faced by the local communication network is that the number of sensing terminals and response terminal devices is huge, the distribution location is scattered, the geographical environment is complex, the number of single communication is small, but the number of concurrent users is large. It is difficult to use the communication structure of the traditional optical fiber private network. The solution is to apply the current low-power wide area communication technology with strong load processing capability, extensive signal coverage capability and low cost, specifically including LoRa, Sigfox, 3GPP and other major standards [7]. The remote communication network is the network medium between the cloud master station and the edge computing terminal. Compared with the load characteristics of the local communication network, the distance between the cloud master station and the edge computing terminal is far, and the service is highly sensitive to delay and reliability. Generally, optical fiber, 4G/5G and other low delay and high reliability communication modes are used.

For regional users, the cost comes from the discomfort caused by purchasing the EN computing service in the corresponding region and adjusting the electricity use behavior. Among them, the purchase cost of computing services should depend on the complexity of the tasks to be processed. The higher the task complexity, the higher the computing resources required to ensure the processing delay. Therefore, the higher the computing cost of this part. Literature believes that task complexity depends on the task content attribute, that is, the total number of CPU cycles to process the task, not just the task length [8]. Because even for businesses with the same task length, the content of the algorithm required for each unit length task is generally different, and the corresponding calculation intensity is also different, this paper needs to combine the comprehensive analysis of the task length and calculation intensity of each business, and use the calculation load to measure the purchase and calculate the service cost, namely:

$$\|y - \theta_i\| = \min(\|y - \theta_i\|) \tag{1}$$

Each edge node provides computing services for its local regional users at a certain price. As the leader in the Stackelberg game with the local regional users, the edge node needs to allocate its internal computing resources reasonably according to the computing load of the arriving service, and determine its task unloading and uploading strategies, as well as the optimal computing load pricing strategy. Therefore, the internal operation optimization of edge nodes can be expressed as:

$$I(s_1, s_2, .., s_n) = -\sum_{i=1}^{m} p_i \log_2(p_i) \tag{2}$$

$$Info(S) = -\sum_{i=1}^{m} p_i \log_2(p_i) \tag{3}$$

In order to study the changes of applications of different computing loads in computing resource transactions, based on the setting that the maximum value of EN1-EN3 computing loads increases in turn, this paper increases the computing loads of each bit of APP1-APP4 in each EN in turn, namely 500 cycles/bit, 800 cycles/bit, 1000 cycles/bit, and 1500 cycles/bit. Figure 5 shows the comparison of calculated load adjustment changes during pre dispatching and re dispatching. From a vertical perspective, with the increase of the maximum value of the computing load in turn, the elastic computing load distribution from the pre scheduling stage to the re scheduling stage tends to be more flat (such as EN3), while the time distribution of computing load in the regions with rich computing resources (ENI) has basically not changed. It can be seen that in the area with a large amount of calculated load, the impact of elastic calculation load on its net income is greater, so the adjustment range of calculated load is more obvious [9]. In the figure, EN1 calculates that the adjustment range of the total load is about 0% in the pre dispatching stage and 0.18% in the re dispatching stage; EN2 calculates that the adjustment range of the total load is 1.57% in the pre dispatching stage and 1.62% in the re dispatching stage; EN3 calculates that the adjustment range of the total load is 3.13% in the pre dispatching stage and 4.98% in the re dispatching stage. From a horizontal perspective, the greater the computing load of elastic business APP, the greater the range of change during computing resource trading. Taking the three elastic APPs in EN3 as an example, APP2-APP4 calculates the total load adjustment range of 3.82%, 5.83% and 2.29% in the pre scheduling phase, and 4.37%, 6.2% and 6.24% in the re scheduling phase [10]. It can be seen that the larger the calculation load is in the rescheduling stage, the larger the adjustment range of the APP will be, especially in areas with large total calculation load.

4 Conclusion

In the mode of distributed resource provision, this paper proposes the optimal allocation of cloud computing resources taking into account the cooperation of multi-agent tasks. From the perspective of multiple agents, the role they play in the operation process and the existing cooperation mechanism are analyzed: from the perspective of computing service users, a user computing load demand response mechanism is established, that is, users can reduce the cost of computing services by adjusting their own business arrangements; From the perspective of computing service providers, a computing service provider cooperation mechanism of multi-agent task collaboration has been established, in which edge computing terminals can trade computing services to the cloud through the network between cloud edges according to the dynamic change of their own computing load, and can also trade computing services to members of the consortium, forming the problem of maximizing net income. According to the interaction and collaboration among computing service users, computing service providers and multilateral alliances, a three-layer collaborative optimization model is established.

References

1. Lou, G., Cai, Z.: A cloud computing oriented neural network for resource demands and management scheduling. Int. J. Netw. Secur. **21**(3), 477–482 (2019)
2. Qing, Z.: Evaluation and optimization for the operation of wind farm (2017)
3. Wang, Y., Qin, D., Liu, J., et al.: User utility optimization of cloud computing resource scheduling algorithm based on production function. Appl. Res. Comput. (2017)
4. Xie, Z.: Resource allocation of cloud computing based on Fourier transform and support vector machine. Microcomput. Appl. (2017)
5. Subhash, L.S., Udayakumar, R.: Sunflower whale optimization algorithm for resource allocation strategy in cloud computing platform (2021)
6. Liu, F., Liu, Y., et al.: Design of information management platform for laboratory construction based on cloud computing (2018)
7. Sabuj, S.R., Asiedu, D.K.P., Lee, K.J., et al.: Delay optimization in mobile edge computing: cognitive UAV-assisted eMBB and mMTC services. IEEE Trans. Cogn. Commun. Network. **2**, 8 (2022)
8. Zhong, S.Q., Gong, X.F.: Cloud workflows scheduling based on improved particle swarm optimization algorithm. Electron. Des. Eng. (2019)

9. Wang, Y.: Design of cloud computing resource intelligent scheduling experiment based on SLA. Research and Exploration in Laboratory (2017)
10. Kuang, Z., Xie, S.: Load balancing scheduling of virtual machine resource in cloud computing based on cat swarm optimization algorithm. J. Jilin Univ. (Sci. Ed.) **54**(5), 1117–1122 (2016)

Design and Implementation of Operation and Maintenance Monitoring Platform Based on Data Mining Algorithm

Jian Chao Guan, Qiang Sun, Ting Li(✉), and Yu Lu

State Grid Anhui Electric Powerco.Ltd.Information and Telecommunication Branch, Hefei 23000, Anhui, China
xjtuliting@163.com

Abstract. The project is to design and implement a monitoring system based on data mining algorithm, which will be able to monitor the performance of the operations and maintenance (O&M) activities performed by operators to detect problems that may occur during these operations. The goal is also to provide managers, engineers and technicians with effective tools that they will use as support in their daily work. The monitoring system must be able to collect relevant information about O&M activities, analyze them using appropriate algorithms and provide suggestions for improvement. In addition, it should have a graphical interface so that users can easily access all necessary information.

Keywords: Data mining · Operation and maintenance monitoring · Platform design

1 Introduction

Traditional network management software usually cannot fully meet the different requirements of complex integrated network space and different units for the network. Yunna innovates the product concept, adheres to the modularization, building block platform structure, and integration "topology I, monitoring, security management, display, assets, orders, and logs, and develops and constructs an integrated full stack operation and maintenance monitoring platform, which is intelligent and dynamic, and is suitable for unified connection and intelligent monitoring.

Things has promoted the construction of smart cities by leaps and bounds. With the cities across the country, urban traffic equipment: electric police checkpoints, signal lights, video surveillance, traffic guidance screens, etc. [1]; Municipal infrastructure: intelligent light pole, intelligent well cover, intelligent trash can, intelligent fire hydrant; Community equipment: AI face recognition terminals, intelligent door locks and other urban infrastructure equipment have soared, forming a situation of large number, variety and manufacturers, bringing challenges to urban managers. How to manage and maintain these facilities and equipment in the city has become an urgent problem for the development of smart cities.

J. C. Hung et al. (Eds.): FC 2023, LNEE 1131, pp. 167–172, 2024.
https://doi.org/10.1007/978-981-99-9299-7_23

In order to change the management confusion, high operation and maintenance cost, efficiency and quality faced by traditional operation and maintenance, Zhicheng Technology's independently developed big data operation and maintenance platform is positioned at the city level or industry level operation and maintenance management platform, and uses to realize the operation of monitoring equipment, comprehensively digitize the operation and maintenance work, mine data efficiency, and truly improve the efficiency of operation and maintenance work. Based on this, this paper studies the design and implementation of operation and maintenance monitoring platform based on data mining algorithm.

2 Related Work

2.1 Algorithm Implementation of Data Mining Platform

Through the efficient integration technology and cloud computing, computer diffusion analysis activities can be efficiently carried out in machine clusters with infinite scale. In the framework based on Map/Reduce, all request activities can be calculated, analyzed and processed. When conducting data mining, specific data mining operations can be divided into different tasks. On this basis, distribution activities can be carried out and different machines can be used to complete relevant tasks. After completing relevant tasks, move to the next data mining step, merge the above documents, and carry out output activities for enterprises. In the data platform system, the circular application algorithm can be used to process the generated Reduce tasks and convert them into data mining target files to meet specific needs. Designing the data mining platform through data mining platform algorithms can make data information more flexible, improve data mining efficiency, strengthen data mining quality, and improve work efficiency [2].

The platform focuses on the innovative operation and maintenance mode. Through the unified monitoring management as the center, it realizes the centralized monitoring and management of the IT infrastructure of the data center, provides the monitoring and management of the basic software and hardware resources of the data center and various applications, builds a unified and integrated system resource monitoring platform, deeply excavates the processes of each link, and establishes the equipment failure model in combination with the self-developed algorithm, At the used to realize the operation of intelligent monitoring equipment, aiming and improve work efficiency. At the same time, combined with the statistical multi-dimensional data, it can not only help managers understand the overall and facilities, but also provide important data support for the assessment of the operation and maintenance work of various manufacturers.

2.2 System Development Technology

The Spring Boot was developed by the Pivota team to simplify the Spring development project. Different from the development method of Spring reconfiguration, Spring Boot is based on the principle that convention takes precedence over configuration. Developers do not need to configure the frameworks and dependencies in detail when using Spring Boot, because Spring Boot provides default configurations for most integrated frameworks. For personalized information, users can configure it in the propcerties file or yaml file. For the integrated framework, Spring Boot also provides the default dependent version, which solves the version conflict and stability problems when jar packages are transferred between projects [3].

In terms of deployment, unlike the traditional J2EE development mode, Spring Boot has Tomcat built-in, so the final project packaging form is a jar package consisting of project files and web containers. This makes the Spring Boot project deployment extremely convenient and can be run directly using the java jar command. Figure 1 below shows the object relationship mapping.

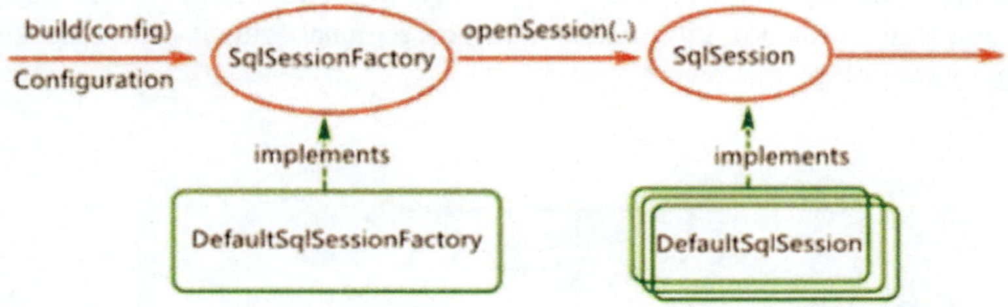

Fig. 1. Object relationship mapping

In terms of development, Spring Boot uses a large number of annotations to simplify code and minimize configuration. Spring's loC and AOP can be implemented through annotations. For example, the @ Bean annotation indicates that the object returned by the method is a bean and is handed over to Spring for management@ The aspect annotation indicates that the class is a facet class [4].

3 Design and Implementation of Operation and Maintenance Monitoring Platform Based on Data Mining Algorithm

Operation the core function of the system, which is realized by real-time monitoring of equipment status on the basis of unified access platform. The traditional manual inspection often has problems that are not found in time, the status inspection is not comprehensive, and the data measurement is not in-depth. These problems lead to the failure can be solved only when the equipment is damaged and the power supply is affected. In the long run, this not only increases the cost, but also affects the power supply quality. Through real-time online monitoring, the can timely handle the abnormal information of the equipment. Operation and maintenance management functions include equipment monitoring, alarm management, map monitoring, repair management and other sub functions [5].

Daily monitoring and reporting of equipment status will generate a large amount of data. Statistical analysis of these data will help administrators understand equipment operation status and assist in the formulation of operation and maintenance plans. For example, by calculating the total power of the distribution room and combining with the historical operation data, the load situation of the next stage is predicted, so as to arrangement of the equipment maintenance plan, so as to improve the operation economy of the distribution network. The alarm information when the equipment status exceeds the threshold value and various status information also provide guidance for maintenance personnel during maintenance, which is the technical basis for fault diagnosis and repair. For example, when the voltage value of a transformer fluctuates violently and exceeds the threshold value, its ambient temperature is often higher than the normal value, so we can find the impact of temperature on its stable operation. Therefore, the system provides users with various forms of data statistics and analysis, so that users can clearly and intuitively feel the changes of data and the internal relationship between attributes through visual display. The data statistics and analysis module involves historical operation status statistics, alarm information statistics, regional statistics and other sub [6]. The logic view in Fig. 2 below.

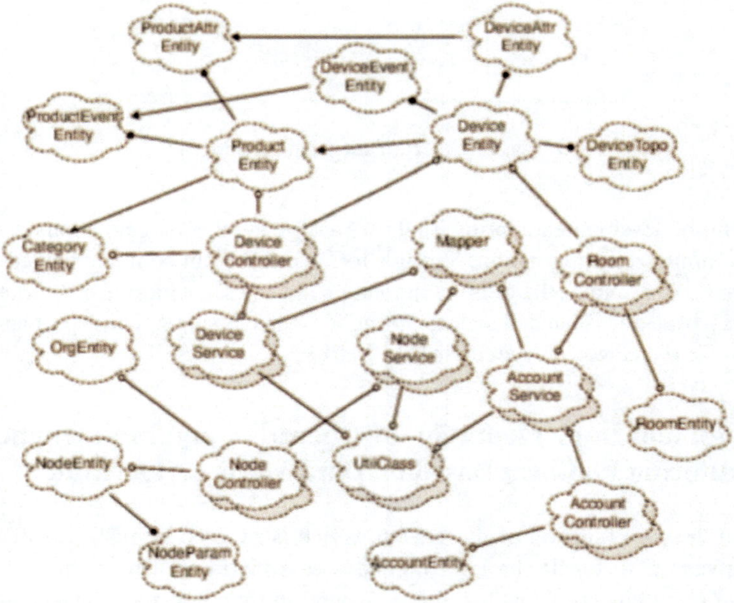

Fig.2. Object relationship mapping

The sensing layer consists of communicable power distribution devices and gateways, which constitute the edge nodes of the system and interact with the application layer through the network layer using MQTT protocol. In the network layer, the main facility is the MQTT message server, which is responsible for receiving and forwarding messages. The application layer is the cloud platform of the system, through which equipment management, permission management, data analysis and other operations can be carried out [7].

The system platform is developed using MVC layered architecture and communicates with the gateway and equipment through MQTT Broker. Under MVC hierarchical architecture, the system can be roughly. For this system, it can be divided into six layers, from top to bottom: view layer, controller layer, service, model layer, data persistence and database entity. The structure of each layer.

4 Operation

1. Switchable patrol inspection: the big data operation and maintenance platform takes equipment as the core, and can comprehensively control the operation status of each core equipment through 7X24 hour zero manpower full volume high-frequency monitoring and link real-time monitoring and tracking. It can also generate manual active patrol inspection tasks through template selection, which lays a data foundation for business and system performance analysis, and achieves the goal of operation and maintenance assessment [8]. Thus, the key indicators of operation and maintenance MTTR and MTBF are reduced by more than 78%, and the online rate of equipment is guaranteed to be more than 95.6%.

2. Accurate fault judgment can be achieved: based on the fault scenario, tmachine learning and other technical means, integrates scientific algorithms such as linear regression, realizes fault prediction in advance through prediction models, learns accumulated experience from the fault database, provides reliable and referential fault solutions, reduces learning costs, and reduces 20.12% of fault repair time.

3. Visualization maintenance can be realized: the supports most public protocols such as SNMP and HTTP. At the same time, through the API open interface, it can interface with the operation and maintenance monitoring and ITOM products such as Zabbix to obtain equipment parameters in real time, and then display the equipment distribution status and operation situation on the GIS engine in real time, so that the see clearly at a glance, and find problems at the first time through the fault root cause location and dynamic baseline [9], In addition, abnormal conditions shall be timely sent to relevant personnel through the system and SMS messages to effectively reduce MTTA and MTTR.

Fourthly, operation and maintenance assessment can be digitalized: record the operation and maintenance situation of all brands of equipment and operation and maintenance manufacturers throughout the process, provide multi-dimensional operation and maintenance effect analysis, and realize the digitalization and refinement of operation and maintenance assessment [10].

V. Globalization of asset management: The big data operation and maintenance platform will automatically record the data information of equipment from purchase, warehousing, outbound, installation, and scrapping, and supervise the whole process, so that it can be implemented to people and seconds, realizing the management mode of combining dynamic data and static data of assets, and tracing the source of asset equipment with flexible and configurable workflow and multi-dimensional report analysis [11], Effectively reduce labor management cost by 14.3%.

5 Conclusion

The sensing layer consists of communicable power distribution devices and gateways, which constitute the edge nodes of the system and interact with the application layer through the network layer using MQTT protocol. In the network layer, the main facility is the MQTT message server, which is responsible for receiving and forwarding messages. The application layer is the cloud platform of the system, through which equipment management, permission management, data analysis and other operations can be carried out.

The system platform is developed using MVC layered architecture and communicates with the gateway and equipment through MQTT Broker. Under MVC hierarchical architecture, the system can be roughly. For this system, it can be divided into six layers, from top to bottom: view layer, controller layer, service, data model layer, data persistence and database entity. The structure of each layer.

References

1. Li, D.-W.: Development and application of intelligent operation and maintenance monitoring platform based on multidimensional data. Electron. Comp. Inf. Technol. (2018)
2. Zhao. J., Chen, Z., Jing, B., et al.: Design of upgrade of remote operation and maintenance platform for metering terminals monitoring. Microcomput. App. (2019)
3. Chen, Z.X., Liang, A.B.: Design of information collection and operation monitoring system based on large data. Electron. Des. Eng. (2018)
4. Hou, Z., Chen, H., Chang, S., et al.: Internet plus distributed photovoltaic power generation operation monitoring platform. Elect. Measur. Instrum. (2016)
5. Ren, G., Shu, M.: Improvement of Aprior algorithm based on Hadoop and its application in operation and maintenance of EMU. Railway Comput. App. (2018)
6. Gao, F., Wang, Y., Yang, L.: Equipment condition monitoring and diagnosis based on IoT and operation big data. Proc. Autom. Instrum. (2018)
7. Bian, G., Huang, H., Liu, D., et al.: Intelligent on-line detection of airport operation and maintenance equipment based on internet of things. Mech. Eng. (2016)
8. Han-Cong, H., Xiao, Z.D.: Data mining and analysis of two equipment defects based on Apriori algorithm. Electron. Des. Eng. (2019)
9. Xie, C., Cao, Z., Wen, D., et al.: State evaluation of distribution transformers based on real-time operation data mining. Zhejiang Elect. Power. (2017)
10. Pei, Z., Wang, L., Chong, L., et al.: Operation indicators monitoring and performance analysis of database server based on power user electric energy data acquire system. Hebei Elect. Power (2016)

Civil Aviation Logistics Transportation Route Optimization Algorithm Based on A+ Algorithm

Fenchou Han(✉)

Sanya Aviation and Tourism College, Hainan 572000 Sanya, China
fc_han13700479787@163.com

Abstract. The role of transportation route optimization in civil aviation logistics is significant, but there is a problem with poor optimization. The route planning method cannot solve the problem of multi-route selection in civil aviation transportation, and the rationality. Therefore, this paper proposes an to construct a logistics route optimization model. Firstly, transportation knowledge is used to classify the transportation route, and the transportation route is selected according to the length of the route to realize the standardized processing of time. The transportation knowledge then classifies the time into a path optimization collection and iteratively analyzes the scored content. MATLAB simulation shows that the A+ algorithm's optimization degree and optimization time of A+ algorithm are better than that of the path planning method under a certain area range.

Keywords: transportation knowledge · Time · A+ algorithm · Optimize the results

1 Introduction

Path optimization is one of the important contents of logistics and transportation and plays a very important role in applying the route optimization system [1]. However, in selecting the logistics transportation route [2], the selection result has the problem of long total distance, affecting logistics transportation and passage. Some scholars believe that applying the A+ algorithm to civil aviation transportation can effectively analyze distance and time, and provide corresponding support for route optimization verification [3]. This paper proposes an A+ algorithm to measure the optimization effect of civil aviation logistics transportation route and verify the model's effectiveness.

2 Related Concepts

2.1 Mathematical Description of the A+ algorithm

The A+ algorithm uses the relationship between the route, the overall distance and the time correlation to optimize the transportation route, and according to the civil aviation transportation, Multidimensional metrics that find outliers in path optimization and form

J. C. Hung et al. (Eds.): FC 2023, LNEE 1131, pp. 173–179, 2024.
https://doi.org/10.1007/978-981-99-9299-7_24

a path table. By integrating the path optimization results, the correlation of the election results is finally judged. The A+ algorithm combines transportation knowledge, uses the path relationship and A+ algorithm, optimizes the time results, and can improve the time Recognition rate.

Hypothesis 1: The transportation route is x_i, the set of path optimization results is $\sum x_i$, the professional distance is y_i, and the judgment function of the selection result is $f(x_i)$ as shown in Eq. (1).

$$f(x_i) = \oint_{y_i} \sum x_i | \xi \tag{1}$$

ξ Adjust the coefficient for the selection of the results to reduce the impact of ambiguity and pass rate.

2.2 Selection of Selection Method

Hypothesis 2: The path optimization method selection function is $F(x_i \rightarrow y_i)$ and the time weight coefficient is κ_i, then the path optimization method selection is shown in Eq. (2).

$$F(x_i)=z_i^2 \cdot f(x_i^2|y_i) \Rightarrow \kappa_i \pm \xi \tag{2}$$

2.3 Time, Single Processing

Before the A+ algorithm analysis, the standard analysis of the time and single time in the selection results should be carried out, and the transportation route should be mapped to the selection table to determine the semantic anomaly of content. First, the transportation route is comprehensively analyzed, and the thresholds and weights of the transportation route are set to support the accurate analysis of the A+ algorithm [4]. The transportation route needs to be standardized, and if the processed results conform to the non-standard distribution, the treatment is valid, otherwise, the standard treatment is re-processed. In the accuracy of the A+ improve the selection level, the scheme should be selected, and the specific method selection is shown in Fig. 1.

Through the transportation route display, the A+ algorithm analysis is uniform, which is in line with the objective facts. The selection method is not directional, indicating that the A+ algorithm analysis has strong accuracy, so it is used as research on the optimization of civil aviation logistics transportation path [5]. The selection method meets the mapping requirements, mainly the transportation knowledge adjusts the selection method, removes duplicate distances, and revises the pass rate so that the whole selection of transportation routes is strong.

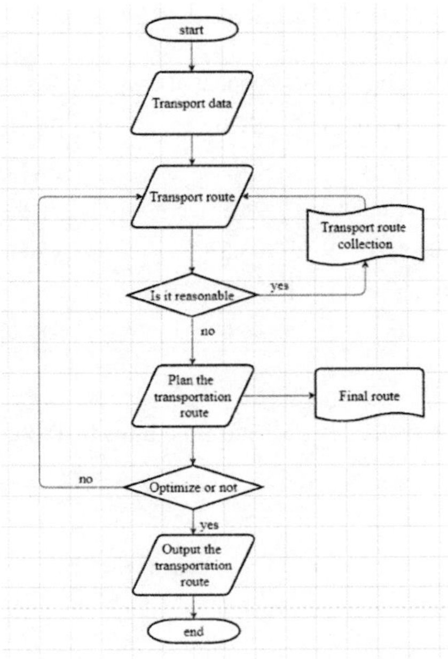

Fig. 1. Selection results of A+ algorithm analysis

2.4 Multi-line Associativity

The A+ algorithm adopts the accurate time judgment and adjusts the corresponding distance relationship to realize the optimization method of civil aviation logistics transportation route. The A+ algorithm divides the optimization selection of civil aviation logistics transportation routes into different data volumes and randomly extracts different methods. In the iterative process, the path length of different data amounts is matched with the selection method. After the matching processing is completed, different methods are compared to optimize the transportation route of civil aviation logistics, and the time results with the highest accuracy are recorded.

3 Practical Examples of Logistics and Transportation Route Optimization

3.1 Select Content Situation

In the analysis of the system scoring, the number of test orders is 2 421, and the time is four based on different types of transportation routes 50 articles, as shown in Table 1.

Table 1. Characteristics of the test path

Path scheme	Path point	Amount of data	Dispersion	threshold
Short distance	starting point	55.36	25.00	48.21
	Endpoints	87.50	78.57	83.93
Long distance	starting point	78.57	33.93	23.21
	Endpoints	103.57	67.86	46.43
Random distance	starting point	73.21	26.79	58.93
	Endpoints	44.64	28.57	51.79

The pass-through rate processing for multiple lines in Table 1 is shown in Fig. 2.

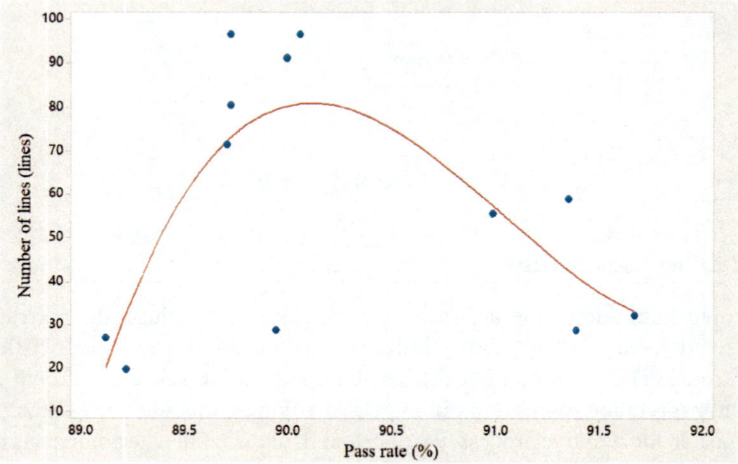

Fig. 2. The processing process of the pass rate

It from Table 1 that e path programming method, the selection result of the A+ algorithm is closer to the actual time, in terms of civil aviation logistics transportation path optimization selection pass rate, selection rate, accuracy, etc., A+ algorithm path planning method. From the change in pass rate in Fig. 4, it can be seen that the accuracy of the A+ algorithm is better, and the judgment speed is faster. Therefore, the time speed, time, and optimization degree of the A+ algorithm are better.

3.2 Reduction Ratio of Distance

Path distance includes distance, pass rate, speed. After the standard threshold screening of A+ algorithm, the preliminary time obtained, and the time obtained to analyze the effect more accurately, select different distances and calculate the overall time of logistics transportation, as shown in Table 2.

Table 2. Overall situation of path optimization

The distance is reduced by the proportion	Regional scope	Outlier recognition rate
25%	25.00, 23.23	33.93
50%	64.29,68.52	46.43
70%	33.93,42.6	42.86
mean	75.00,78.62	71.43
X2	60.271	50.230
P = 0. 012		

3.3 Time and Accuracy of Path Optimization

The optimization time and accuracy are compared with the path planning method, and in Fig. 3.

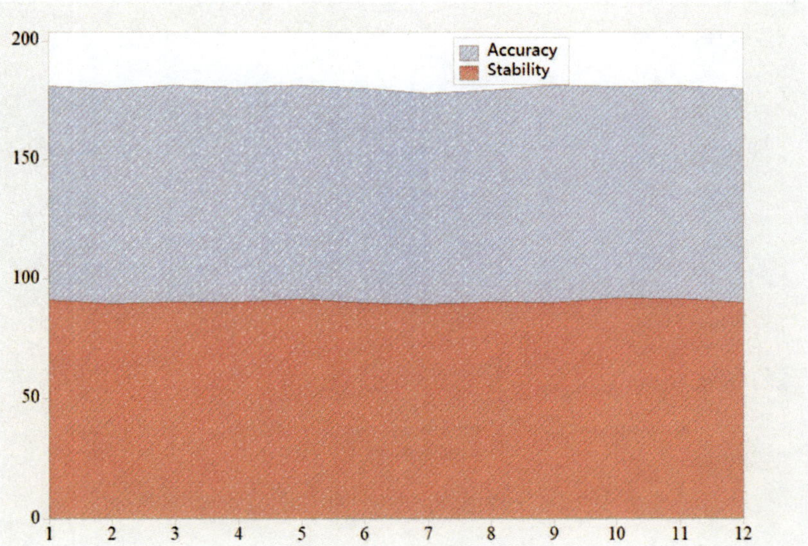

Fig. 3. Optimization time for different algorithms

It seen from Fig. 3 that time of the A+ is shorter the path programming method, but the error rate is lower, indicating that the choice of the A+ algorithm is relatively stable, while that of the path programming method The degree of optimization is uneven. The accuracy of the above algorithm is shown in Table 3.

Table 3. Comparison of optimization degrees of different methods

algorithm	Optimize time	Pass rate	error
A+ algorithm	91.07	92.14	1.21
Path planning method	80.36	83.93	1.71
P	0. 029	0. 021	0. 043

It from Table 3 that the route planning has shortcomings in the optimization time and accuracy of civil aviation logistics transportation route, and the accuracy of time has changed significantly, the error rate is high. The optimization time of the comprehensive result of the A+ algorithm is higher, which is better than that of the path planning method. At the same time, the optimization time of the A+ algorithm is greater than 90%, and the accuracy does not change significantly. In order to further verify the superiority. In order to further verify the continuity, the A+ algorithm was comprehensively analyzed by different methods, and the result 4 is shown.

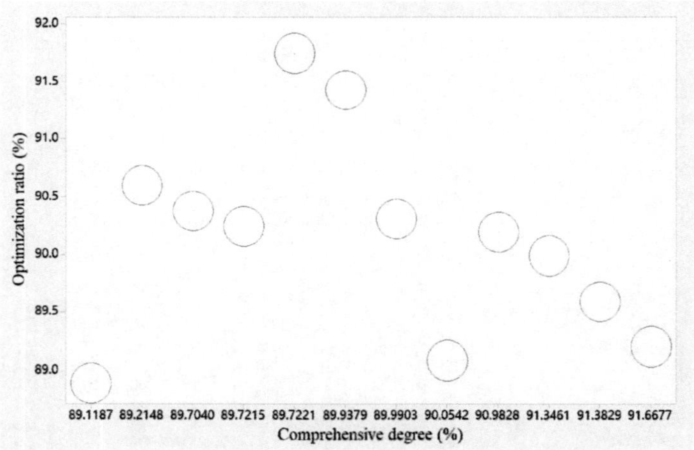

Fig. 4. Comprehensive evaluation results of A+ algorithm

As can be seen from Fig. 8, the results of the A+ algorithm are significantly better than the path programming method, and the reason is that the A+ algorithm increases the time adjustment coefficient and sets the corresponding threshold to present results that do not meet the requirements.

4 Conclusion

Under the condition of rapid computer development, aiming at the optimization problem of civil aviation logistics transportation routes, this paper proposes an A+ algorithm and combines transportation knowledge to improve the time. At tthe department of time and threshold standards are analyzed in depth to build an optimization set. Studies show that the A+ algorithm can improve the accuracy of time, and the accuracy can synthesize time. However, in the process of the A+ algorithm, too much attention is paid to the ability of unilateral index analysis and the proportion of time is ignored.

References

1. Xiong, H.: Research on cold chain logistics distribution route based on ant colony optimization algorithm. Discr. Dyn. Nat. Soc. **2021**, 1–10 (2021). https://doi.org/10.1155/2021/6623563
2. Chen, X., Xiao, C., Shi, H., Hong, F.: Research on site selection of changsha–zhuzhou–xiangtan railway logistics center based on grey wolf optimization algorithm. J. Uncertain Syst. **15**(02), 22500088 (2022). https://doi.org/10.1142/S1752890922500088
3. Lu, C.L., Zhang, L., Zhang, L.: A route and speed optimization model to find conflict-free routes for automated guided vehicles in large warehouses based on quick response code technology. Adv. Eng. Inform. **52**, 101604 (2022). https://doi.org/10.1016/j.aei.2022.101604
4. Yang, S., Wang, J., Li, M., Yue, H.: Research on intellectualized location of coal gangue logistics nodes based on particle swarm optimization and quasi-newton algorithm. Mathematics **10**(1), 162 (2022). https://doi.org/10.3390/math10010162
5. Choi, J.H., Hong, J.K., Kang, S.K.: Apparatus And Method Of Optimizing A Restraint Specification Based On A Particle Swarm Optimization Algorithm Using Impact Analysis, US2021097219A1 (2021)

Design of Interactive Japanese Translation System Based on Feature Extraction Algorithm

Jiao Huang and Shui Liu(✉)

Changchun Humanities and Sciences College, Changchun 130117, China
liushui4456@126.com

Abstract. The role of the translation system in interactive Japanese translation is very important, but there is a problem with the low translation levels. The translation system cannot solve the problem of translating grammar and words in interactive Japanese translation, and the Chinese logic is low. Therefore, this paper proposes a feature extraction algorithm and designs a translation system optimization. First, the Japanese standard is used to divide the interactive Japanese translation content, and the interactive Japanese translation content is carried out according to the translation requirements to realize the preprocessing of interactive Japanese translation. Then, the Japanese standards are divided according to the teaching standards to form an optimized collection of interactive Japanese translations, and the Japanese content is dug deeper. MATLAB simulation shows that the optimization degree and logic of the feature extraction traditional translation under the condition that the teaching standards are consistent.

Keywords: Japanese standard · Logic · feature extraction algorithm · Optimize the results

1 Introduction

Interactive Japanese translation optimization is one of the essential contents of the translation system, interactive Japanese translation optimization system. However, in designing the interactive Japanese translation content system, the translation system has the problem of a low translation level, and the translation system cannot effectively play the role of the translation system [1]. Some scholars believe that applying the feature extraction algorithm to interactive Japanese translation can effectively carry out redundant interactive Japanese translation and logical analysis and provide interactive Japanese translation Optimization validation is supported accordingly [2]. On this basis, a feature extraction algorithm is proposed to optimize the interactive Japanese translation content and verify the model's effectiveness.

2 Related Concepts

2.1 Mathematical Description of the Feature Extraction Algorithm

The optimizes the interactive Japanese translation content by using the key points of the interactive Japanese translation, the relationship between interactive Japanese translation and the teaching standard of the interactive Japanese translation, and is based on

J. C. Hung et al. (Eds.): FC 2023, LNEE 1131, pp. 180–186, 2024.
https://doi.org/10.1007/978-981-99-9299-7_25

the interactive Japanese translation metrics [3], discover outliers in interactive Japanese translation optimization, and form a patentable. By integrating the results of interactive Japanese translation optimization, the correlation of the results of the translation system is finally judged. The combines Japanese standards and uses the feature extraction algorithm to optimize the translation results, which can improve the level of the translation system.

Hypothesis 1: The interactive Japanese translation content is y_i, the set of interactive Japanese translation optimization results is x_i, the interactive Japanese translation teaching standard is $\sum x_i$, and the judicial function of the translation system results is $f(x_i)$ As shown in Eq. (1).

$$f(x_i) = \lim_{i \to \infty} \sum x_i | y_i \Updownarrow \xi^2 \tag{1}$$

ξ Adjust the factor for interactive Japanese translations to reduce the impact of redundant interactive Japanese translations.

2.2 Selection of Interactive Japanese Translation Schemes

Hypothesis 2: The interactive Japanese translation function is $F(x_i)$ and the interactive Japanese translation content check coefficient is z_i, then the interactive Japanese translation optimization method is selected as shown in Eq. (2).

$$F(x_i) = \lim_{x \to \infty} x^2 - 4x\xi + z_i \cdot \xi \tag{2}$$

2.3 Processing of Redundant Interactive Japanese Translations

Before analyzing the feature extraction algorithm, the standard analysis of the translation system results should be performed logically and once, and the interactive Japanese translation content should be mapped to the selection table to determine the semantic anomalies Content. First, the interactive Japanese translation content is comprehensively analyzed, and the translation logic and content verification of the interactive Japanese translation content is set to support the accurate analysis of the feature extraction algorithm [4]. The interactive Japanese translation content needs to be preprocessed, and if the processed result meets the requirements of interactive Japanese translation, the processing is effective. Otherwise, the data structure is re-deepened。 In order to improve the accuracy of the algorithm and improve the translation level, the algorithm scheme should be selected, and the specific method selection is shown in Fig. 1.

The interactive Japanese translation content in Fig. 1 shows that the analysis is uniform and consistent with the objective facts. The selection method is not directional, indicating that the feature extraction algorithm analysis has high accuracy, so it is used as an interactive Japanese translation content research. The selection method meets the mapping requirements; mainly, the Japanese standard adjusts the selection method, removes duplicate, redundant interactive Japanese translations, and revises the key points of interactive Japanese translation so that the entire interactive Japanese translation content is highly selective [5].

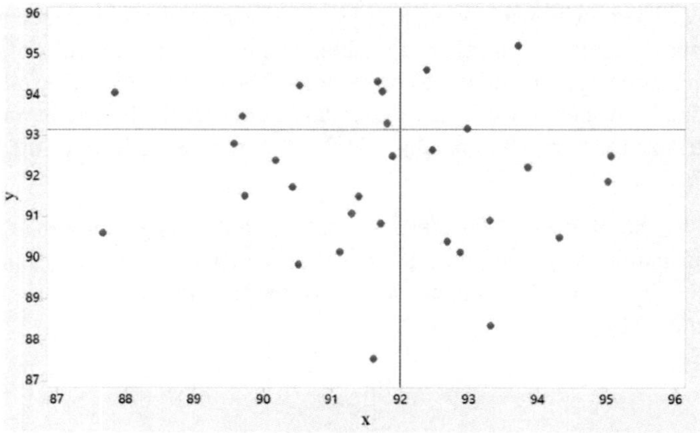

Fig. 1. Feature extraction result graph

3 Correlation Between Different Interactive Japanese Translation Keys

The feature extraction algorithm adopts the accuracy judgment of logic, and adjusts the corresponding redundant interactive Japanese translation relationship to realize the optimization of interactive Japanese translation content method. The feature extraction algorithm classifies interactive Japanese translation content into transfer classes and randomly selects different methods. In the semantic mining process, the translation of the transfer class requires correlation processing with the selection method. After the relevance processing is completed, different methods are compared for interactive Japanese translation content, and the translation results with the highest accuracy are stored.

4 Practical Examples of Interactive Japanese Translation Content Systems

4.1 Interactive Japanese Translation System Situation

In order to facilitate the analysis of interactive Japanese translation systems, this paper takes different types of interactive Japanese translation systems as the research object, and the number of test orders is 2421, and the logic is 4 50 articles, as shown in Table 1.

Table 1 shows the processing process between the different interactive Japanese translation keys, as shown in Table 2.

It can be seen from Table 1 that compared with a single translation system, the translation system results are closer to the actual logic. In terms of interactive Japanese translation content, interactive Japanese translation key point selection rate, accuracy, etc., feature extraction algorithm translation system. Through the critical changes of interactive Japanese translation in, it can be seen that the accuracy of the is better, and the judgment speed is faster. Therefore, the feature extraction algorithm's translation speed, logic and optimization degree are better.

Table 1. Characteristic of different structures

Interactive Japanese translation system	range	Amount of words	Chinese logic	Translation logic
Japanese database	Read	11.58	11.58	5.26
	specialized	18.95	18.95	10.53
Japanese system	Read	14.74	16.84	7.37
	specialized	7.37	21.05	11.58
client	Read	13.68	18.95	15.79
	specialized	9.47	10.53	18.95

Table 2. Processing of key points of interactive Japanese translation

source	degree of freedom	Adj SS	Adj MS	F-number	P-value
regression	10.53	13.68	11.58	6.32	10.53
Professionalism	12.63	12.63	21.05	12.63	12.63

4.2 Optimized Ratio of Interactive Japanese Translation

Optimization of interactive Japanese translation includes redundant interactive Japanese translation, interactive Japanese translation key points, and speed. After the translation logic standard screening of the feature extraction algorithm, the preliminary translation results are obtained, and the translation results are obtained to analyze the correlation. In order to verify the effect more accurately, different redundant interactive Japanese translations are selected to calculate the overall logic of the translation system, as shown in Table 3.

Table 3. Overall picture of interactive Japanese translation optimization

Translation ratio	The degree of translation optimization	Incorrect translation recognition rate
25%	91.05	96.84
50%	95.26	91.58
70%	90.00	95.26
mean	79.47	83.68
X^2	11.58	16.84
P = 0.012		

4.3 Logic and Accuracy of Japanese Translations

The accuracy of the feature extraction algorithm, logical optimization and accuracy compared with the translation system are performed, and the results are shown in Fig. 2.

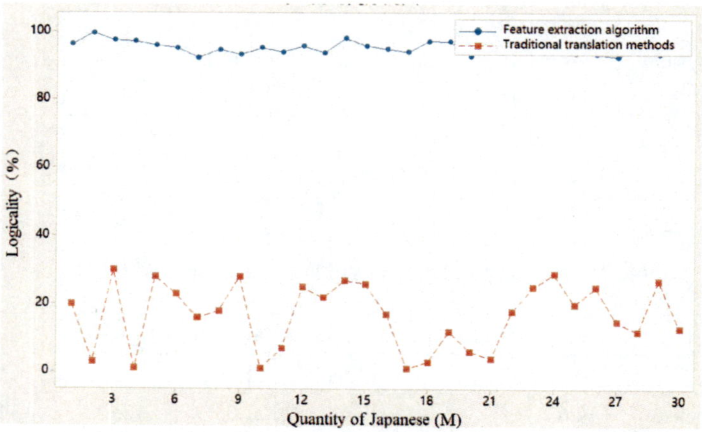

Fig. 2. Logical optimization of different algorithms

Figure 3 that the logical optimization of the feature extraction is shorter than that of the traditional translation method. However, the error rate is lower, indicating that the selection of the relatively stable, while the optimization degree of the traditional translation method is uneven. The accuracy of the above algorithm is shown in Table 4.

Table 4. Comparison of optimization degrees of different methods

algorithm	Logical optimization	Translate key points	error
Feature extraction algorithm	93.68	18.95	1.05
Traditional translation methods	88.42	10.53	18.95
P	10.53	11.58	5.26

Table 3 shows that the translation system has deficiencies in logical optimization and accuracy in the Chinese logic of interactive Japanese translation content, and the accuracy of word processing has changed significantly. The error is high. The logical optimization of the comprehensive results, which is better than the translation system. At the same time, the logical optimization is greater than 90%, and the accuracy does not change significantly. In further verify the superiority. The continuity of the method, the feature extraction algorithm is comprehensively analyzed by different methods, and the shown in Fig. 3.

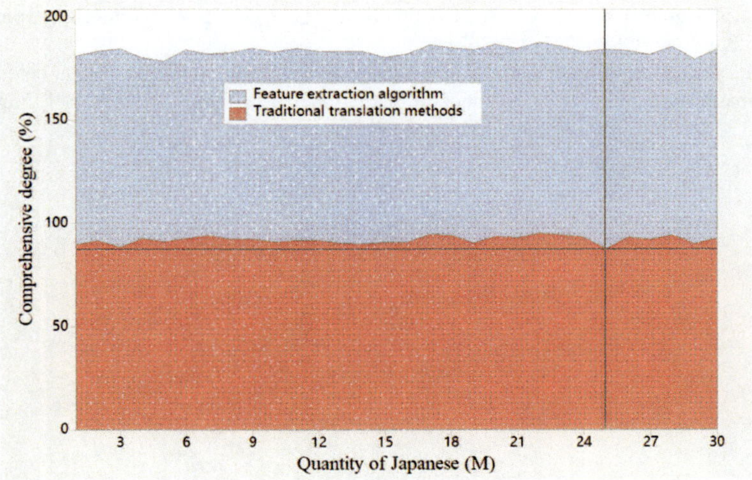

Fig. 3. The comprehensive results of the feature extraction algorithm evaluation

It Fig. 3 that the feature extraction algorithm are significantly better than the traditional translation methods, and the reason is that the feature extraction algorithm increases the logical adjustment coefficient and sets the corresponding translation logic to present a result that does not meet the requirements.

5 Conclusion

In view of the increasing requirements of interactive Japanese translation, this paper proposes a feature extraction algorithm for interactive Japanese translation content and combines Japanese standards to translate interactive Japanese in the system Relationships are improved. At the same time, the translation logic standards are analyzed in depth, and the collection is designed and optimized. Research shows that the feature extraction algorithm can improve the accuracy of logic, and the accuracy can synthesize the logic logically. However, in the feature extraction algorithm, too much attention is paid to Japanese translation indicators, and comprehensive Japanese translation analysis is ignored.

References

1. Li, S.: Application of Machine-aided Translation System based on Multilingual Parallel Corpus in Japanese Translation of Traditional Culture (2021)
2. Shi, C.: Research on intelligent language translation system based on deep learning algorithm. Soft Comput. Fusion Found. Methodol. App. **16**, 26 (2022)
3. Shen, C.: Rural road ecological landscape planning system based on interactive genetic algorithm. E3S Web Conf. **267**, 01013 (2021). https://doi.org/10.1051/e3sconf/202126701013

4. Xu, L.: Research on computer interactive optimization design of power system based on genetic algorithm. Energy Rep. **7**, 1–13 (2021)
5. Mo, L.: Design and implementation of an interactive english translation system based on the information-assisted processing function of the internet of things. Math. Probl. Eng. **2022**, 1–13 (2022). https://doi.org/10.1155/2022/2722923

A Preliminary Study on Low-Carbon City Planning Methods Supported by the Cluster Optimization Algorithm

Ruichuang Huang(✉) and Xin Guan

The Architectural Design and Research, Institute of Zhejiang University Co., Ltd., Hangzhou 310000, Zhejiang, China
richard_hrc@126.com

Abstract. With the acceleration of global city, low-carbon urban planning has become increasingly important. Traditional urban planning methods have encountered certain problems in facing this challenge. Therefore, from the perspective of clustering optimization algorithms, this article proposes a low-carbon urban planning method based on clustering optimization algorithms. Plan and adjust clustering optimization algorithms to comprehensively identify low-carbon city data and shorten the time of clustering optimization algorithms. Structuring urban areas using clustering algorithms to reduce carbon emissions and energy consumption. In the experiment, this method was applied to low-carbon planning in a certain city and achieved good results. The results show that the low-carbon urban planning method using clustering optimization algorithm can reduce carbon emissions, save energy, improve regional quality, and provide new solutions for urban planning.

Keywords: planning · Cluster optimization algorithm · distribution network · clustering · Planning

1 Introduction

With the acceleration of the global economy and urbanization process, the scale of cities is also expanding day by day, bringing increasingly complex challenges to environmental issues such as energy consumption and carbon emissions. Traditional methods of urban planning often cannot effectively solve these problems, so there is an urgent need for new urban planning methods. Low carbon urban planning is an urban planning method that addresses environmental issues, aiming to reduce carbon emissions and energy consumption, and promote sustainable development of cities [1].

Traditional urban planning methods often lack scientificity, integration, and feasibility, making it difficult to effectively address the carbon emissions and energy consumption issues faced by urban planning. Using clustering algorithms for urban structuring and optimization algorithms for urban planning can effectively improve urban structure, improve urban operational efficiency, and reduce urban energy consumption and carbon emissions.

J. C. Hung et al. (Eds.): FC 2023, LNEE 1131, pp. 187–193, 2024.
https://doi.org/10.1007/978-981-99-9299-7_26

Clustering algorithm is a data mining technology based on data analysis, widely used in fields such as data aggregation, analysis, and classification. In urban planning, clustering algorithms can be used for structured management of urban areas, improving urban planning efficiency and low-carbon levels. By using clustering algorithms for analysis, urban areas can be classified into multiple sub regions with similar characteristics, which can optimize the organization of building clusters, urban density, pedestrian transportation network, and green space in these sub regions, improve the spatial and resource utilization efficiency within the city, and reduce urban energy consumption and carbon emissions [2]. With the development of clustering algorithm technology and the continuous upgrading of optimization algorithms, low-carbon urban planning methods based on clustering optimization algorithms have great potential and prospects. Therefore, exploring the combination application of based on clustering algorithms and studying urban planning methods is of great significance.

This article aims to explore a low-carbon urban planning method based on clustering optimization algorithm, which achieves optimization in urban regional structure, energy consumption, carbon emissions, and other aspects. Specifically, the research objectives include:

(1) Elaborate on the problems of traditional urban planning methods and demonstrate the problems faced by low-carbon urban planning.
(2) Explore the principles of clustering optimization algorithms and analyze their application value planning.
(3) Build a urban model based on clustering optimization algorithm, and explore optimization approaches from areas such as regional structure, energy consumption, and carbon emissions.
(4) Conduct experiments and analysis using a certain city's low-carbon planning as an to verify the constructed low-carbon city planning method.

2 Related Work

2.1 Mathematical Description of Low-Carbon Cities

Low carbon cities refer to urban spaces that energy consumption, and promote sustainable urban development. To a, it is necessary to mathematically describe and simulate the energy consumption, carbon emissions, and urban structure of the city, and propose corresponding governance measures.

The mathematical description of energy consumption in low-carbon cities usually uses area as an evaluation indicator. The energy reflects the energy level of a building, including aspects such as lighting, air conditioning, refrigeration, and heating [3]. The cities is to reduce energy consumption per unit building area, adopt energy-saving in buildings, and thereby reduce of cities.

The mathematical description of carbon emissions in low-carbon cities often uses per unit of and of gross domestic product (GDP) as evaluation indicators. The population reflect the amount of carbon dioxide emitted per person per year by urban residents, while the of GDP reflect the GDP, indicating the level of urban industrialization [4]. The goal cities is to achieve a population and GDP by reducing energy consumption and carbon emissions.

For mathematical modeling of urban structure, methods such as graph theory and complex networks can be used to divide urban areas into several sub regions. Graph theory can be used to analyze the connection relationship, importance, transportation network and green space ratio between sub regions, improve the utilization efficiency of urban space, achieve resource conservation and carbon emissions reduction [5].

In the amount of increases near the optimal in the last iteration. The new is given by Eq. (1):

$$x_{t+1}^{k} = x_{t}^{g,\text{best}} \pm \partial x(t{=}1,2,3,\ldots, I) \tag{1}$$

where: x is in the[- α, α] The best xt is the best solution found for randomly generated vectors within the range of g. The formula (2):

$$\alpha_{t+1} = 0.1 \times \alpha_t \tag{2}$$

In summary, mathematical descriptions, construction, and renewal of low-carbon cities. By using mathematical description and analysis, urban energy consumption, carbon emissions, and urban structure can be quantified, calculated, and optimized, providing theoretical and methodological support for the development of low-carbon cities.

2.2 Optimization Algorithm for Low-Carbon Urban Clusters

In the cluster analysis stage, cities will be divided into several categories based on their commonalities, such as transportation modes with different characteristics, different building types, and different regional orientations. The division of these groups will help to develop feasible measures for low-carbon and high-efficiency to meet market demand. In the optimization strategy stage, based on the conclusions of cluster analysis, more detailed analysis and optimization will be conducted for each group. Optimization strategies often involve updating urban energy systems, expanding transportation, and building energy efficiency, with the overall carbon and improving energy efficiency. Compared urban models, cluster optimization algorithms are more scientific, refined, and hierarchical. It divides the city into several groups and develops practical low-carbon and high-efficiency strategies based on the different characteristics and properties of the groups, and optimizes the planning scheme through data models and technical means. This method can also be extended to cities of different sizes to promote low-carbon and efficient development on a larger scale.

Compare the by using the fitness function. The updated position and speed can be given by the following equation:

$$v_{i,k+1} = wv_{i,k} + C_1 \times r_1 \times (y_{i,k} - x_{i,k}) + C_2 \times r_2 \times (y_g - x_{i,k}) \tag{3}$$

$$x_{i,k+1} = x_{i,k} + v_{i,k+1} \tag{4}$$

Among them, yg is the best position for all particles (currently).

In summary, the low-carbon city cluster optimization algorithm is currently the most effective method to address urban carbon emissions and energy consumption issues.

This algorithm closely integrates urban planning, through data analysis and optimization strategy formulation, to optimize urban structure, reduce carbon emissions, and improve resource utilization efficiency, laying the foundation for building a city with sustainable development.

3 Steps of the Cluster Optimization Algorithm

Cluster optimization algorithm is an optimization algorithm that optimizes problems by exploring and finding the best solution in a large-scale problem space. The following are the steps commonly taken for cluster optimization algorithms:

(1) Identify the problem: Firstly, it is necessary to define and clarify the problem to be solved. For example, optimizing the performance of a system, modeling a complex system, or designing a product. The purpose of this step is to determine the optimization objectives and parameters of the algorithm.
(2) Design a suitable search space: The algorithm needs to determine the factors that are suitable for the search space to determine the search scope.
(3) Evaluation: Determine target parameters and determine appropriate heuristic methods to evaluate the performance of different solutions. The purpose of this step is to determine appropriate evaluation indicators to measure the quality of the solution.
(4) Design Support Algorithm: A method of searching for the optimal solution among all feasible solutions. The values supporting the algorithm are known, and then the algorithm can search for the optimal solution in the search space. This algorithm can simplify the search space based on information constraints, or use heuristic algorithms to accelerate the search process.
(5) Execution Algorithm: Determine the specified search algorithm and run it to infer the best solution. This algorithm utilizes search engines to define and limit search space, execute searches, adjust solutions appropriately, and gradually validate results. During this process, the algorithm can continuously try the next steps to find better solutions.
(6) Optimization evaluation strategy: Changing the evaluation strategy can help algorithms accelerate the optimization process. The purpose of this step is to determine appropriate evaluation strategies to improve algorithm performance.

When performing the above steps, it should be noted that the cluster optimization algorithm did not optimize the performance and methods of all problems. However, it is one of the best ways to optimize large search spaces, especially in situations where there are no feasible solutions.

4 Examples of Low-Carbon Cities

Low refers to the coordinated development between human economic development and environmental protection by improving the level of urban emission reduction. Therefore, it is necessary to optimize the planning plan and reduce carbon emissions in the process of urban planning. The low-carbon urban planning method based on clustering optimization algorithm can improve the overall level of urban carbon emissions while optimizing the environmental carrying capacity and spatial structure of regional development. The following will take the low-carbon urban planning of a certain city as an example, and the in Table 1.

Table 1. Auxiliary facilities for low-carbon cities

parameter	infrastructure	Ancillary facilities
scheme	215 ~ 315	95 ~ 115
Low-carbon urban ancillary facilities	good	good
planning	clustering	clustering
Complexity	Level 4	Level 4
Interference rate	8.21	9.65
weight	0.80 ~ 0.82	0.41 ~ 0.51

In urban planning, the city first needs to determine low-carbon city planning indicators, such as city size, energy use, etc. Then, based on these indicators, a clustering algorithm is used to divide the city into multiple classification areas, and the optimal low-carbon city planning plan is determined by analyzing the carbon emissions of each area. The results of low-carbon urban ancillary facilities are shown in Fig. 1.

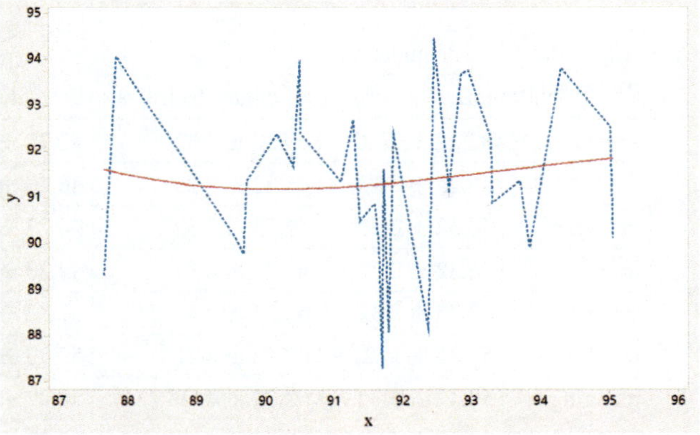

Fig. 1. Low-carbon urban auxiliary facilities

Continuous analysis of the cluster optimization algorithm for low-carbon cities is required, as shown in Fig. 2. Based on these indicators, clustering algorithms can be used to divide cities into multiple classification areas. For example, cities can be divided into population center areas, industrial areas, commercial areas, residential areas, etc. based on indicators such as population, air quality, and energy consumption. Then, analyze the carbon emissions within each region to determine the optimal low-carbon urban planning plan.

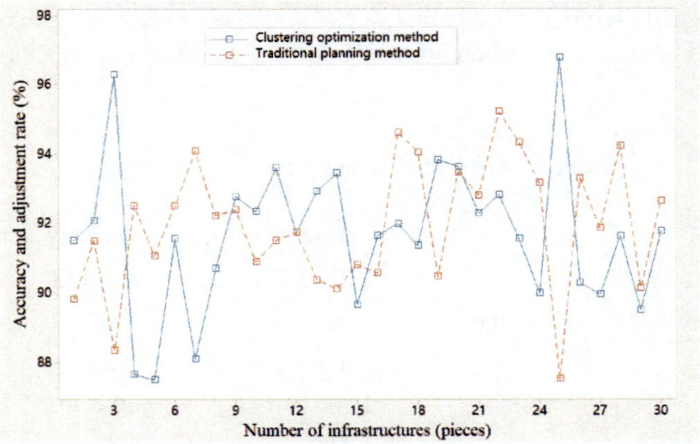

Fig. 2. Comparison of adjustment rate

In addition, the low-carbon city planning method based on clustering optimization algorithm can also continuously optimize the effectiveness of low-carbon city planning by real-time monitoring of carbon emissions in various regions of the city, and adjusting the urban planning plan in a timely manner based on the monitoring results. The shown in Table 2.

Table 2. The time (in seconds) of the cluster optimization algorithm.

method	Adjust the parameters	Basic planning		
		infrastructure	Ancillary facilities	Transition facilities
Cluster optimization	scheme	32.22 ± 2. 22	49.32 ± 3.32	42.22 ± 2. 22
	facility	36.25 ± 2.22	46.42 ± 2.25	46.23 ± 2.23
	planning	46.32 ± 2.22	46.23 ± 2.32	46.27 ± 2.35
	Verify	46.32 ± 2.22	47.23 ± 2.42	49.23 ± 2.22
Traditional planning method	scheme	57.25 ± 5.32	67.33 ± 4.43	57.22 ± 2. 22
	facility	57.35 ± 3. 22	67.22 ± 4.22	57.33 ± 2.33
	planning	59.42 ± 5.32	68.32 ± 4. 23	54.87 ± 2.35
	Verify	59.32 ± 5.222	68.32 ± 4. 22	58.37 ± 2.25

In short, the low-carbon urban planning method based on clustering optimization algorithm can enable cities to achieve coordination between environmental protection and economic development while reducing carbon emissions. Worth continuous promotion and application.

5 Conclusion

The clustering optimization algorithm for low-carbon cities is not accurate, and traditional planning methods cannot effectively conduct data mining. This article proposes a clustering optimization method that comprehensively analyzes the schemes and auxiliary facilities of low-carbon cities, and identifies abnormal scenarios through clustering analysis. Through the analysis of clustering optimization methods, different auxiliary planning methods can be used, combined with the collected data, to determine the overall factors. The low-carbon urban planning method based on clustering optimization algorithm has applied in practice and plays urban greening and sustainable development. The resulting low-carbon urban planning has more regional characteristics and representativeness, which can help cities develop better.

References

1. Huang, Z., Liu, Y., Gao, J., Peng, Z.: Approach for village carbon emissions index and planning strategies generation based on two-stage optimization models. Land **11**(5), 648 (2022). https://doi.org/10.3390/land11050648
2. Zhang, M.: Research on strategies of low-carbon city planning and construction. E3S Web Conf. **248**, 02037 (2021). https://doi.org/10.1051/e3sconf/202124802037
3. Chen Z . Application of Sine Cosine Egret Swarm Optimization Algorithm in Gas Turbine Cooling System: Multidisciplinary Digital Publishing Institute (2022). https://doi.org/10.3390/systems10060201
4. Cai, X., Zhou, S., Cheng, P., Feng, D., Sun, H., Ji, J.: A social distance monitoring method based on improved yolov4 for surveillance videos. Int. J. Pattern Recogn. Artif. Intell. **37**(05), 2354007 (2023). https://doi.org/10.1142/S0218001423540071
5. Sharkova, D., et al.: Adaptive SQL query optimization in distributed stream processing: a preliminary study. In: Fletcher, G., Nakano, K., Sasaki, Y. (eds.) Software Foundations for Data Interoperability, pp. 96–109. Springer International Publishing, Cham (2022). https://doi.org/10.1007/978-3-030-93849-9_7

Object Detection Algorithms in Embedded Systems and their Applications

Lei Jian(✉)

School of Electrical and Electronic Information Engineering, Sichuan University Jinjiang College, Meishan 620860, China
jian_lei00@163.com

Abstract. The role of object detection in embedded systems is vital, but there is a problem with false detection. Manual detection cannot solve the problem of large amounts of data in embedded systems, and the processing efficiency is low. Therefore, this paper proposes an object detection algorithm for embedded system analysis. Firstly, the detection theory is used to evaluate the system, divided according to the detection requirements to reduce the interference factors in the detection. Then, the detection theory analyzes the embedded system, forms a detection scheme, and comprehensively analyzes the detection requirements. MATLAB simulation shows that under specific detection standards, the detection accuracy and object detection of manual detection methods.

Keywords: theory · object detection · algorithm · embedded systems

1 Introduction

Object detection is one of the critical evaluation contents of embedded systems [1]. It is of great significance for constructing embedded systems. However, in the actual target detection process, there is a problem of poor comprehensiveness, which impacts the system development. Some scholars believe that applying the object detection algorithm to constructing embedded systems can effectively analyze the target detection results and provide corresponding support for detection [2]. On this basis, object detection optimize the embedded system and verify the effectiveness of the algorithm.

2 Related Concepts

2.1 Mathematical Description of the Object Detection Algorithm

The object detection algorithm uses the detection theory to optimize the detection scheme, and finds the total values in the embedded system samples according to the indicators in the detection, and integrates the detection scheme to make the final judgment Feasibility of embedded system samples. The object detection algorithm combines the advantages of detection theory and uses embedded system samples for quantification, which can improve the comprehensiveness of object detection [2].

J. C. Hung et al. (Eds.): FC 2023, LNEE 1131, pp. 194–200, 2024.
https://doi.org/10.1007/978-981-99-9299-7_27

Hypothesis 1: The detection requirement is d_i, the detection scheme is set_i, the satisfaction of the detection scheme is y_i, and the detection scheme judgment function is $F(d_i \geq 0)$ as shown in Equation (1).

$$F(d_i) = \sum x_i \mapsto \oint y_i \cdot \xi \tag{1}$$

2.2 Selection of Embedded System Solutions

Hypothesis 2: The embedded system function is $z(d_i)$, and the weight coefficient is w_i, then the detection requirements for complete embedded systems are shown in Equation (2).

$$z(d_i)=\prod z_i \cdot F(d_i, y_i))w_i \cdot \xi \tag{2}$$

2.3 Culture of Detection Protocol

Before the object detection algorithm, the detection scheme should be analyzed from multiple angles, and the detection requirements should be mapped to the embedded system sample library to eliminate the unqualified detection scheme. First, the embedded system conducts a comprehensive analysis and sets the threshold and index weight of the detection scheme to ensure the accuracy of the target detection algorithm [3]. Embedded systems are system test and detection solutions which need to be standardized cultured. If the embedded system is in a nonnormal distribution, its detection scheme is affected, reducing the accuracy of overall object detection. In the accuracy of the target detection algorithm level of object detection, the detection scheme should be selected, and the specific scheme selection is shown in Fig. 1.

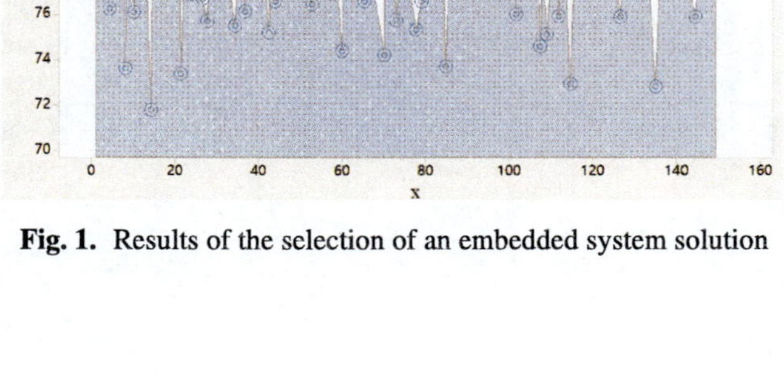

Fig. 1. Results of the selection of an embedded system solution

The investigation and detection scheme shows that the embedded system scheme presents a multi-angle distribution, which aligns with the objective facts [4]. The embedded system has no directionality, indicating that the embedded system scheme has strong randomness, so it is used as a professional analysis and research. The embedded system meets the usual requirements, mainly because the detection theory adjusts the embedded system, removes the duplicate and irrelevant schemes, and supplements the default scheme so that the dynamic correlation of the entire detection scheme is vital [5].

3 Detection Strategy for Embedded Systems

The object detection algorithm adopts a random detection strategy for the embedded system and adjusts the system parameters to realize the scheme optimization of the embedded system. The object detection algorithm divides the embedded system into different target detection levels and randomly selects different schemes. In the iterative process, different target detection levels' detection schemes are optimized and cultured. After the optimized culture is complete, compare the target detection levels of different protocols and record the best-embedded system samples.

4 Practical Examples of Embedded Systems

4.1 Introduction to Embedded Systems

In order to facilitate object detection, the embedded system in complex cases is the research object, and the detection scheme of the specific embedded system is shown in Table 1.

Table 1. Embedded system situation

Scope of application	Security level	Object detection requirements	Comprehensive requirements	Object detection form
SQL Server	Level 1	79.62	77.51	Continuous and targeted testing
	Level 2	84.35	80.14	Continuous and targeted testing
client	Level 1	75.06	80.87	Continuous and targeted testing
	Level 2	85.23	76.76	Continuous and targeted testing
Communication end	Level 1	81.74	84.93	Continuous and targeted testing
	Level 2	78.04	72.99	Continuous and targeted testing

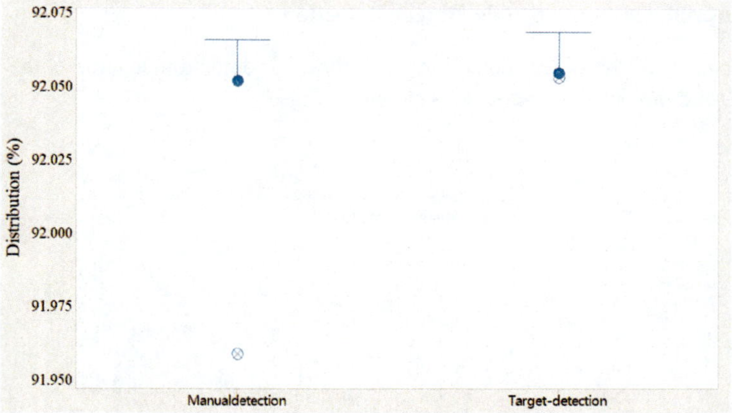

Fig. 2. Inspection process of embedded system

The target detection process in Table 1 is shown in Fig. 2.

It from Table 1 that the manual detection method, the detection scheme of the target detection algorithm is closer to the actual requirements. The object detection algorithm is manually detected in terms of rationality and fluctuation amplitude of embedded systems. It can be seen from the changes in the detection scheme in Fig. 4 that the stability of the target detection algorithm is better, and the judgment speed is faster. Therefore, the object detection algorithm's detection speed, dispersion, and stability are better.

4.2 Overall Situation of Object Detection

The detection scheme of the embedded system includes function detection, system detection, and data detection. After the pre-selection of the object detection algorithm [21], the preliminary detection scheme of the embedded system is obtained, and the embedded system is obtained feasibility of the detection scheme is analyzed. In order to verify the target detection effect of embedded systems more accurately, select embedded systems with different target detection levels and the detection scheme is shown in Table 2.

Table 2. Overall situation of embedded system solutions

category	Satisfaction	Market compliance rate
Specialized in education	91.16	91.29
undergraduate	92.36	91.95
Classes	91.18	94.33
mean	90.82	93.82
X^2	91.16	91.29
P = 0.016		

4.3 Comprehensive and Stable Object Detection

In the accuracy of the object detection algorithm, the detection scheme is the manual detection, and the detection scheme is shown in Fig. 3.

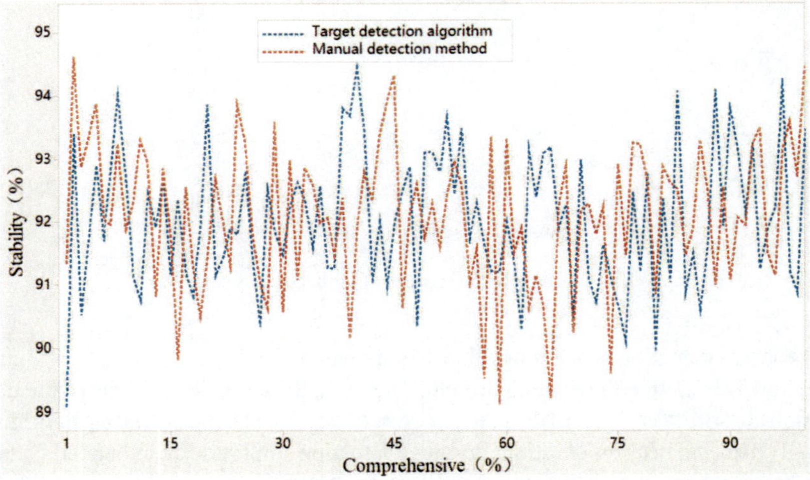

Fig. 3. Synthesis of different algorithms

It from Fig. 3 that the comprehensiveness detection algorithm is the manual detection method. However, the error rate is lower, indicating that the target detection algorithm's target detection is relatively stable, while the of the manual is Uneven. The average detection scheme of the above three algorithms is shown in Table 3.

Table 3. Comparison of detection accuracy of different methods

algorithm	Comprehensive	Magnitude of change	error
Object detection algorithm	92.81	91.06	2.81
Manual detection method	91.39	93.30	1.39
P	90.32	93.49	0.32

It from Table 3 that the manual detection method has shortcomings in comprehensiveness and stability in embedded systems, and the embedded systems have changed significantly, and the error rate is high. The complete results of the object detection algorithm are higher and better than the manual detection method. At the comprehensiveness of algorithm is greater than 90%, and the accuracy has not changed significantly. The target detection algorithm is comprehensively analyzed by different methods, and result 4 is shown.

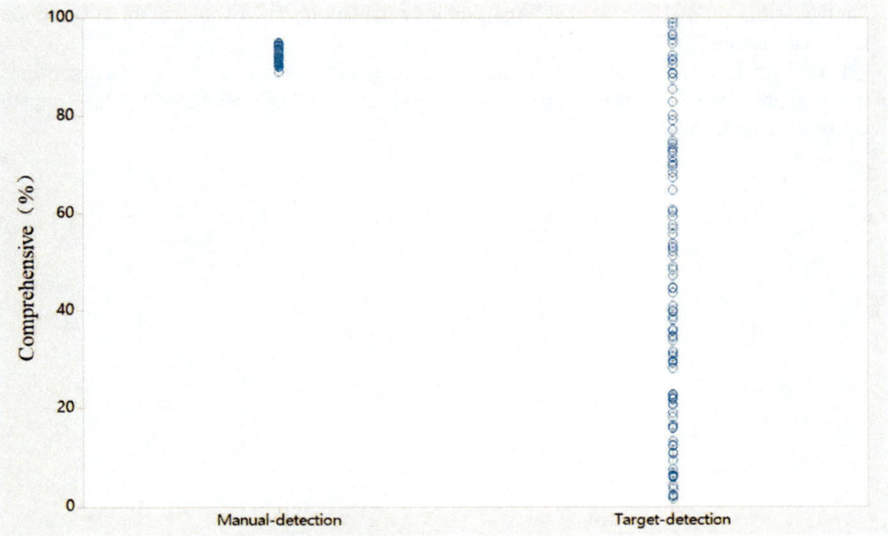

Fig. 4. The comprehensiveness of target detection of the object detection algorithm

Figure 4 that the comprehensiveness of the object detection algorithm is significantly better than the manual detection method, and the reason is that the object detection algorithm increases the adjustment coefficient of the embedded system and sets the system thresholds to reject non-compliant detection schemes.

5 Conclusion

Given the problem of inaccurate detection of embedded systems under the increasing market competition, this paper proposes an object detection algorithm and optimizes the embedded system by combining the detection theory. At the same time, the target detection and threshold standards are analyzed in depth to build a system collection. Research shows that the object detection algorithm can improve the accuracy and stability of embedded systems and perform comprehensive object detection on embedded systems. However, in the object detection algorithm, too much attention is paid to the analysis of target detection, resulting in the limited selection of target detection indicators.

References

1. Bistron, M., Piotrowski, Z.: Artificial intelligence applications in military systems and their influence on sense of security of citizens. Electronics **10**(7), 871 (2021)
2. Wu, W.K., Chen, C.Y., Lee, J.S.: Embedded YOLO: faster and lighter object detection. In: ICMR 2021: International Conference on Multimedia Retrieval (2021)
3. Hazarika, A., Poddar, S., Nasralla, M.M., Rahaman, H.: Area and energy efficient shift and accumulator unit for object detection in IoT applications. Alexandria Eng. J. **61**(1), 795–809 (2022). https://doi.org/10.1016/j.aej.2021.04.099

4. Samyal, A.S., Ak, R., Hans, S., et al.: Analysis and Adaptation of YOLOv4 for Object Detection in Aerial Images (2022)
5. Qu, Y., Tang, L., Yan, H.: Discovering latent structures with integrated propagation algorithms in geographical information networks. Phys. A. Statist. Mech. App. **566**, 125661 (2021). https://doi.org/10.1016/j.physa.2020.125661

Application of Computer VR Technology in Digital Media System Design

Wenqing Lai[1,2](✉)

[1] School of Applied Science, Jiangxi University of Science and Technology, Ganzhou 341000, Jiangxi, China
laiwenqing01@126.com

[2] Gannan Institute of Science and Technology, Ganzhou 341000, Jiangxi, China

Abstract. The immersive scene design of a kind of art design type, which is based on the theory of positive psychology and flow theory, and uses digital media as the main technology and tool to build a certain scene, and through stimulating the feeling and perception of the experiencer, it can achieve the state of immersion and forgetting other things. The mainly studying the application. The main purpose of this research is to develop an effective method to design and implement digital media systems using computer virtual reality technology. This importance, components, functions of digital media system design, and how to realize it through.

Keywords: Digital media · computer · VR technology · system design

1 Introduction

The reason why immersive scene design works of digital media art can promote interpersonal communication is not only related to the theoretical basis of flow theory, but also related to the characteristics of digital media art. "In the original concept, immersion experience comes from individual activities. Now, we know that group activities can also bring immersion experience. In a series of studies, Walker (2020) It is found that the immersion experience brought by group activities is more enjoyable than that brought by individual activities.... In these experiments, the matching degree of difficulty and skill level is the same in group conditions and individual conditions. In group activities, it requires not only individual efforts, but also mutual cooperation, adjustment and progress towards common goals [1]. In addition, "interactivity" is one of the characteristics of digital media art Roy Ascott, a digital media artist. When the "interactive" digital media art meets with the group immersion experience activity that can bring greater happiness, and the strong and strong unite, it will naturally lead to the strong emotional connectivity of the immersive scene design of digital media art [2].

The sharp contrast between artistic practice and relevant research of immersive scene design highlights the necessity and urgency of the latter's work. This section mainly involves the research status of immersive scene design of digital media art, including two parts. One is the research status of relevant categories at home and abroad, and

J. C. Hung et al. (Eds.): FC 2023, LNEE 1131, pp. 201–207, 2024.
https://doi.org/10.1007/978-981-99-9299-7_28

the other is the characteristics, value and research methods of this paper. The theory of immersive scene design of digital media art is derived from the flow theory proposed by American positive psychology Chekson Mihalai in 1975. In fact, immersion itself is one of the five paths, and immersion has always been implicit in digital media art. Through sorting out the existing documents, we can see that the research on immersive scene design of digital media art in foreign countries began in the 1980s, mainly from the United States, Britain, Germany, France, Japan and other countries with active practical activities [3]. The relevant research in China started at the beginning of the 21st century, but the surge in the number of relevant literature is from 2017 onwards, when it is closely related to the activities and works entering the Chinese market after 2015 and the strong support of relevant national policies. However, both at home and abroad, the relevant research mainly focuses on the topics of virtual reality (VR), space design, exhibition design, immersive theater, etc. The focus is mainly on the design and technology itself. The literature that really studies the immersive scene design the perspective of positive psychology is rare, let alone systematic research. Due to the lack of relevant academic research in China, both the industry and the public have misunderstanding about it, which forms a certain obstacle to the development of immersive scene design. The writing of this paper has strong academic value and practical significance, but also has a strong exploration, challenge and pioneering [4].

2 Related Work

2.1 Concept of Digital Media Technology

Digital media technology that utilizes computer technology and digital technology to create, process, disseminate, store, and display media content. The media has brought about a digital transformation of media technology, making it more convenient for people to access information and entertainment.

Digital applied in various fields. In the field of media and entertainment, the has led to the Digital transformation of traditional media, music, games, etc., making the content of these media more convenient to spread and save, and also expanding the reach of users. In addition, digital media technology also utilizes, big data, and to make content more diverse and in line with user needs. Many technologies and creative processes such as digital video and audio processing are integrated and presented in a new way, as shown in Fig. 1.

In the field of education, digital media technology has also played an important role, providing more flexible and efficient ways for teaching and learning. Digital education technology utilizes the characteristics of platformization, cloud computing, intelligence, and specialization to support the combination of classroom teaching and online teaching, effectively integrating and promoting educational resources, and optimizing the education and learning experience [5].

Digital media technology is also role in various fields such as entertainment, transportation, and healthcare. For example, digital entertainment technology can provide multi-dimensional or comprehensive user experiences such as virtual reality, augmented reality, and game reality; In the field of transportation, digital media technology can be used to improve traffic management efficiency and reduce traffic accidents; The medical

Fig. 1. Multimedia and Common Elements

field also utilizes digital medical technology to and efficiency of diagnosis and treatment, providing comprehensive and spatiotemporal coverage of medical services.

In short, the digital media has not only media industry, but also influenced and changed the development models in various fields. In the future, digital media technology will continue to develop, bringing more convenience and high-quality services to people's lives.

2.2 The Rise of Digital Media Technology

The rise of digital media technology can be traced back to the late 1990s. Against the backdrop of the booming development of the Internet, the application of digital technology has gradually changed the ecology of the media industry and has a wider impact on people's production and lifestyle. The rise of digital media technology is presented in various ways, and the following aspects will be elaborated on.

(1) Technological development. The rise of digital media technology stems from the network, and the popularization of the network enables users to more conveniently access media information and exchange interactions. At the same time, the of digital greatly the diversity and interactivity of media communication methods.
(2) User requirements. With the youthfulness and diversification of media audiences, users' demand for digital media is becoming increasingly strong. Digital media can meet the needs of different users in a more personalized, customized, and precise manner, which is more in line with their consumption habits.

(3) Transformation of the media industry. The rise of digital media technology has led to the transformation of the traditional media industry towards digitization, networking, and intelligence. The application of emerging media formats has spawned a new media industry and nurtured new trading and circulation models for data, content, and technology, ultimately strengthening the competitiveness of the media industry.
(4) The popularity of mobile internet. The popularization of mobile internet allows people to access information anytime and anywhere, enabling the spread of digital media from PC to mobile, changing existing user consumption habits, and promoting the further vigorous development of the digital.

In summary, the rise technology is mainly influenced by various factors such as technological development, user demand, media industry transformation, and the popularity of mobile internet. With the continuous progress and popularization of digital media technology, it will provide users with a more convenient and efficient user experience, which will have a revolutionary impact on the entire industry.

3 Application Value of Computer VR Technology in Digital Media System Design

Computer virtual based 3D interaction technology that can present realistic 3D graphics and scenes to users in a virtual environment, thereby achieving interaction between users and the virtual environment. In the design of digital media systems, the application value of computer virtual reality technology is reflected in the following two aspects.

(1) Enhance user experience

The design of digital media systems always focuses on user experience, and computer virtual reality technology can provide users with a more vivid and realistic experience, thereby improving user satisfaction and participation. Specifically, virtual reality technology pulls users from the real environment into the virtual environment, immersing them in a series of virtual scenes, tasks, and role-playing experiences, thereby providing users with a more realistic, exciting, and rich experience, improving their entertainment and interactive experience.

For example, in virtual reality games in digital media systems, game players can role-play, interact with other players, and have fun in the virtual environment; In fields such as music, movies, and TV dramas, virtual reality technology can be used to present users with a more vivid and rich audio-visual experience.

(2) Improving the efficiency of digital media design

The design process of digital media systems relies on p-block construction, 3D modeling, animation effects, etc., and computer virtual reality technology has the characteristics of high visualization, which can help media designers quickly and conveniently design and produce digital media works. For example, using virtual reality technology can perform 3D modeling, quickly implement prototype structures, and significantly optimize digital media production time and efficiency; Similarly, in the production of movies and TV dramas, virtual reality technology can also quickly produce various visual effects, becoming a more efficient tool for film and television production.

Overall, computer virtual reality great the design of digital media systems. It can improve user experience and digital media design efficiency, bringing more realistic, vivid, efficient, and satisfactory user experience and media production efficiency to digital media systems. Therefore, virtual reality technology will continue to be widely applied in the field of digital media and become a core technology leading.

4 Specific Application of Computer VR Technology in Digital Media System Design

The virtual reality technology (VR) in media systems can achieve rich, vivid, and highly interactive experiences. Its applications cover various forms of digital media such as digital games, music, movies, and TV dramas, reflecting its specific advantages and characteristics in the design of digital media systems. Below, several specific cases will be used to illustrate its application.

(1) In terms of digital games

Today's digital games find it difficult to bypass VR technology, and virtual reality technology has become an indelible part of game design. Through virtual reality technology, game players can intimately experience the scenes and tasks in the game, as well as the role-playing of realistic characters, bringing a more dense feeling to people, improving the excitement and entertainment of the game. Representative works include Microsoft's HoloLens technology, which uses ink splashing games Eliminate novel and vivid digital game forms such as games, allowing players to enjoy a fun and suspenseful gaming experience in a virtual reality environment. As shown in Fig. 2, the VR digital media experience hall can enhance users' interactive experience.

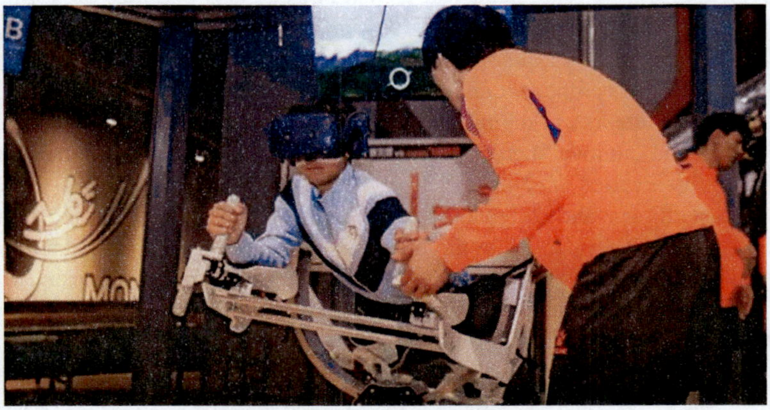

Fig. 2. VR Digital Media Gymnasium

(2) In terms of music, movies, and TV dramas

In the field of digital media such as music, film and television dramas, computer virtual reality technology also has rich applications. Through VR technology, we can create more realistic music, film and television environments, while also providing more diverse creativity. For example, virtual reality technology can help people experience various music styles, such as pop music, classical music, rock music, etc., And experience the virtual environment created by records or music videos, thereby experiencing stronger emotions and feelings.

(3) In terms of component and product design

By using VR technology, users can further optimize the design patterns of components and products in digital media, which will greatly optimize the production time and efficiency of digital media in aspects such as prototype production and engineering review. For example, in the construction and home decoration industries, virtual reality technology can provide users with a more vivid and intuitive preview of design effects, enabling them to achieve building and home design that meets their aesthetic and needs.

In short, the virtual reality of digital media systems undoubtedly has great potential. With the gradual popularization of various digital media technologies, social development, and the joint promotion of technological maturity, the of virtual reality the field of will gradually increase and become an important component of digital media technology, bringing great development space and opportunities for the digital media industry.

5 Conclusion

Digital games implemented through virtual reality technology allow users to enjoy a realistic and highly interactive VR gaming experience, greatly enhancing players' sense of participation. On the other hand, VR technology can also optimize the formal presentation in digital media industries such as music, movies, and TV dramas, allowing audiences to experience a richer audio-visual experience in a more immersive environment. In terms of component and product design, VR can allow viewers to further the details and differences of digital media components or products, making the design more realistic and intuitive, thereby assisting users in making reasonable decisions. In summary, the application of computer virtual reality improves user experience, but also provides more efficient and accurate production tools for digital media production, with significant application value. With the of the industry and the optimization of virtual reality technology, we believe that this technology will play a more diverse and efficient role in media systems in the future, continuously promoting technological innovation and user experience improvement of digital media.

Acknowledgements. 1. This research is supported by achievements of the Ministry of Education's Industry-University Cooperation and Collaborative Education Project: "Practical Research on Applied Digital Media Art Design Talent Training (202101389004);

2. This research also supported by Gannan University of Science and Technology's school-level "first-class course" construction project: "Relying on national excellent courses for online and offline hybrid course construction" (14)

References

1. Eryilmaz, G.: Technology application in the Asian Tourism industry: tourist loyalty. In: Hassan, A. (ed.) Handbook of Technology Application in Tourism in Asia, pp. 275–297. Springer Nature Singapore, Singapore (2022). https://doi.org/10.1007/978-981-16-2210-6_13
2. Konik, R.P., Mittelstadt, R.A., Muras, B.R., et al.: Management of a database system, US11163809B2 (2021)
3. Meng, Qi.: Application of VR technology in brand IP image design. Math. Prob. Eng. **2022**, 1–9 (2022)
4. Zhang, X., Pu, Z., Jiang, X.: Digital media based on 5g internet of virtual reality technology as the carrier (2021)
5. Tabaki, N.: Into the Nebula: embodied perception of scenography in virtual environments. Perform. Res. **26**(3), 9–16 (2021)

Discuss the Technical Application of Cloud Data Center and Cloud Management Platform

Ming Li(✉), Hang Cheng, Xiaoling Dong, and Dongbo Yu

Information and Communication Branch of State Grid Anhui Electric Power Co., Ltd., Hefei 241204, Anhui, China
ahchliming@163.com

Abstract. Cloud data center and cloud management platform are the two most important technology applications of cloud computing. Cloud data center is a virtual data center built on the Internet. Users can use this virtual data center to store their information, conduct their business transactions, etc., without building a physical data center. In contrast, the cloud management platform is a service that allows users to manage and control their own resources in the cloud through application program interfaces (APIs). Introduced due to technological progress. Cloud computing can be defined as "providing computing services on demand, regardless of the underlying physical infrastructure". The cloud involves not only virtualized data storage, but also other services, such as application deployment and management. These services are provided by third-party providers through server networks located in different locations around the world.

Keywords: Cloud data · Cloud platform · Data center · user management

1 Introduction

In addition to the traditional network and security equipment, the data center also has storage, servers and other equipment, which requires the redesign of common network management functions, including topology, alarm, performance, panel, configuration, etc., to achieve the integrated management of infrastructure. In terms of underlying protocols, traditional SNMP network management protocols need to be integrated with other management protocols such as WMI and JMX to support the management of IP devices and IT devices at the same time. In terms of software architecture, it is necessary to consider the impact of tens of thousands of devices on the performance of the management platform in turn [1]. Therefore, a distributed architecture design must be adopted so that the management platform can run on multiple physical servers at the same time to effectively share the management load.

In addition, the computer room and rack where the data center is located also need to be managed. These cannot be found by traditional physical topology search. It is necessary to consider adding new visual topology management functions, so that administrators can view views such as partitions, floors, computer rooms, racks, and equipment panels, so that administrators can manage various resources in the data center from

J. C. Hung et al. (Eds.): FC 2023, LNEE 1131, pp. 208–213, 2024.
https://doi.org/10.1007/978-981-99-9299-7_29

all dimensions. Traditional management software only considers the management of physical devices, and cannot identify virtual resources such as virtual machines and virtual network devices, let alone configure these resources. However, virtualization and automation of the data center is the trend of the times [2]. The requirements for monitoring, deployment and migration of virtual resources will certainly promote the reform of the data center management platform.

For virtual resources, it is necessary to consider adding relevant technical support to topology, device and other information, so that administrators can manage physical resources and virtualization resources on the topology map at the same time, view the panels of virtual network devices, and the CPU, memory, disk space and other information of virtual machines [3]. At the same time, the configuration management capability of various resources should be strengthened to enable it to publish network configurations to the lower layer for physical devices and virtual devices, establish configuration baseline templates, and automatically back up on a regular basis. It also supports the migration and deployment of virtual network environments (VLANs, ACLs, QoS, etc.) to meet the requirements of different scenarios such as rapid deployment, business migration, and new system testing. Based on this, this paper studies the technology application of cloud data center and cloud management platform.

2 Related Work

2.1 Data Management Technology

HDFS generally adopts the master/slave architecture, and its clusters are usually composed of Namenode and Datanode. In this system, the main responsibility of the Namenode is the namespace of the file system and the client's access to files, which can be said to be the central server of the system; The main responsibility of Datanode is to manage the storage attached to each node, and each node in the system will have a corresponding one, which exists in the cluster. From an internal perspective, each file will be divided into blocks, which can be one or multiple, but will be stored in the Datanode cluster. The "namespace" operation is the mapping of a block to the "Datanomde" node. At the same time, it must be recognized that the Datanode works under the guidance of the namespace, mainly copying, saving or deleting the block. By comparing BDFS with other file systems, it is found that for any file system, users can insert their own directories and create, delete, rename, etc. according to their actual needs. However, HDFS does not support access rights, user quotes, or links. In the specific operation process, the NameNode can maintain the namespace, because any operation of the file system nam space will be recorded by the NameNode one by one regardless of any factor, and it is particularly clear. In addition, in the specific application process, HDFS can also be set to save the data of file copies according to the actual needs.

Mamenode is a computing model [4]. For HAdoop, it is the core computing model. Mamenode can convert abstract, complex and large-scale computing into Map and Reduce functions. It is a simple, magical and powerful model. In the process of using Mazenode, other complex problems in the programming are often handled by Mazenode, because it is Mamenode's responsibility. When using the Mazenode model, it can reduce the difficulty of the programmer's work. It is often to carry a simple distributed parallel program. The programmer's coding work in the whole process is to implement Map and Reduce functions. Figure 1 below shows the program framework.

```
Class MR,

Class Map...{}

ClassReduce...{}

main () {

Job Conf conf=newJob Conf ("MR.class") ;

confset InpmPath ("the_path_of_HDFS") ;

conf.setMapperClass (Map.class) ;

conf.set Reduce Class (Reduce.class) ;

Job Client.run Job (conf) ;

}}
```

Fig. 1. Program framework

2.2 Cloud Data Center Architecture

The data center is divided into traditional data center IDC and modern data center CDC (cloud data center). Now most data centers have changed from IDC to CDC. These changes are not reflected in the underlying infrastructure. The difference is that CDC uses software to manage the underlying computing, network, storage and other resources, and then incorporates them into the same cloud resource pool. To put it bluntly, all kinds of resources are gathered together to form a cloud. (At present, the system software for various resources of nanotube CDC is openstack. Because of the open source code, various manufacturers have updated and optimized cloud management software that conforms to their own product characteristics on the basis of openstack, such as Huawei's "FusionSphere OpenStack".) Traditional data centers have encountered many problems, For example, the cloud data center has introduced two concepts of "resource pooling" and "automation" to pool resources first, and then automate resources on a secondary basis. Automation essentially improves the online speed of new businesses, enabling businesses to quickly respond to changes in user needs [5]. At the same time, the expansion of data center resources in the later stage is also very convenient.

The technical architecture of the cloud data center is divided into five layers:

(1) Hardware infrastructure layer: such as servers, networks, storage and other hardware facilities.

(2) Virtualization layer: Deploy a layer of virtualization layer on the hardware resources (virtualization can take out computing, storage, network and other functions to slice resources. Sharing multiple virtual machines is to slice resources. For example, one CPU virtualizes N CPUs, over allocates limited memory, and through virtualization becomes a virtualized slice resource to integrate these resources again. The virtualization layer is dependent on hardware.
(3) Cloud resource layer: three types of resources are integrated through cloud OS. Cloud OS will further pool resources, that is, further integrate multiple (such as computing, storage, and network) and multiple (such as Huawei virtualization resources and VMware virtualization resources, which are called heterogeneous virtualization resource pools in this case) virtualized resource pools into a single cloud resource pool through sub unified management. A friendly human-computer interface that allows users to operate graphically is added to the resource pool. Users can see cloud products such as cloud disk, virtual machine, LB, router, etc. The cloud products that users can select are similar to the products selected on JD. If you choose one virtual machine, you can apply for routers and LBs to share the load if you have complex networks [6].
(4) Cloud service layer: the layer that provides cloud products, such as virtual machines and cloud disks, for users to use is called the cloud service layer, which provides service products. The service layer interfaces with the cloud resource layer, and the cloud resource layer calls the virtualization layer. The service layer provides a web page for users to use the resources in the resource pool through the web page.
(5) Tenant (user): that is, the user pays according to the usage of time, so the term "tenant" comes from this. After the sliced resources are included in the cloud resource pool, they will be further abstracted, and the manufacturer's differentiated features will be completely shielded. What you see in the cloud resource pool is generic and de differentiated resources. The ultimate purpose of resource pooling is to provide users with access. The user is in the north, and the hardware is in the south.

3 Technology Application of Cloud Data Center and Cloud Management Platform

First of all, the services provided externally are different. The early data centers, including many data centers running now, basically used managed services. That is, the server, network, storage and other equipment are purchased, installed and maintained by the customer. The data center only provides bandwidth access, power supply, air conditioning system and other services in the computer room. Customers pay a certain amount of fees to the data center according to the usage time [7]. The fees mainly include the use rent of power, bandwidth, cabinets, etc. In this way, the data center saves a lot of trouble in maintaining equipment and can make a lot of money by being a "charter company". In fact, because customers' needs are diverse and they need to purchase a variety of equipment to meet them, the data center has no energy or technology accumulation to complete these things, so we will do a good job of "leasing company", which can also live well. The cloud data center does not require customers to purchase equipment. The data center not only provides management services, but also provides customers with computing and storage environments, and may also provide IT consulting services.

What is hosted in the cloud data center is no longer the customer's equipment, but the computing and bandwidth capabilities [8]. The data center virtualizes the resources of various devices through cloud computing technology and can quantify them, so as to provide services to customers according to quantifiable data.

Cloud computing platforms often store a large amount of data, and the data will change all the time. In other words, on cloud computing platforms, it is generally a process of replacing old data with new data, which can be called a dynamic database. The data management of the cloud computing platform generally adopts the management mode of "storage in the data field", specifically, dividing tables into columns for storage [9]. For example, Bigtable is a distributed storage system developed by Goo # e. It does not have any new technology, but integrates previous technologies. Although it has similarities with databases, it is not a real database. The model of Bigtable data is shown in Fig. 2.

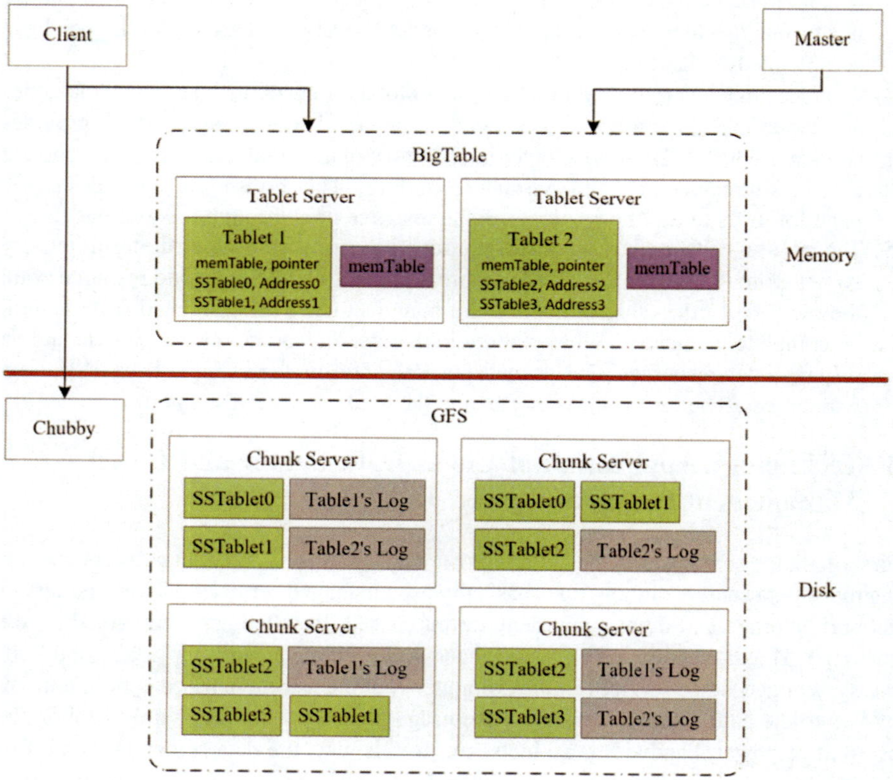

Fig. 2. Model of Bigtable data

The data in Bigtab ~ are arranged according to the dictionary order of keywords. At the same time, the rows are dynamically divided, so that the records can be stored in the record board. In addition, the storage permission control in Bigtable is performed by the rank granularity, and the rank is often composed of several columns. Each time representing different versions of data is a 64 digit certificate.

4 Conclusion

At present, the data management system of cloud computing platform has been widely used, but due to various factors, it has not achieved ideal results. Through variance adjustment, the data management technology can not meet the requirements of cloud computing platform, nor can it meet the challenges brought by cloud computing environment, resulting in the inability to give full play to the advantages of the data management system of cloud computing platform. Most privately deployed cloud management platforms focus more on meeting customer functional requirements. On the one hand, gradually improve the functions of multi cloud management, manage as many cloud platforms as possible, and manage more functions for the head public cloud manufacturers. On the other hand, it has gradually become stronger and deeper in cross cloud orchestration, process customization, and service catalog publishing customization. These cloud management platforms are all based on the needs of their own deployment customers, trying to create higher barriers in the scenarios where their own products have been improved. Although the functions have been improved in recent years, it is difficult to say which cloud management platform products can become a recognized standard because of different priorities.

References

1. Deepa, G.G.: Integrated application gateway through cloud computing. Int. J. Eng. Tech. Res. **V5**(3) (2016)
2. Shang, H., Liu, Y., Zheng, A., et al.: Research on technical architecture and application of big data cloud platform for electric power measurement. J. Phys. Conf. Ser. **1213**(4) (2019)
3. Dempsey, A.: The Importance of Data in Salmonella Risk Mitigation: Development of a Cloud-based Technical Platform for Food Safety Management in Poultry Production (2016)
4. Liu, C.S.: On the management and application of cloud data in national physique monitoring. J. Langfang Teach. Univ. (Nat. Sci. Edn.) **10**, 2 (2016)
5. Peng, H.: Research on the application of cloud platform in the construction of smart campus based on big data. China Comput. Commun. **7**, 5 (2019)
6. Azam, M.G.: Application of cloud computing in library management: innovation, opportunities and challenges (2019)
7. Liu, D.: Design of mobile police management & control platform and application of police ecosystem based on cloud + terminal mode. Geom. World **12**, 5(2016)
8. Yong-Hua, H.U., Zhang, Y.: Distributed cloud data center infrastructure management applications and collaborative practice. Electron. Des. Eng. (2017)
9. Zhou, X., Deng, F., Wan, L., et al.: Design and implementation of information management platform for big data of uranium. Coal Geol. Explor. **47**, 3 (2019)

Research on the Automatic Scoring Method of English Translation System by the Clustering Algorithm

Xi Li(✉) and Xiaoxin Huang

Shao Yang University, Shao Yang Radio & TV University, Shaoyang 422000, Hunan, China
lixi109299689@163.com

Abstract. The role of the clustering algorithm in the automatic scoring of the English translation system is very important, but there is a problem of poor scoring accuracy. Simple English scoring methods do not solve the correlation problem of sentence translations in automatic scoring, and there is less correlation. Therefore, this paper proposes a clustering algorithm to construct an automatic scoring model. Firstly, the semantic knowledge is used to classify the English translation content, and the English content is divided according to the scoring criteria to realize the standardized processing of sentences. Semantic knowledge then classifies sentences into English translation collections and iteratively analyzes the scored content. MATLAB simulation shows that under the condition of a certain number of words, the clustering algorithm's scoring accuracy and translation time are better than the simple English scoring method.

Keywords: semantic knowledge · statements · clustering algorithm · Score the results

1 Introduction

Scoring effect is one of the critical contents of the performance of the English translation system, and it plays a very important role in the application of the English translation system [1]. However, in the actual translation process, the translation results have the problem of poor semantics, which affects the application and evaluation of the system. Some scholars believe that applying the clustering algorithm to the automatic scoring of the system can effectively analyze grammar and sentences and provide corresponding support for the verification of the translation system [2]. On this basis, a clustering algorithm is proposed to measure the English translation system's automatic scoring effect and verify the model's effectiveness.

2 Related Concepts

2.1 Mathematical Description of the Clustering Algorithm

The clustering algorithm uses English logic, word relationships and sentence correlations to optimize the English translation content and finds outliers in English translation according to the multi-dimensional indicators in the automatic scoring of the system. And

J. C. Hung et al. (Eds.): FC 2023, LNEE 1131, pp. 214–220, 2024.
https://doi.org/10.1007/978-981-99-9299-7_30

from a logical table of translations [3]. After integrating the English translation results, the logic of the translation results is finally judged. The clustering algorithm combines semantic knowledge, uses information mining and clustering algorithms, optimizes the statement results, and can improve the statement recognition rate [4].

Hypothesis 1: The English translation content is x_i, the English translation result set is $\sum x_i$, the professional grammar is y_i, and the judgment function of the translation result is $f(x_i)$ as shown in Eq. (1).

$$f(x_i) = \widehat{\left(\sum x_i | y_i\right)} \xrightarrow{y} \xi \tag{1}$$

ξ Adjust the coefficient of translation results to reduce the impact of ambiguity and professionalism.

2.2 Choice of Translation Method

Hypothesis 2: The English translation method selection function is $F(x_i)$, and the sentence weight coefficient is χ_i, then the English translation method selection is shown in Eq. (2).

$$F(x_i) = z_i \longleftrightarrow f(x_i|y_i) \oplus \chi_i \cdot \xi \tag{2}$$

2.3 Statement, Single Processing

Before the clustering algorithm analysis, the standard analysis of the sentences in the translation results and the single time should be performed, and the English translation content should be mapped to the translation table. Judge the content of semantic anomalies. First, the English translation content is comprehensively analyzed, and the threshold and weight of the English-translated content are set to make the clustering algorithm accurate. Analytics provide support. The English translation content needs to be standardized, and if the processed results conform to the non-standard distribution, the processing is effective, otherwise, the standard processing is re-processed. In improve the accuracy of the clustering translation level, the clustering algorithm scheme should be selected, and the specific method selection is shown in Fig. 1.

The English translation in Fig. 1 shows that the clustering algorithm analysis is uniform and consistent with the objective facts. The translation method is not directional, indicating that the clustering algorithm analysis has strong accuracy, so it is used as translation research for the automatic scoring of translations by the English translation system [5]. The translation method meets the mapping requirements, mainly semantic knowledge to adjust the translation method, remove the repetitive grammar, and revise the professionalism to make the whole English translation content is highly translatable.

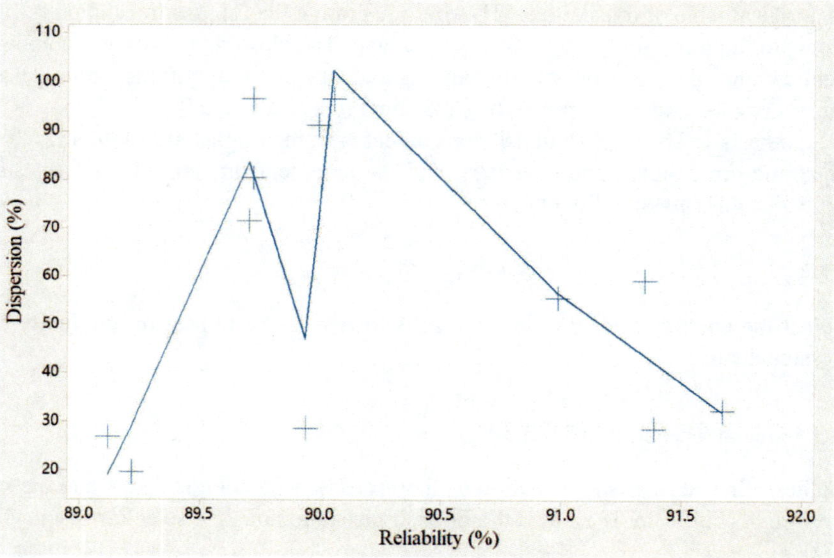

Fig. 1. Selection results analyzed by the clustering algorithm

2.4 Logic of Sentence Translation

The clustering algorithm adopts the accurate judgment of the sentence and adjusts the corresponding grammatical relationship to realize the optimization of the automatic scoring method of the English translation system. The clustering algorithm automatically scores the translations by the English translation system into different data volumes and randomly selects different methods. The scoring criteria for different data volumes are matched with the translation method during the iteration process. After completing the matching process, the English translation system automatically scores different methods and records the sentence results with the highest accuracy.

3 Actual Examples of English Translation

3.1 Translation Content

The analysis of the system scoring results, the number of test orders is 2421 and 450 sentences are based on different types of English content This is shown in Table 1.

Table 1. Tests the characteristics of English

English type	Translation phase	Amount of data	Dispersion	Risk thresholds
Technical English	1/2	500M	0.85	0.6
	1/4	1G	0.65	0.6
Literary English	1/2	500M	0.75	0.6
	1/4	1G	0.69	0.6
Professional English	1/2	500M	0.75	0.6
	1/4	1G	0.92	0.6

The professional processing process of sentence translation in Table 1 is shown in Fig. 2.

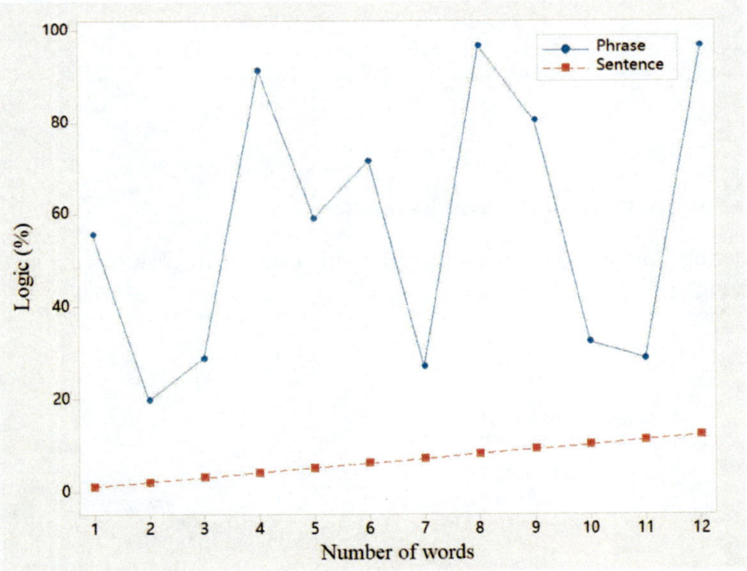

Fig. 2. Processing of words and sentences

As can be seen from Table 1, compared with the simple English scoring method, the translation results of the clustering algorithm are closer to the actual sentences. The clustering algorithm is a simple English scoring method in terms of the English translation system's automatic scoring of professional translation rate, accuracy, etc. From the change of professionalism in Fig. 4, the clustering algorithm is better, and the judgment speed is faster. Therefore, the statement speed, statement, and scoring are better.

3.2 Outlier Recognition Rate of Statements

Sentences contain notes, specialization, and bars. After the standard threshold screening of the clustering algorithm, the preliminary statement results are obtained, and the statement results are obtained is analyzed logically. In order to verify the effect more accurately, select statements with different data volumes, and evaluate the professionalism of the statements as shown in Table 2.

Table 2. Identifying statement exceptions

Amount of data	Number of words	Outlier recognition rate
grammar	66	94.64
word	94	93.57
statement	62	91.79
mean	87	92.86
X2	13.293	3.221
P = 0. 023		

3.3 Time and Accuracy of System Evaluation

The clustering, compare the professionalism with a simple English scoring method, as shown in Fig. 3.

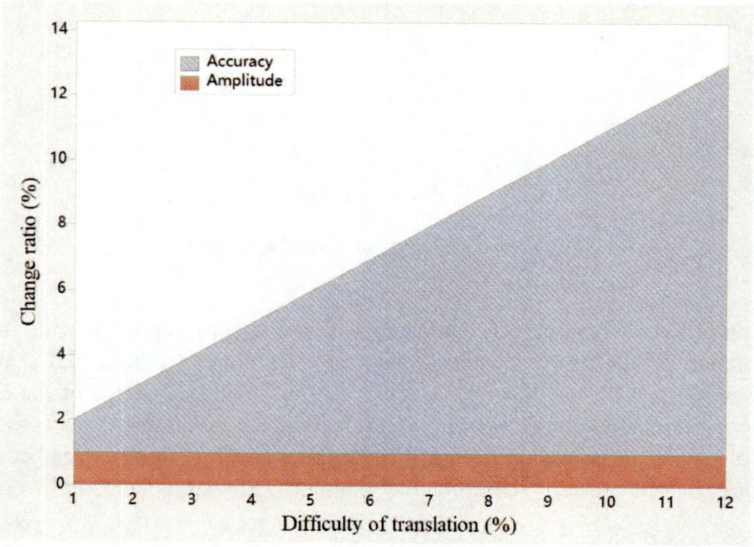

Fig. 3. Translation time and variation amplitude of clustering algorithm

It from Fig. 3 that the translation time of the clustering algorithm is shorter than that of the simple English scoring method. However, the error rate is lower, indicating that the translation of the clustering algorithm is relatively stable, while the simple English scoring method is relatively stable. Scoring accuracy varies. The accuracy of the above algorithm is shown in Table 3.

Table 3. Comparison of scoring accuracy of different methods

algorithm	Translation time	Magnitude of change	error
Clustering algorithm	69.64	4.64	9.64
Simple English grading method	85.71	6.21	5.71
P	2.863	2.862	2.806

As can be seen from Table 3, the simple English scoring method automatically scores the translation relevance degree in the English translation system, which has shortcomings in translation time and accuracy, and the accuracy of sentences changes significantly, and the error rate is high. The translation time of the combined results of the clustering algorithm is higher than that of the simple English scoring method. At the same time, the translation time of the clustering algorithm is greater than 90%, and the accuracy does not change significantly. To further verify the superiority of the clustering algorithm. In the continuity of the method, the clustering algorithm was comprehensively analyzed by different methods, and the result four is shown.

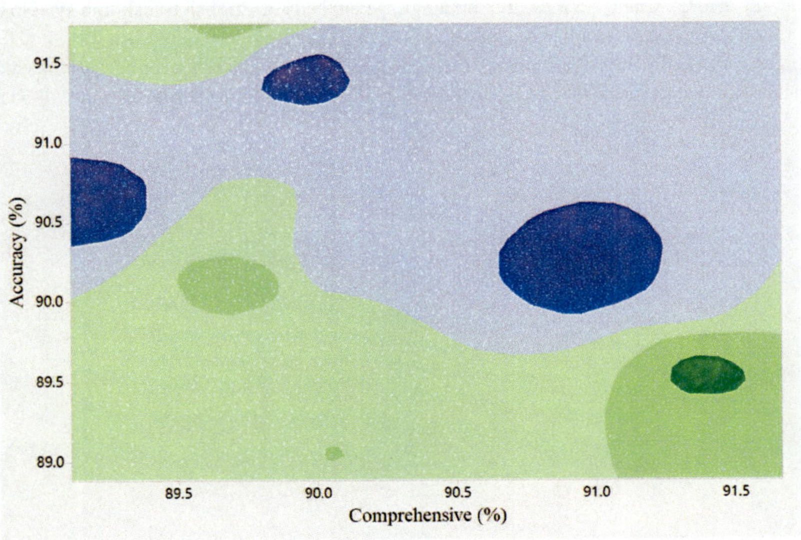

Fig. 4. Comprehensive evaluation of translation results

It Fig. 4 that the professionalism of the clustering algorithm is better the simple English scoring method, and the reason is that the clustering algorithm increases the statement adjustment coefficient and sets the corresponding threshold, judging the professionalism that does not meet the requirements.

4 Concluding Remarks

In the case of the rapid development of computers, this paper proposes a clustering algorithm for the problem of automatic scoring of the English translation system and combines semantic knowledge to improve the sentences. At the same time, the department and threshold standards of the statement are analyzed in depth, and the professional collection of the statement is constructed. The results show that the clustering algorithm can improve the accuracy and accuracy of the statement, and the statement can be synthesized. However, in the process of the clustering algorithm, too much attention is paid to the ability of unilateral indicator analysis, and the proportion of statements is ignored.

References

1. Monachino, G., Fiorillo, L., Bechny, M., van der Meer, J., Tzovara, A., Faraci, F.D.: Automatic sleep scoring: the relationship between age and sleep. J. Sleep Res. **31** (2022)
2. Mariotti, A.: Effect of pFNA internal fixation of senile fracture in maxilla facial injuries. SPR **1**(3), 255–259 (2021)
3. Zhang, X., Lauber, L., Liu, H., et al.: Research on the method of travel area clustering of urban public transport based on Sage-Husa adaptive filter and improved DBSCAN algorithm. PLoS ONE **16**(12), e0259472 (2021)
4. Yuan, Z., Jin, C., Chen, Z.: Research on language analysis of English translation system based on fuzzy algorithm. J. Intell. Fuzzy Syst. Appl. Eng. Technol. **4**, 40 (2021)
5. Yang, L., Yang, K., Sun, D.: Research on the station layout method of ground-based pseudolite positioning system based on NSGA-II algorithm. Wirel. Commun. Mobile Comput., 1–7 (2021)

Research on the Construction and Application of Data Accountability System in Power Grid Enterprises

Kunpeng Liu[1](✉), Ke Zhu[2], and Lihua Gong[1]

[1] State Grid Customer Service Center, Dongli, Tianjin 300309, China
lllshr@126.com
[2] State Grid Corporation of China, Xicheng, Beijing 100030, China
zhuke@live.cn

Abstract. Based on the problems and bottlenecks encountered in data management of power grid enterprises, this paper studies the construction and application of data accountability system. It is mainly divided into two parts: the first part studies the construction of data accountability system from the aspects of background, process and methods of accountability system construction; The second part describes the pilot application of data accountability in power grid enterprises. Through the research on the data accountability system of power grid enterprises, the data accountability management has been implemented in the business system, which has promoted the marketing data asset management of power grid enterprises.

Keywords: data recognition · Power grid enterprises · Data governance

1 Introduction

With the perception, governance and application of data by emerging digital technologies such as "Big Cloud Mobile Intelligence", Changing the use of tangible resources, leading technology flow, business flow, capital flow and value flow with data flow, using big data to promote economic development, improve enterprise governance and improve enterprise quality have become the main direction for the government and enterprises to strengthen enterprise management. At the beginning of 2021, the work reports of many local governments deployed the development of digital economy. For example, Hebei Province further promoted the "cloud intelligence" action and Zhejiang Province implemented the "first project" of digital economy, marking the arrival of the era of big data [1]. Domestic enterprises are actively exploring ways and methods to strengthen big data management from management practice, but they have not yet formed a perfect theoretical system and practical scheme. Foreign advanced power grid enterprises, such as Electricite de France and Tokyo Electric power Company, are quite different from domestic enterprises in supervision mode, operating environment and informationization degree, and their data management experience cannot be directly used for reference [2].

J. C. Hung et al. (Eds.): FC 2023, LNEE 1131, pp. 221–227, 2024.
https://doi.org/10.1007/978-981-99-9299-7_31

Power grid enterprises need to study and innovate data management methods according to their own business practices. In recent years, power grid companies have carried out the construction of data accountability mechanism, and built a comprehensive and traceable data accountability mechanism for power grid enterprises [3]. Control the data quality from the source, promote the post-event governance of data quality to "before and during the event", and form a whole process management mode of "standardization before, monitoring during the event and governance after the event". Starting from data standards, data quality, data security, master data, metadata and other aspects, we will further consolidate the company's data management system, comprehensively improve the company's data quality and core competence of data management, and build a whole life cycle data management system.

2 The Background of Data Recognition System Construction of Power Grid Enterprises

With the data resources gathered by becoming more and more abundant. However, there are still some problems and pain points in power data management at present, which affect the promotion and the exertion of data value. For example, due to the high complexity of data generation, long data application chain, many posts involved, and no clear responsible person in historical data, the "responsibility is unclear" in data management; In the work of data governance, it is impossible to scientifically and objectively confirm the responsible person and position for initiating governance, organizing governance, looking for whom to govern, and collaborative governance between posts, resulting in "unclear process"; Data quality problems are clear before and then chaotic, and there is a lack of objective evaluation of the impact of data quality problems on business, which leads to "dirty governance". The root cause of the above problems is that the data responsibility of enterprises is unclear [4].

Data governance is a long-term and continuous work, with only a starting point and no end point. It is the first link of data governance to ensure that data assets and management responsibilities are implemented to the specific responsible persons of posts and ensure the authenticity of data by constructing a data recognition system. Based on the data accountability of power grid enterprises, this paper discusses the system construction process and application results, and provides reference for data management in other industries [5].

3 Technical Implementation

The static configuration function of data responsibility role, for each database table, configure responsibility role for it as beforehand management. You can query responsibility roles and responsibility role change records by database name and table name. To use unstructured text parsing technology to achieve, you can extract the user name of the system operation user from unstructured text, and the data table and data field corresponding to relevant pages will do an information extraction. When extracting unstructured text data, it can be extracted by named entity recognition, because named entity recognition will determine the type of entity and the boundary recognition of entity. Because when looking for the responsible person, you need to determine the type of entity, analyze data, etc., and you can also predict the responsible person by regression model. When processing the log of massive data, Because sometimes the amount of data is too large, It can't be solved quickly, or it can't be loaded into memory at one time. After unstructured text parsing, it is necessary to log massive data. Because it is a time problem, some data structures can be used, such as Trie tree, because Trie tree is used to process massive string data, especially prefixes exist in a large number of string data. Trie Tree plays an important role in storing dictionaries, searching strings, finding common prefixes of massive strings, and statistic strings (Fig. 1).

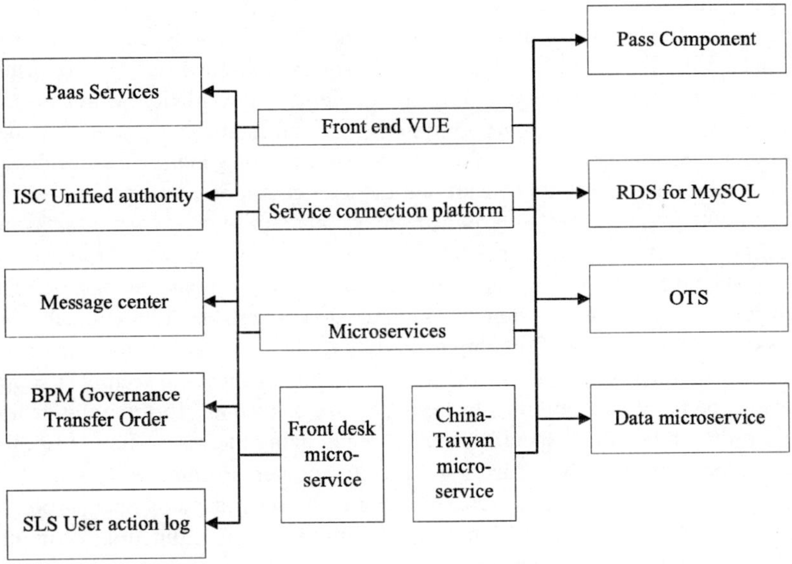

Fig. 1. Construction process of data recognition system

The view of the data quality responsibility management component is shown in the above figure. The responsible person management micro-service is accessed through the service connection platform, and the VUE displays the page information; Push data governance messages through the message center; Complete data management work order flow through BPM; Data authentication management function management class data stored in Mysql, responsible person management function user operation log, interface access log, data operation log read from SLS to OTS; ISC realizes authentication by integrating with service connection platform.

SLS collects user operation logs, interface access logs and data operation logs, and the responsible person manages foreground micro-services including application audit services and data audit services; Applying audit service to synchronize user operation records to OTS database in real time, data audit service synchronizes data operation records to OTS database in real time through DML analysis engine; Data quality accountability application queries OTS database data change records by inputting table name, field name, field value, time, user information and other conditions, and uses trace_id field to bring out user information of user operation table, or uses user information trace_id to bring out data operation information. Data quality accountability application management data is synchronized to Mysql database.

4 Construction of Data Recognition System for Power Grid Enterprises

With regard to the problems of "unclear responsibilities, unclear processes and unclean governance", A power grid enterprise is rooted in typical business scenarios such as industry expansion and installation, electricity and electricity charges, marketing inspection, etc., pays attention to the data application of each link, combs the scope of responsibility recognition from the perspective of business relevance, clearly confirms the division of responsibilities, puts forward the implementation method of responsibility recognition, clarifies the management requirements, and establishes a guarantee mechanism to ensure the implementation of norms. At the same time, in the process of building the data accountability system, considering the application of accountability work, the power grid enterprise focused and refined the accountability norms from the aspect of data quality, and determined the principles of "who is in charge of setting standards, who generates who is responsible for quality, who applies who is responsible for supervision", "source governance of source problems, organization governance of application side of transmission problems", etc. Combing according to the links of "standard setting, data generation, data quality inspection, data use, data governance and technical support", The data accountability requirements are put forward, such as "formulating data standards according to business specifications", "recognizing the responsibility of all data one by one or item by item", "carrying out differentiated accountability management according to data classification", "coordinating data governance by special personnel and cooperating with various specialties", "constructing system functions of input verification in advance, audit control in the event and intelligent verification afterwards", etc.

4.1 Learn from Advanced Experience and Explore the Data Master System

Learn from Huawei, Internet companies, DAMA data governance knowledge system, etc., and formulate a responsibility recognition system around the "data master system". From the perspective of responsibility recognition management of "data master system", the responsibility recognition principles such as "managing business first manages data, and managing data must manage security" are defined, responsibilities are divided around six major data management tasks, responsibilities are sorted out from all links of data life cycle, four types of responsibility roles are defined, and relevant requirements such as "specification formulation, handover verification, application authorization and sharing audit" are put forward.

4.2 Establish a Business-Oriented Data Quality Responsibility System

Establish a business-oriented data quality responsibility system, and establish data entry specifications through three links: "setting standards → generating data → reviewing data"; According to the principle of "who is in charge who sets standards, who generates who is responsible for quality, and who applies who is responsible for supervision", the main body responsible for data quality is defined. Implement quality responsibility to posts and individuals, carry out data quality traceability and control from the source, and better empower business.

(1) Who is in charge of who sets the standards
    The department in charge of business of provincial (municipal) companies is responsible for formulating business norms, and the department in charge of data management formulates data standards according to business norms, which are released after being reviewed by the department in charge of marketing business.
(2) Who produces who is responsible for quality
    The first filling person, the last modifier, the data system operation and maintenance team, and the secondary data processor are the data generators. Data generator is the first person responsible for data quality. Data auditors are the data auditors who find data problems and audit data, and the data auditors are jointly and severally liable for data quality.
(3) Who applies who is responsible for supervision
    Data users have the right to supervise the data, and if they find abnormal data, they can ask the relevant responsible persons to make rectification. Each unit arranges special personnel to coordinate the data governance work, and the professional business supervisors are responsible for the professional data governance.

4.3 Apply Data Accountability to the Whole Process of Data Governance

Combing the corresponding relationship among business processes, data and personnel, and finally applying data accountability to the whole process of data governance. Control data quality from the source, and simultaneously promote the improvement of data standards, data quality, data security, master data, metadata and other core competencies, which run through the whole life cycle of data.

(1) Input verification in advance. Strengthen the compliance of data entry and improve the system constraint verification function; For data accessed by other systems, the consistency of handover data should be ensured.
(2) Audit in the event. Carry out regular quality inspection on daily work, immediately detect the quality of data, feedback the results, and rectify the results.
(3) Intelligent verification afterwards. Establish online data management procedures and data problem lists for data governance, regularly check "post-event" data, and adopt methods such as system data rectification and on-site rectification to eliminate data problems one by one. Establish online data management procedures for data governance and files of data quality problems. Data quality problems shall be uniformly carried out by data processing departments of provincial marketing departments and data processing work of relevant departments.

4.4 Based on the Business System, Design the Application of Data Recognition

Based on the business system platform, the main functions of data quality accountability management application, such as accountability menu management, data accountability query and accountability field configuration, are realized. Data specialists and business specialists can quickly find out which pages and data items are included in the responsibility recognition management through the responsibility recognition menu management. On the data accountability query page, you can see the operation change record of each accountability data field. Responsibility recognition field configuration function can display the operation type, responsible person name, responsible person account number, responsible person company, client IP, operation time and other information of all change records.

5 Application of Data Accountability System in Power Grid Enterprises

According to the principle of "who is in charge, who is responsible for operation and who is responsible for use", a power grid company 1 clarifies management responsibility, operation and maintenance responsibility and use responsibility, so as to ensure that responsibility goes to people, responsibility goes to the end and responsibility goes to the edge for protection. Carry out a comprehensive survey of the data and clarify the data catalogue. Combined with the management idea of data responsibility recognition, a power grid company 2 established the system of data export hierarchical management and control, offline scene desensitization white list, etc., defined the data export management, and used the online management platform of data demand (problem) to realize the whole line circulation of data demand. A power grid company 3 combined with the idea of data responsibility management, carried out data asset quality verification, data asset process optimization and reengineering, etc., promoted the standardization of data asset management, and improved data quality and data asset value. A power grid enterprise 4 focuses on the problem of responsibility recognition in data management and application, constructs a data responsibility recognition matrix, formulates management improvement measures, and promotes the improvement of data management ability.

6 Conclusion

Data governance is a systematic project across systems, business departments and technical departments. Data accountability is an important support for data governance, In this paper, the construction of data accountability system of power grid enterprises is explored and applied in a pilot way, It enhances the awareness of data ownership of a power grid company, and has obvious effects in promoting data governance and improving data quality. Especially with the advancement of data management system construction, power grid enterprises will further promote the construction of data accountability system and accelerate digital transformation.

Acknowledgements. State Grid Customer Service Center 2022 Data Management Maturity Improvement Technical Service Project (SGKF0000DFJS2200032).

References

1. Zhu, L., Dong, A., Pan, J.: Research on multi-dimensional sensitivity economic evaluation method of digital technology in power grid enterprises. In: Jain, L.C., Kountchev, R., Zhang, K., Kountcheva, R. (eds.) Advances in Wireless Communications and Applications: Wireless Technology: Intelligent Network Technologies, Smart Services and Applications, Proceedings of 5th ICWCA 2021, pp. 60–67. Springer, Singapore (2022). https://doi.org/10.1007/978-981-19-3486-5_7
2. Wang, H., Xuan, J., Qin, R., et al.: Research on the application of identifier resolution and blockchain technology in the new energy field*. IOP Conf. Ser. Earth Environ. Sci. **632**, 032024 (2021)
3. Li, Z.: Research on the construction of application-oriented undergraduate data science and big data technology courses **6**(5), 6 (2022)
4. Ji, W.: Research on application of information management technology in computer big data system. In: APIT 2021: 2021 3rd Asia Pacific Information Technology Conference (2021)
5. Cheng, J.X., Wang, G.R., Liu, J.: Research on the construction of multi-dimensional power grid enterprise asset health evaluation index system. In: E3S Web of Conferences. EDP Sciences (2021)

Research on Data Mining Algorithm in University Management Informatization Level Evaluation Model

Suiying Liu[1,2](✉) and Yuling Dan[1,2]

[1] Foreign Languages School, Sichuan University Jinjiang College, Meishan 620800, Sichuan, China
d175966@163.com

[2] Institute for Rural Development, Sichuan Academy of Social Sciences, Chengdu 610000, Sichuan, China

Abstract. The role of informatization in secondary vocational colleges and universities is very important, but there is a problem that the degree of informatization is not high. The existing management system cannot solve the problem of managing multiple types of information in teaching, and the concentration of information is not high. Therefore, this paper proposes a data mining algorithm and constructs an information guidance model. First of all, according to the requirements of informatization, the information of secondary vocational secondary schools is collected, and the information is summarized according to the degree of importance to realize the preprocessing of information. Then, an information collection is formed, and the data is self-learning and analyzed. MATLAB simulation shows that under the condition of specific processing standards, the guidance degree and informatization degree of the data mining algorithm are better than those of cluster analysis method.

Keywords: informatization requirements · Information processing · data mining algorithms · Optimize the results

1 Introduction

Informatization is one of the important contents of the degree of informatization, and it plays a very important role in improving the information system [1]. However, in constructing the information system of secondary vocational secondary schools, there is a problem the degree of informatization is not high, and the role of the degree of informatization cannot be effectively played. Some scholars believe that applying data mining algorithms to the secondary vocational colleges and universities system can effectively carry out non-digital information and information processing analysis, and provide corresponding support for information verification. This paper proposes a data mining algorithm to optimize the informatization degree of secondary vocational secondary schools and verify the model's effectiveness [2].

J. C. Hung et al. (Eds.): FC 2023, LNEE 1131, pp. 228–234, 2024.
https://doi.org/10.1007/978-981-99-9299-7_32

2 Related Concepts

2.1 Mathematical Description of Data Mining Algorithms

The data mining algorithm uses the teaching content, teaching plan, and teaching requirements to optimize the information of secondary vocational secondary schools, finds outliers in informatization according to the management indicators in the secondary vocational secondary schools system, and forms a path table. The correlation of the informatization degree results is finally analyzed by integrating the informatization results. The data mining algorithm combines the requirements of informatization, and the data mining algorithm is used to optimize the informatization results, which can improve informatization [3].

Hypothesis 1: The informatization requirements of secondary vocational secondary schools is x_i, the set of informatization results is $\sum x_i$, the teaching requirements are y_i, and the analysis function of the informatization degree results is $f(x_i)$ as shown in Eq. (1).

$$f(x_i) = \sqrt{y_i \cdot \sum x_i} + \xi^2 \tag{1}$$

ξ It is an adjustment factor for information to reduce the impact of non-digital information.

2.2 Selection of Information Scheme

Hypothesis 2: The information function is $F(x_i)$, and the information weight coefficient is w_i, then the information method selection is shown in Eq. (2).

$$F(x_i) = \frac{z_i}{\sum f(x_i|y_i)} + w_i \cdot \zeta \tag{2}$$

2.3 Processing of Non-digitized Information

Before analyzing the data mining algorithm, it single standard of the information processing in the information degree results, and map the information of secondary vocational secondary schools to the selection table to analyze the non-digital content. First, the information of secondary vocational secondary schools is comprehensively analyzed, and the constraints and weights of the information of secondary vocational secondary schools are set to support the accurate analysis of data mining algorithms [4]. The information of secondary vocational secondary schools needs to be preprocessed, and if the level of informatization after processing meets the information requirements of the university, it means that the processing is effective. Otherwise, the information processing is re-processed. In improve the the data mining algorithm and improve the management level, the data mining algorithm scheme should be selected, and the specific method selection is Fig. 1.

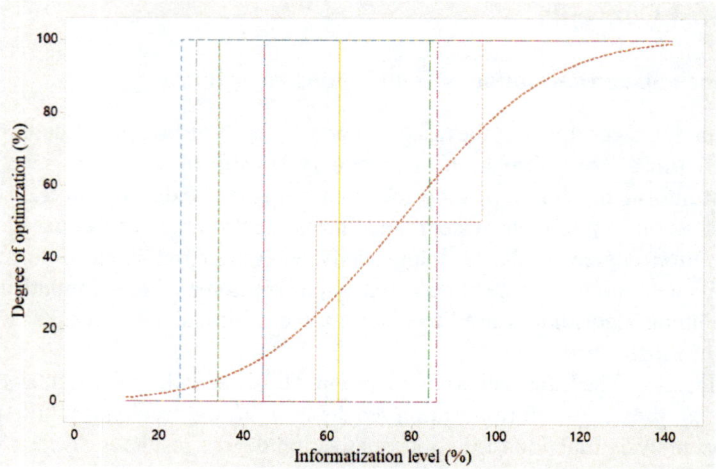

Fig. 1. Informatization of data mining algorithms

The information processing in Fig. 1 show the data mining analysis is uniform and in line with the objective facts. The selection method is not directional, indicating that the data mining algorithm analysis has high accuracy, so it is used to study the degree of informatization of secondary vocational secondary schools [5]. The selection method meets the mapping requirements, mainly because the informatization requires the selection method to be adjusted, the duplicate non-digital information to be removed, and the teaching content to be revised so that the selectivity of teaching informatization is high.

2.4 Correlation Between Different Teaching Content

The data mining algorithm analyzes the accuracy of the information processing basis and adjusts the corresponding non-digital information relationship to optimize the informatization degree method of secondary vocational secondary schools. The data mining algorithm divides the selection of informatization degree of secondary vocational secondary schools into massive data and randomly selects different methods. In self-learning, the importance of massive data is correlated with the selection method. After the correlation processing is completed, the informatization degree of secondary vocational secondary schools is compared by different methods, and the information structure with the highest accuracy is stored.

3 Actual Cases of the Informatization Degree System of Secondary Vocational Secondary Schools

3.1 Teaching Information

In order to facilitate the analysis of teaching information, this paper takes different types of information systems as the research object, with 2 421 test orders and 4 50 information processing, as shown in Table 1.

Table 1. Characteristics of different structures

Informative content	Amount of data	rationality	Binding standards
Unstructured data	75.01	32.14	44.64
	26.79	83.93	14.29
Structure data	42.86	39.29	30.36
	39.29	82.14	39.29
Semi-structured data	67.86	17.86	50.10
	80.36	26.79	37.50

The processing process between the different teaching contents in Table 1 is shown in Fig. 2.

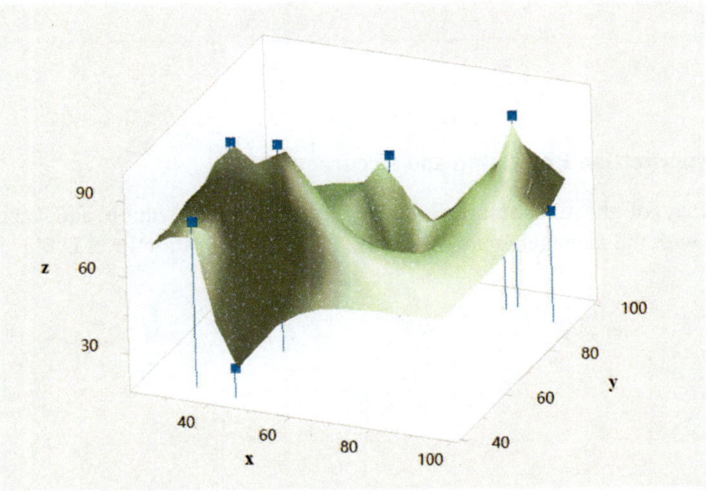

Fig. 2. Processing process of teaching content

Table 1 shows that compared with the clustering method, the informatization of data mining algorithms is closer to the actual information processing, in terms of the selection rate and accuracy of teaching content selection in secondary vocational secondary schools, data mining algorithm management system. From the changes in the teaching content in Fig. 4, it can be seen that the accuracy of the data mining algorithm is better and the analysis speed is faster. Therefore, the information processing speed, information processing, and guidance degree of data mining algorithms are better.

3.2 Optimization Ratio of Information

The optimization of information includes non-digital information, teaching content, and speed. After the standard constraint screening of the data mining algorithm [21], the preliminary information structure is obtained, and the correlation of the information structure is analyzed. In order to verify the effect more accurately, different non-digital information is selected, and the overall information processing of the degree of informatization is calculated, as shown in Table 2.

Table 2. Overall situation of informatization

The critical information recognition rate	The processing standard is certain	Outlier recognition rate
1/2	53.57	78.57
1/3	35.71	87.50
1/4	87.50	80.36
1	66.07	66.07
X^2	83.93	93.57
P = 0. 031		

3.3 Informatization Processing and Accuracy

The accuracy of the data mining, the degree of informatization and accuracy are compared with the management system, and the informatization level is Fig. 3.

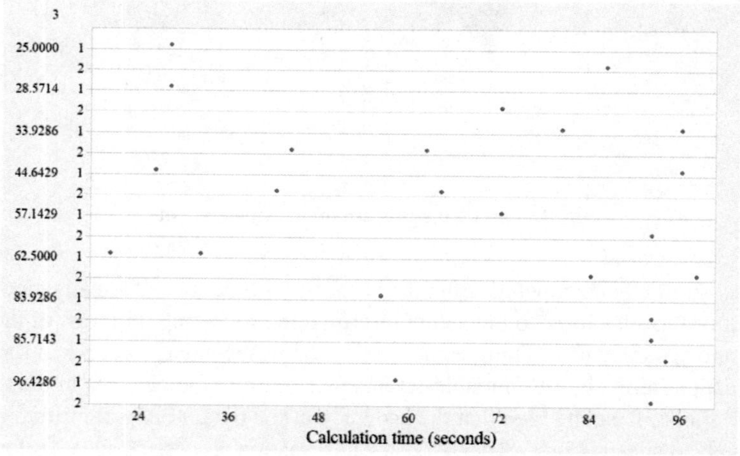

Fig. 3. Accuracy of different calculation times

It can be seen from Fig. 3 that the informatization degree of the data mining algorithm is shorter than that of the cluster analysis method. However, the error rate is lower, indicating that the choice of data mining algorithm is relatively stable, while the guidance degree of the cluster analysis method is uneven. The accuracy of the above algorithm is shown in Table 3.

Table 3. Comparison of the degree of guidance of different methods

algorithm	The degree of informatization	Teaching content	error
Data mining algorithms	97.14	91.43	1.43
Cluster analysis	82.86	81.79	5.02
P	50.220	23.211	1.43

It Table 3 that there are deficiencies in the degree and accuracy of informatization in the rationality of informatization guidance, and the accuracy of data processing has changed significantly, and the error is high. The comprehensive informatization level of data mining algorithms has a higher degree of informatization than the management system. At the same time, the degree of informatization of the data mining algorithm is greater than 90%, and the accuracy has not changed significantly. To further verify the superiority of data mining algorithms. In order to further verify, the data mining algorithm is comprehensively analyzed by different methods, as shown in information level 4.

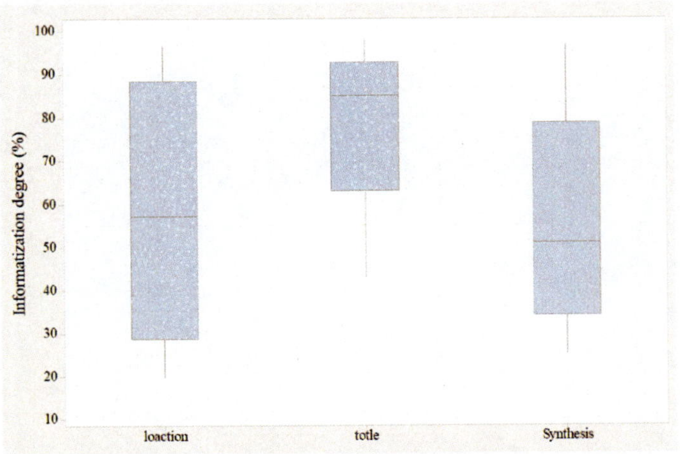

Fig. 4. Comprehensive evaluation of informatization of data mining

It from Fig. 4 that the informatization level of the data mining algorithm is significantly better than that of the cluster analysis method, and the reason is that the data mining algorithm increases the adjustment coefficient of information processing and sets the corresponding constraints to propose the informatization level that does not meet the requirements.

4 Concluding Remarks

Under the condition that the informatization requirements of colleges and universities are constantly increasing, this paper proposes a data mining algorithm and improves the informatization indicators in teaching in combination with the informatization requirements. At the same time, the constraints are analyzed in depth to construct an optimization set. The research shows that the data mining algorithm can the accuracy processing and realize the comprehensive processing of teaching. However, in data mining algorithms, too much attention is paid to the preprocessing ability of information, and the relationship between different information structures is ignored.

References

1. Zhang, H., Fang, M.: Research on the integration of heterogeneous information resources in university management informatization based on data mining algorithms. Comput. Intell. **37**, 1254–1267 (2020)
2. Hui, Z.: Research on vocational college course evaluation based on data mining (2018)
3. Wu, F., Liu, X., Wang, Y., et al.: Research on evaluation model of hospital informatization level based on decision tree algorithm. Secur. Commun. Netw., 1–9 (2022)
4. Yin, Y.: Research on ideological and political evaluation model of university students based on data mining artificial intelligence technology. J. Intell. Fuzzy Syst. **40**(6), 1–10 (2020)
5. Chang, R.: Evaluation model of enterprise lean management effect based on data mining. Discrete Dyn. Nat. Soc. **2022** (2022)

Embedded Network Equipment Fault Remote Monitoring System Based on Internet of Things

Weixue Liu(✉)

Collage of Mathematics and Compute Science, Chifeng University, Chifeng 024000, Inner Mongolia, China
13739938768@163.com

Abstract. The embedded network equipment fault remote monitoring is an embedded system that uses the Internet to connect and communicate with other equipment. It can be used for industrial automation, medical treatment, home security, etc. The monitoring platform mainly includes three parts: data receiving and transmitting module, database and client. The data receiving and sending module is not only responsible for receiving data and storing it in the database, but also responsible for forwarding the data to the client. The database is responsible for storing and maintaining data, and providing links to clients for querying historical data. The client obtains the real-time data of remote sensors and controllers through the data receiving and transmitting module connected to the server, performs status display and remote monitoring, and accesses the database for historical data query and analysis. Aiming at the two key technologies of signal acquisition and information transmission in the monitoring system, this paper studies and designs a monitoring system applied to the remote status of equipment failures. This system mainly focuses on the function realization of the lower computer, which can complete the accurate acquisition of equipment status signals and reliable transmission through LoRa wireless technology.

Keywords: Remote monitoring system · Equipment failure · Embedded Network

1 Introduction

With progress in the field of industrial control, emerging technology wireless sensor network has entered the field of industrial equipment monitoring, becoming another research hotspot in the field of industrial equipment fault monitoring, which is a new technology after fieldbus. The real-time, low power consumption, low cost and remote transmission of today's emerging technology wireless sensor network make this contemporary hot technology become one of the important technologies in the field of industrial monitoring [1]. Although wireless sensor technology continues to develop in the field of industrial control monitoring, most modern enterprises are still using inefficient manual inspection methods and high cost wired monitoring methods to monitor the operating state of equipment, which are not conducive to industrial modernization. Intelligent development. It also restricts the future development of enterprise construction.

J. C. Hung et al. (Eds.): FC 2023, LNEE 1131, pp. 235–240, 2024.
https://doi.org/10.1007/978-981-99-9299-7_33

The advantages of wireless sensor network technology make up for the shortcomings of the current manual inspection and wired monitoring methods, and become an indispensable key technology in the industrial monitoring fault system. Compared with wired network [2]. The cost of wireless network is relatively low; Simple installation without wiring; In case of node failure, it is easy to find the cause and maintain; Good mobility; Flexible networking to reduce construction amount; Wide coverage; Strong scalability, easy to expand nodes.

As China's industrial equipment moves towards automation. Complicate. With the rapid development, enterprises have increasingly high requirements for fault monitoring methods and technologies of their equipment, and traditional monitoring technologies and methods can no longer meet the development needs of modern industry. Especially for some production enterprises, if the equipment failure monitoring is not timely, once a production accident occurs, it will cause serious consequences [3]. Therefore, the research on improving the equipment fault monitoring efficiency and improving the equipment management mechanism can not only improve the production efficiency of enterprises and prolong the service life of equipment, but also promote the development of industry.

2 Related Work

2.1 Equipment Fault Research

Industrial production equipment is mainly CNC machine tools, and CNC machine tools are highly automated and very complex mechanical processing equipment with integrated machine, electric, hydraulic, pneumatic and other technologies. Due to their high degree of automation and control, failure or pre failure will have a serious impact on the production of enterprises. Therefore, the industrial equipment monitored by this system is mainly CNC machine tool operability equipment, and the monitoring of the running state of such equipment is particularly important. The real-time monitoring system can not only optimize the management of production equipment, timely repair and repair faulty equipment, but also increase the production efficiency of enterprises [4].

According to the field survey, the production equipment failures are mainly divided into two types: one is the equipment operating system and electrical component failures, mainly including electronic hardware failures, control component failures and software system failures; The other is mechanical failure of equipment, mainly motor, hydraulic press, pneumatic motor and other mechanical failures. The manual operator judges the running state of the equipment, presses the key in which state the current equipment is in, and then the sensor node will collect the status information in real time [5]. The collection cycle is queried by the central processor, and then the central processor encodes the status information of the equipment into a data frame, and adds the synchronization frame header and check bit according to the wireless communication protocol to form a data packet, Then the control radio frequency module sends data packets, and transmits them to the remote monitoring management center through the wireless gateway, and the management center schedules the maintenance department to provide an optimized equipment management scheme.

2.2 Monitoring Network Topology

Network topology refers to the wireless communication network formed by sensor nodes in a self-organized manner through network protocols, which is mainly divided into two-dimensional topology and three-dimensional topology. At present, the research and application of three-dimensional spatial topology is still in the initial stage, and the application of relevant theories is relatively few, so this section mainly analyzes two-dimensional network topology. Two dimensional spatial network topology can be divided into star network topology, cluster network topology and mesh network topology, and each topology has its own characteristics.

First of all, star network topology is a relatively simple network topology. It can not only be point-to-point communication, but also point to multipoint communication. The is shown in Fig. 1. This topology is generally composed of a sink node and multiple sensor nodes.

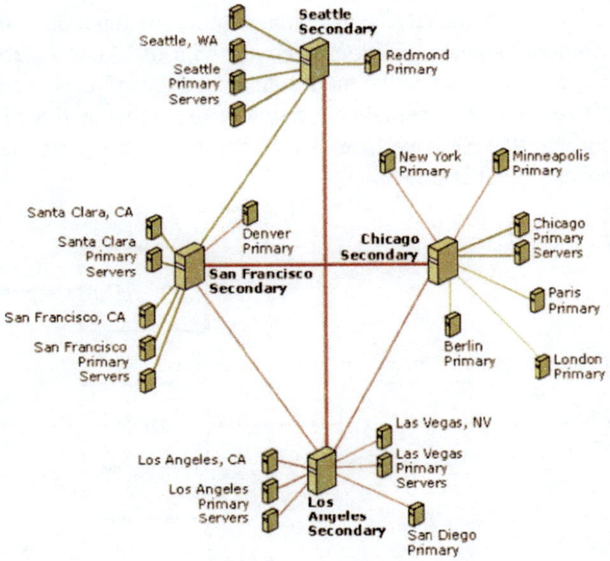

Fig. 1. Star network topology

When the star network topology network starts, it is first initiated by the sink node to send a start command to the sensor nodes, and establish a monitoring network. When sending the startup command, the sink node will set its own network ID, and the sensor nodes in the network will judge whether to send the network access request to the sink node according to this network ID. After receiving the network access application, the sink node will check the accuracy of the network access device information. If the network access is allowed after the inspection, it will start to listen to the device status information monitored by the node.

3 Embedded Network Equipment Fault Remote Monitoring System Based on the Internet of Things

This unit is responsible for signal conditioning of analog signals collected by the monitoring information measurement unit, and then converting analog signals into digital signals through analog to digital conversion circuits and outputting them to the next level. This paper mainly studies the small analog signals that are difficult to process. Other analog signals with relatively large amplitude only need to be properly reduced and then transmitted to the analog-to-digital converter.

The signal conditioning circuit can also be divided into two parts: the gain control circuit and the filter circuit. The gain control circuit is responsible for amplifying some of the collected small analog signals and suppressing the introduced common mode interference. The filter circuit is mainly responsible for suppressing the RF interference at the front end and the noise coupled to the analog circuit at the rear end. The circuit design is flexible and can be tailored according to needs. The A/D conversion circuit includes ADC drive circuit and ADC conversion circuit. The function of the ADC drive match the the front-end signal conditioning circuit with the ADC in terms of voltage size, input signal form, etc., while the main function of the ADC conversion circuit is to be responsible for analog to digital conversion of the signal, and then transmit it to the next level through the SPI interface. As shown in Fig. 2 below, the arrangement framework of power circuit is shown.

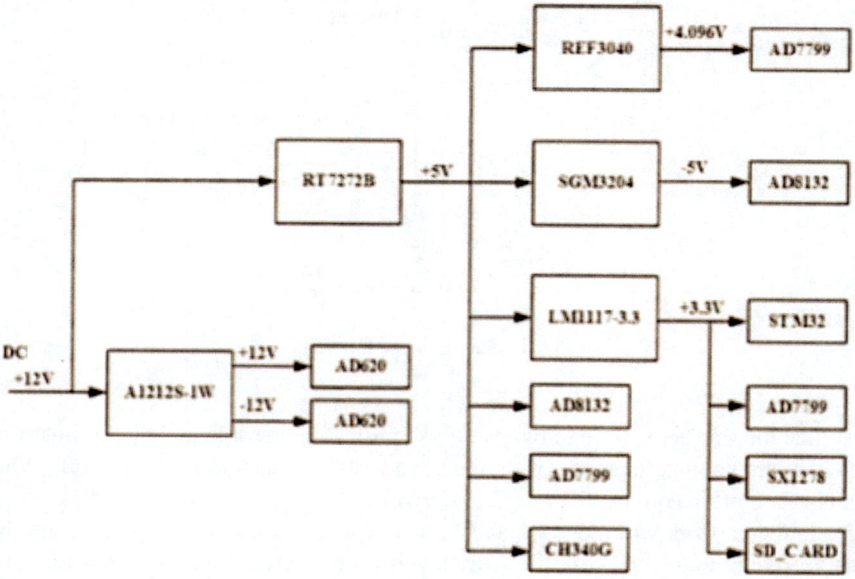

Fig. 2. Arrangement framework of power circuit

The equipment monitoring system network can use the broadcast mechanism to release monitoring tasks, that is, using the Flooding protocol. The broadcast communication mechanism has different control and transmission modes. The sink node is used to issue monitoring tasks to sensor nodes in the whole monitoring area, which is called global broadcast communication mechanism. However, any node communicates with its neighbor nodes by multi hop transmission, which is called cluster head broadcast communication mechanism.

This section combines the theoretical knowledge of MAC protocol to design the communication protocol based on the network topology of the system to achieve wireless communication of the monitoring network. However, because different enterprises have different requirements for monitoring the operation status of production equipment, the communication protocol designed in this system is one of the ways to realize the networking of the entire monitoring system, and specific problems should be analyzed according to different enterprise needs.

Assume that the design is based on an enterprise that needs of equipment. The system uses TDMA technology to design a communication protocol. The base station uses a broadcast mechanism to send time-sharing commands. Each sensor node calculates the activity time based on the time-sharing command received, and completes time synchronization and waits for the activity time, Sensor nodes will only send monitoring status information in their own time domain to achieve time-division debugging.

4 Simulation Analysis

In the wireless monitoring system network, the types of dynamic wake-up MAC protocols can be divided into fixed allocation, random competition, and on-demand allocation. In the equipment fault monitoring network, nRF24LO1 has a limited wireless transmission distance in the workshop and coverage of the whole workshop, it is necessary to divide the workshop monitoring area into several small micro networks to cross cover the monitoring to avoid monitoring blind areas. However, due to the competition mechanism of MAC protocol, each sensor node randomly allocates resources. In a micro network, when multiple data streams arrive at the sink node at the same time, collision conflicts are very likely to occur, which seriously affect the effectiveness and reliability of the ad hoc network. This can be resolved by TDMA technology time-sharing modulation (the data collision conflicts section in the next section will be described in detail). In the same way, in the cross region of two micronets, sensor nodes may also receive data streams broadcast by different micronet sink nodes, which will also cause collision conflicts. According to the relevant theoretical knowledge of MAC protocol, channel resources are allocated reasonably, and a fixed network identifier is set for each micro network to distinguish the cross monitoring area network.

Time synchronization mechanism is an important guarantee for the normal, stable and reliable operation of a wireless monitoring system network. For a monitoring network, the time synchronization of nodes in the network is an important basis for data transmission. However, each node is independent of each other and cannot communicate with each other. There are always some network delays and timing errors, which make it difficult to ensure real-time synchronization between each node. Aiming at the time synchronization problem, the system adopts the instantaneous synchronization scheme, which can ensure the instantaneous synchronization of the entire network.

Each micro monitoring network takes the time of the sink node as the benchmark. The sink node sends a frame pilot time command to the sensor node in the network, and the sensor node will wait for its own active time domain after receiving the command. The time error of timer timing is relatively small during this period of time, which has little impact on time synchronization and realizes instantaneous synchronization of monitoring network.

5 Conclusion

The embedded network equipment fault remote monitoring a solution that allows you to remotely monitor the equipment in your home or office. It is built on the existing infrastructure and uses standard protocols, such as SNMP, HTTP, HTTPS, etc. The data collected by such devices can be viewed through applications or Web portals, which enables you to view the status of various devices. This type of device has been used for many years, and because it provides great flexibility in use, it has become very popular.

References

1. Zhou, C., Wenhua, H.E., Xiong, S., et al.: Development of transformer remote monitoring system based on internet of things. Modern Machinery (2018)
2. Zhang, X., Wang, Y.: Research on prepaid account financing model based on embedded system and internet of things. Microprocess. Microsyst. **82**(VI), 103935 (2021)
3. Han, C., Liu, X., Liu, B., et al.: Design of embedded remote software update system based on FPGA+ARM. J. Interconnect. Netw. **22**, 2143040 (2022)
4. Zhang, J.: Design and implementation of sewage monitoring system based on internet of things technology (2021)
5. Bilandi, S.T., Ramezani, M., Nouranian, S., et al.: Method, Apparatus and System for Monitoring a Condition Associated With Operating Heavy Equipment Such as a Mining Shovel or Excavator: US20210262204A1[P] (2021)

Dynamic Monitoring Algorithm of Online College Physical Education Student Behavior in Complex Background

Li Lu(✉)

College of General Education, Xi'an Eurasian University, Xi'an 710065, Shaanxi, China
luli@eurasia.edu

Abstract. The role of dynamic monitoring in the behavior of students in college physical education is significant, but there is a problem that the monitoring accuracy is not high. Previous statistical monitoring methods could not solve the problem of accurate monitoring in behavioral monitoring, and there were few monitoring indicators. Therefore, a dynamic monitoring method is proposed to construct a behavior monitoring model. Firstly, the big data mining theory is used to plan behavior monitoring data, and the data collection and division are carried out according to student behavior to reduce the subjective factors in monitoring. Then, the big data mining theory plans the behavior monitoring, forms a data collection of monitoring results, and continuously monitors the data. MATLAB simulation shows that the dynamic monitoring method's evaluation accuracy and monitoring time are better than the previous statistical monitoring methods under the condition of certain monitoring data.

Keywords: big data mining theory · behavioral monitoring · dynamic monitoring · Monitor the results

1 Introduction

Monitoring results monitoring is one of the important evaluation contents of behavior monitoring, which is of great significance to the construction of behavior monitoring. However, in the actual monitoring and management process, there is a problem of poor accuracy of monitoring results, which brings certain economic losses to power enterprises [1]. Some scholars believe that applying intelligent algorithms to the precise monitoring of behavior monitoring can effectively analyze the risks of monitoring results and provide corresponding support for monitors. On this basis [2], a dynamic monitoring algorithm is proposed to optimize behavior monitoring accuracy and verify the model's effectiveness.

J. C. Hung et al. (Eds.): FC 2023, LNEE 1131, pp. 241–247, 2024.
https://doi.org/10.1007/978-981-99-9299-7_34

2 Related Concepts

2.1 Mathematical Description of Dynamic Monitoring

Dynamic monitoring uses dynamic theory to optimize behavioral data and, according to various indicators in behavioral data, find outliers in behavioral monitoring results and make corresponding data. Integrate and ultimately determine the feasibility of behavioral monitoring results [3]. The dynamic monitoring algorithm combines the advantages of big data mining theory and uses behavioral data to quantify the behavioral monitoring results, which can improve the monitoring accuracy.

Hypothesis 1: The data of behavior monitoring is that the behavior monitoring data is d_i, and the monitoring indicator set is $\sum d_i \neq 1$, the behavior monitoring results is $F(d_i \geq 0)$, The data judgment function is shown in Eq. (1).

$$F(d_i) = \sum x_i \to y_i \cdot \xi \tag{1}$$

ξ It is an adjustment factor for behavioral monitoring results, reducing subjective and subjective factors' influence.

2.2 Selection of Monitoring Programs

Hypothesis 2: The monitoring scheme selection function is $z(d_i)$ and the weight coefficient of the monitoring scheme is w_i, then the behavioral monitoring scheme selection is shown in Eq. (2).

$$z(d_i)=\frac{z_i \cdot F(d_i, y_i)}{w_i \cdot \xi}+ \tag{2}$$

2.3 Processing of Behavioral Monitoring Data

Before dynamic monitoring, the monitoring result data should be analyzed discretely, and the data should be mapped to a two-dimensional plane to eliminate abnormal behavior data. First, the behavior monitoring data is comprehensively analyzed, and the threshold and index weights of the data are set to ensure the accuracy of dynamic behavior monitoring. The behavior monitoring data is text test data and needs to be standardized. If behavioral monitoring data is not normally distributed, its monitoring results are affected, reducing the accuracy of overall monitoring [4]. In order to improve the accuracy of dynamic behavioral monitoring and the monitoring level of monitoring results, the behavior monitoring scheme should be selected, and the specific program selection is shown in Fig. 1.

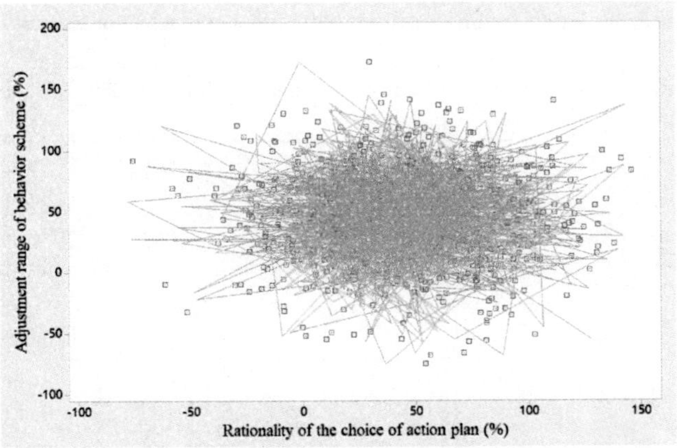

Fig. 1. Results of the selection of a behavioral monitoring programme

Survey data show that the behavior monitoring scheme shows a discrete distribution consistent with objective facts. The monitoring program is not directional, indicating that the behavior monitoring program has strong randomness, so it is used as an analytical study for 8–12 weeks. The monitoring scheme meets the normal requirements, mainly because the big data mining theory adjusts the monitoring scheme to eliminate duplication and irrelevant news and supplement the default schema, so that the entire data is dynamically correlated.

2.4 Strategies for Behavior Monitoring

The dynamic monitoring method adopts a random strategy for behavior monitoring and adjusts the corresponding parameters to optimize the behavior monitoring scheme. The dynamic monitoring method divides behavioral monitoring into different periods and randomly selects different protocols [5]. The behavior monitoring scheme of different time periods is matched in the iterative process. After the matching process is completed, the monitoring results of different protocols are compared, and the best behavior monitoring results are recorded.

3 Practical Examples of Behavior Monitoring

3.1 Introduction to Behavior Monitoring Data

In order to facilitate monitoring, this paper takes the behavior monitoring data of offline college physical education teaching under complex circumstances as the research object, and the monitoring of 12 sports indicators is carried out, and the test time is 12 weeks. The monitoring data for specific physical education are shown in Table 1.

Table 1. Relevant parameters of physical education behavior monitoring

Physical Education Teaching Direction	Time period	Monitor the number of people	Monitor data	Data form
Power	1–8 weeks	96	90.71	Structured data, unstructured data
	8–12 weeks	96	81.79	Structured data, unstructured data
Agility	1–8 weeks	91	98.21	Structured data, unstructured data
	8–12 weeks	97	97.50	Structured data, unstructured data
Continuity	1–8 weeks	53	91.43	Structured data, unstructured data
	8–12 weeks	46	89.29	Structured data, unstructured data

The data processing process for behavior monitoring in Table 1 is shown in Fig. 2.

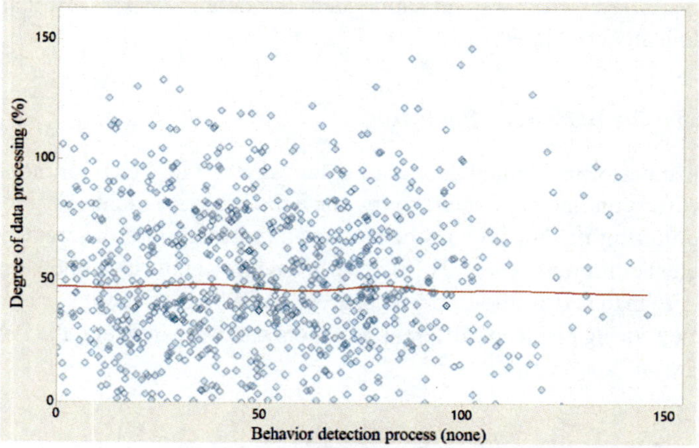

Fig. 2. Processing process of behavior monitoring data

As from Table 1, the monitoring the dynamic monitoring method are closer to the actual monitoring data than the previous statistical methods. In terms of the rationality and fluctuation range of monitoring data selection, dynamic monitoring methods have been used as statistical methods. The data changes in Fig. 4 show that the dynamic monitoring method has better stability and faster judgment speed. Therefore, the monitoring result monitoring speed, behavior monitoring data monitoring results monitoring and summation stability of the dynamic monitoring method are better.

3.2 Behavioral Monitoring

The monitoring results of behavior monitoring include non-structural information, semi-structural information, and structural information. After the pre-selection of the dynamic monitoring method [21], the preliminary monitoring result data of behavior monitoring were obtained, and the behavior was monitored. The feasibility of the monitoring result data is analyzed. In order to verify the effect of behavior monitoring and evaluation more accurately, select behavior monitoring in different periods, and the data evaluation data of behavior monitoring are selected. This is shown in Table 2.

Table 2. Overall picture of behavioral monitoring data

Time period	The rate of simplification of complex factors	Monitor the completeness of the data
1–4 weeks	98.57	85.71
5–8 weeks	93.21	95.6
8–12 weeks	99.64	90.71
mean	96.79	90.00
X^2	55.36 3	9.642
P = 0. 531		

3.3 Accuracy and Stability of Monitoring Data

To dynamic monitoring, the data are compared with previous statistical monitoring methods, and the data are shown in Fig. 3.

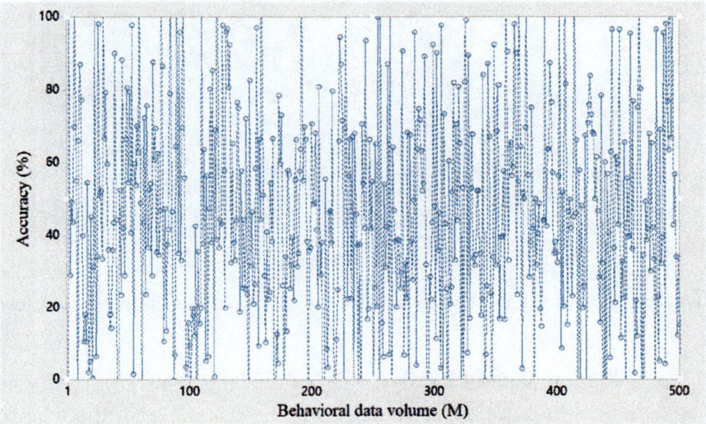

Fig. 3. The accuracy of different algorithms

It seen from Fig. 3 that the accuracy of the dynamic monitoring method the,previous statistical monitoring method, but the error rate is lower, indicating that the monitoring of the dynamic monitoring method is relatively stable Previous statistical monitoring methods have been uneven. The average data of the above three algorithms is shown in Table 3.

Table 3. Comparison of monitoring accuracy of different methods

algorithm	Precision	Magnitude of change	error
Dynamic monitoring methods	96.79	80.36	98.21
Previous statistical monitoring methods	95.71	91.07	91.43
P	0.027	0.010	0.023

It can be seen from Table 3 that the previous statistical monitoring methods had shortcomings in accuracy and stability in behavioral monitoring, the monitoring data changed greatly, and the error rate was high. The complete results of dynamic monitoring methods have higher accuracy than previous statistical monitoring methods. At the same time, the accuracy of the dynamic monitoring method is greater than 90%, and the accuracy has not changed significantly. To further validate the superiority of the dynamic monitoring method. In order to further method, different methods were used to analyze the dynamic monitoring of behavior comprehensively, and the results are shown in Fig. 4.

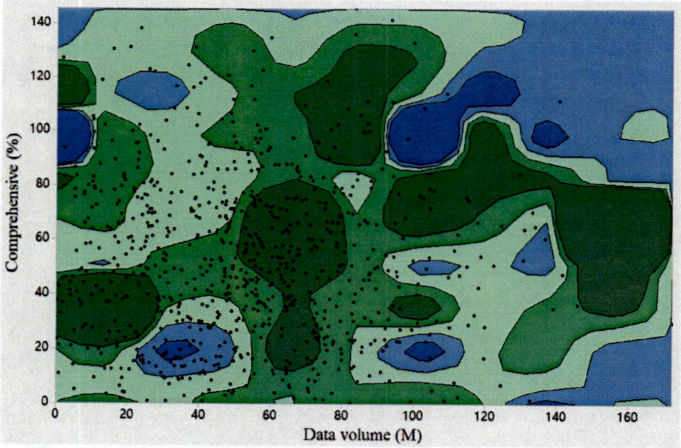

Fig. 4. Dynamic monitoring method to monitor the results of the comprehensive evaluation

It from Fig. 4 that the data of the dynamic monitoring are significantly better than the previous statistical monitoring methods, and the reason is that the dynamic monitoring method increases the regulation of behavior monitoring coefficients and set the corresponding thresholds to exclude data that does not meet the requirements.

4 Conclusion

In the complex background, this paper proposes a dynamic monitoring method for the physical education teaching situation in online colleges and universities and combines the theory of big data mining to improve behavior monitoring. At the same time, the departments and threshold standards of behavior monitoring are analyzed in depth, and the data collection for behavior monitoring is constructed. The research shows that the dynamic monitoring method can improve the accuracy and stability of behavior monitoring and can monitor the comprehensive monitoring results of behavior monitoring. However, in the dynamic monitoring process, too much attention is paid to the analysis of monitoring capabilities, resulting in a relative decline in monitoring accuracy.

References

1. Wu, S., Dai, J., Yang, J.: An Experimental and Algorithm Research on the Influence of OTO Teaching Mode on College Students' PE Learning Interest Based on Cloud Computing. Hindawi Limited (2021)
2. Kim, H., Song, T., Lim, S., et al.: Physical activity engagement outside of college physical education: application of the transtheoretical model. Am. J. Health Behav. **45**(5), 924–932 (2021)
3. Wang, B., Barakati, A.A., Hasan, H.: Characteristics of mathematical statistics model of student emotion in college physical education. Appl. Math. Nonl. Sci. **7**(1), 841–850 (2021)
4. Guo, H., Zhang, L.: Online evaluation system of college physical education students based on big data and artificial intelligence technology. In: International Conference on Frontier Computing, pp. 459–465. Springer, Singapore (2021). https://doi.org/10.1007/978-981-16-0115-6_50
5. Yang, F.: Construction of graduate behavior dynamic model based on dynamic decision tree algorithm. Discrete Dyn. Nat. Soc. **2022**, 1–10 (2022)

Design of ARM-Based Program Automatic Shelling Data Acquisition System

Yanyan Ren(✉) and Rongrong Cui

College of Mathematics and Computer Science, Chifeng University, Chifeng 024000, China
yan0203140203@163.com

Abstract. The role of the ARM program in automatic shelling data collection is very important, but there is a problem with the low accuracy of collecting shelling data. The ant colony algorithm cannot solve the problem of collecting multiple types of automatic hulling data, and the recognition rate is low. Therefore, this paper proposes an ARM program to build an automatic shelling data optimization model. Firstly, the dehulling data collected is classified by the dehulling standard, and the automatic dehulling scheme is selected according to the particle diameter Implement preprocessing of shelled data. Then, according to the degree of shelling, a collection of shelling data is formed, and the parameters are iteratively optimized. MATLAB simulation shows that the ARM program can improve the shelling depth and shorten the shelling time when the shelling diameter is consistent, and the relevant results are better than the ant colony algorithm.

Keywords: shelling criteria · Time · ARM programs · Shelling effect

1 Introduction

The automatic shelling system is an important indicator of the shelling work, and the ARM program the optimization of the shelling data [1]. However, in the process of building the shelling data system [2], the automatic shelling system has the problem of low accuracy of collecting shelling data [3], and cannot effectively play the role of automatic shelling [4]. Some scholars believe that the application of ARM program to the automatic shelling data acquisition system can effectively carry out redundant data rejection and time analysis [5], and provide corresponding support for the optimization of shelling data [6]. On this basis, this paper proposes an ARM program, optimizes the program for automatic hulling data collection, and verifies the model's effectiveness.

2 Related Concepts

2.1 Mathematical Description of the ARM Program

The ARM uses the optimization index, the relationship between the shelling data and the degree of shelling data, to optimize the self-shelling data, and according to the automatic shelling data collection system management indicators to find outliers in the

J. C. Hung et al. (Eds.): FC 2023, LNEE 1131, pp. 248–254, 2024.
https://doi.org/10.1007/978-981-99-9299-7_35

automatic shelling data collection system shelling data optimization and form a path table [7]. By integrating the shelling effect of the shelling data, the correlation of the ARM program results is finally judged. ARM programs combine shelling criteria and use ARM programs to optimize management results, which can improve the level of ARM programs.

Hypothesis 1: The self-shelling data is, the set of shelling effects of shelling data is, the degree of shelling of da_i shelling data is, and the $\sum da_i$ judicial function of the ARM program result $\oint y_i$ is $f(x_i)$ shown in Eq. (1).

$$f(x_i) = \sum x_i|y_i + \sum \xi \tag{1}$$

ξ is the adjustment factor for shelling data to reduce the impact of redundant shelling data.

2.2 Selection of Shelling Data Optimization Scheme

Hypothesis 2: The shelling result's judgment function is and the shelling data's coefficient is, then the optimization method $F(\alpha \cdot x_i)$ of the shelling data o_i is selected as shown in Eq. (2).

$$F(x_i)=z_i \cdot f(x_i|y_i) \rightleftarrows o_i \cdot \xi \tag{2}$$

2.3 Processing of Redundant Shelling Data

Before the analysis of the ARM program, the standard analysis of the time and single time in the results of the ARM program should be carried out, and the self-shelling data should be mapped to the selection table to determine that the standard is not met with particles. First, the self-shelling data is comprehensively analyzed, and the constraints and weights of the self-shelling data set for the ARM program is supported by accurate analysis. The self-shelling data needs to be preprocessed, and the processed results meet the university's shelling data requirements. Explain that the processing is valid. Otherwise, re-deepen the data structure. The ARM program and improve the level of shelling, the data optimization scheme should be selected, and the specific method selection is shown in Fig. 1.

The self-shelling data in Fig. 1 shows that the ARM program analysis is uniform and consistent with objective facts. The selection method is not directional, indicating that the ARM program analysis has high accuracy and can be used as an optimization method for automatic hulling data. The selection method meets the mapping requirements, mainly according to the shelling standard to adjust the selection method, remove duplicate redundant data, and revise the optimization indicators to make the entire self-shelling data is highly selective.

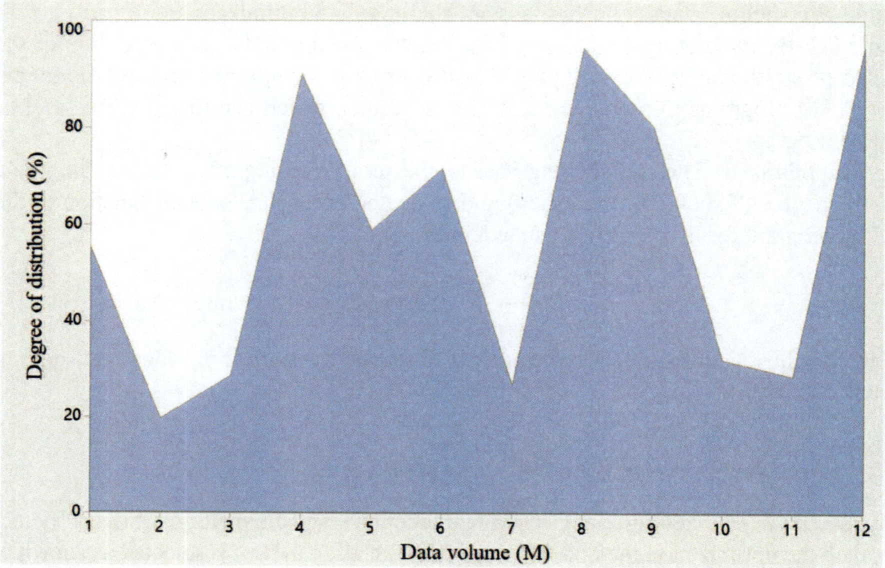

Fig. 1. The results analyzed by ARM programs

2.4 Correlation Between Different Optimization Indicators

The ARM program adopts the accurate judgment of the time and adjusts the corresponding redundant shelling data relationship to realize the optimization of the ARM program method for automatic shelling data collection. The ARM program divides the automatic shelling data collection ARM program selection into massive data and randomly selects different methods. In the self-learning process, the particle diameter of the massive data is correlated with the selection method. After the correlation processing is completed, the ARM program of automatic shelling data collection is compared with different methods to store the management results with the highest accuracy.

3 Optimization Case of Automatic Shelling Data Acquisition System Based on ARM Program

3.1 Shelling Data System Situation

In order to facilitate the analysis of shelling data systems, different types of shelling data systems were used as the research objects, and the number of test orders was 2421, and the time was 4 50, as shown in Table 1.

Table 1. Shelling parameter characteristics

Shelling data	stage	Shedding amount	Recognition rate	Constraints
Basic data	1	55.36	92.86	3.93
	2	19.64	83.93	2.50
Optimize data	3	28.57	85.71	5.00
	4	91.07	94.64	5.71
Verify the data	5	58.93	91.07	6.43
	6	71.43	91.07	7.14

The processing process between different optimization indicators in Table 1 is shown in Fig. 2.

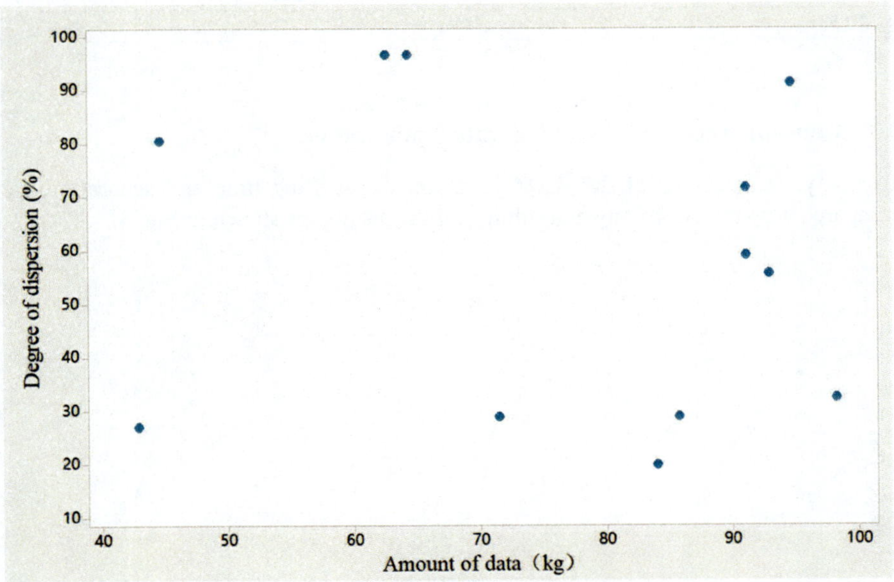

Fig. 2. Optimizing the processing process of indicators

As can be seen from Table 1, compared with the ant colony algorithm, the shelling result of the ARM program of the ARM program is closer to the actual standard. In terms of optimization index selection rate and accuracy based on ARM program automatic shelling data collection selection, ARM program ant colony algorithm. From the changes in optimization indicators in Fig. 4, it can be seen that the ARM program is better and the judgment. Therefore, the processing speed, time, and optimization of the shelling data of the ARM program are better.

3.2 Optimization Ratio of Hulling Data

Optimization of shelling data includes redundant shelling data, optimization metrics, speed. After the constraint criteria screening of ARM program [21], the preliminary management results were obtained, and the correlation of the management results was analyzed. In order to verify the effect more accurately, different redundant shelling data are selected to calculate the overall time of the ARM program, as shown in Table 2.

Table 2. Overall situation of logistics

Shelling ratio	The shelling diameter is consistent	Parameter compliance rate
25%	89.29	41.07
50%	42.86	64.29
70%	96.43	96.43
mean	23.21	98.21
X^2	16.07	32.14
P = 0.531		

3.3 Time and Accuracy of Hulling Data Optimization

To verify the accuracy of the ARM program, the shelling time and accuracy were compared with the ant colony algorithm, and the results are shown in Fig. 3.

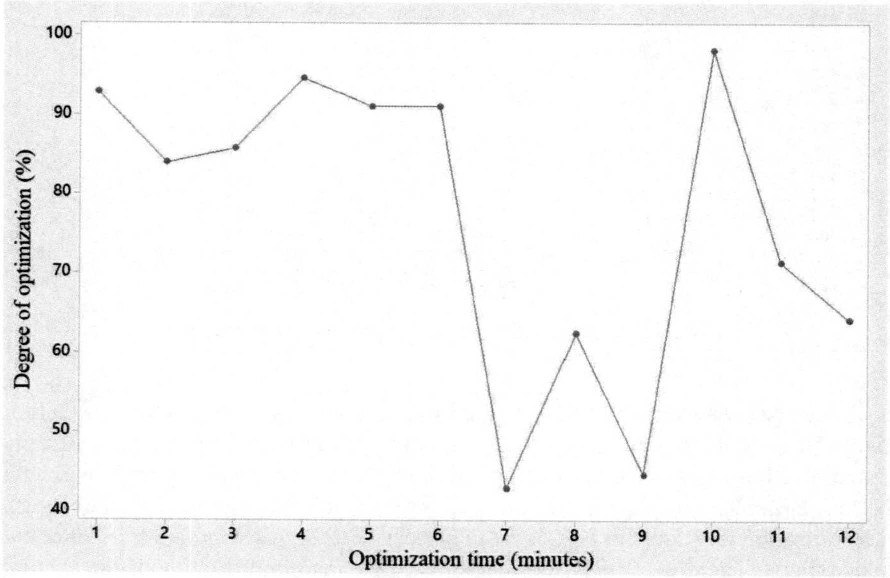

Fig. 3. Shelling time for different algorithms

It from Fig. 3 that the shelling time of the ARM program is shorter the ant colony, but the error rate is lower, indicating that the selection of the ARM program is relatively stable, while the optimization degree of the ant colony algorithm is uneven. The accuracy of the above algorithm is shown in Table 3.

Table 3. Comparison of optimization degrees of different methods

Algorithm	Shelling time	Optimize metrics	Error
ARM program	92.11	95.79	4.74
Ant colony algorithm	70.25	85.26	6.21
P	0.012	0.021	0.023

Table 3 shows that the ant colony algorithm has shortcomings in the dehulling time and accuracy of automatic shelling data collection selection recognition rate based on ARM program, and the accuracy of data processing changes significantly, the error is high. The comprehensive results of the ARM program have a higher shedding time, which is better than the ant colony algorithm. At the same time, the shelling time of the ARM program is greater than 90%, and the accuracy does not change significantly. To further verify the superiority of the ARM program. In to further the continuity, a comprehensive analysis of the ARM program was performed using different methods, as shown in Result 4.

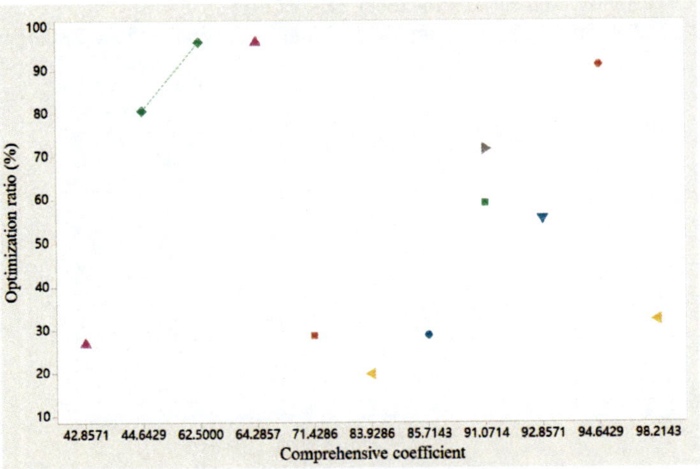

Fig. 4. Comprehensive results of ARM program result evaluation with different methods

As can be seen from Fig. 4, the results of the ARM program are significantly better than the ant colony algorithm, and the reason is that the ARM program increases the time adjustment coefficient and sets the corresponding one Constraints, which propose non-compliant results.

4 Conclusion

In the case of the continuous expansion of rice planting area, because of the problem of automatic hulling data collection, this paper proposes an ARM program method, combined with the dehulling standard Shelling criteria are optimized. At the same time, the constraints of shelling data processing are analyzed in depth to construct a set. Studies have shown that ARM programs can improve shelling accuracy and shorten the total judgment time. However, in the process of ARM program, too much attention is paid to the adjustment of single parameters, and the proportion of different parameters is ignored.

Acknowledgements. Chifeng University application-oriented Demonstration course construction project.

References

1. Luo, X.: Design of efficient search system for English translation terms in massive internet data. Mod. Electron. Technol. **40**(13), 134–136 (2017)
2. Li, F.: Design of high similarity English word selection system. Mod. Electron. Technol. **40**(23), 147–150 (2017)
3. Liu, B., Li, X.: Design and application of meteorological information query system based on Silverlight. Meteorol. Sci. Technol. **44**(3), 369–373 (2016)
4. Li, L., Meng, X., Liu, L.: Design and implementation of computer aided system based on code automatic generation. Aerosp. Control **33**(4), 68–72 (2015)
5. Huang, X.: Reliability Engineering, pp. 2–15. Tsinghua University Press, Beijing (1990)

Research on Management Information System Design Platform of Vocational College with Decision Tree ID3 Algorithm

Hongchun Shen(✉), Jiayan Wu, and Shirui Li

School of Art, Sichuan University Jinjiang College; School of Literature and Media, Sichuan University Jinjiang College; Office of Student Affairs, Sichuan University Jinjiang College, Chengdu 620800, Sichuan, China

xkshendu@163.com

Abstract. The role of system information in vocational colleges is significant, but there is a problem of a low level of information processing. The management system cannot solve the problem of managing multiple types of information in the vocational college system, and the rationality is low. Therefore, this paper proposes a decision tree ID3 algorithm to construct a system information optimization model. First of all, the management standards of colleges and universities are used to classify the information of vocational colleges, and the information of vocational colleges is selected according to the degree of importance to realize the preprocessing of information. Then, the university management standards are classified according to importance, form an information optimization collection, and conduct a self-learning analysis of the scoring content. MATLAB simulation shows that the optimization degree and optimization stability of the decision tree ID3 algorithm are better than those of the ant colony algorithm when the system is fixed.

Keywords: university management standards · stability · Decision Tree ID3 algorithm · Optimize the results

1 Introduction

Information optimization is one of the important contents of system information, which plays a crucial role in improving the information optimization system [1]. However, in constructing the information system of vocational colleges, the system information results have the problem of low information processing level, and the role of system information cannot be effectively played. Some scholars believe that applying the decision tree ID3 algorithm to the vocational college system can effectively carry out redundant information and stability analysis and provide corresponding support for information optimization verification [2]. On this basis, this ID3 algorithm to optimize vocational colleges' system information and verify the model's effectiveness.

J. C. Hung et al. (Eds.): FC 2023, LNEE 1131, pp. 255–261, 2024.
https://doi.org/10.1007/978-981-99-9299-7_36

2 Related Concepts

2.1 Mathematical Description of the Decision Tree ID3 Algorithm

The key points of information, the relationship between information and the importance of information to optimize the information of vocational colleges, and finds outliers in information optimization according to the management indicators in the vocational college system, and forms a path table. By integrating the information optimization results [3], the correlation of the system information results is finally judged. The decision tree ID3 algorithm combines the university management standards, and the decision tree ID3 algorithm is used to optimize the management results, which can improve the system information level.

Hypothesis 1: The vocational college information is x_i, the set of information optimization results is $\sum x_i$, the information importance is $f(x_i)$, and the judicial function of the system information results y_i is as shown in Eq. (1).

$$f(x_i) = \lim_{i \to \infty} \sum x_i | y_i + \sqrt{\xi^2 - 4xy} \tag{1}$$

ξ Adjust the factor for information to reduce the impact of redundant information.

2.2 Selection of Information Optimization Scheme

Hypothesis 2: The information management function is $F(x_i)$, and the information weight coefficient is z_i, then the information optimization method is selected as shown in Eq. (2).

$$F(x_i) = \frac{-x \pm \sqrt{x^2 - 4x\xi}}{2x} + z_i \cdot \xi \tag{2}$$

2.3 Processing of Redundant Information

Before analyzing the decision tree ID3 algorithm, the stability and single standard analysis of the system information results should be carried out, and the vocational college information should be mapped to the selection table to determine the semantic abnormality Content. First, the vocational college information is comprehensively analyzed, and the constraints and weights of the vocational college information are set to support the accurate analysis [4]. Vocational college information needs to be preprocessed, and if the results of processing meet the requirements of university information, the processing is effective Otherwise, re-deepen. The accuracy of the decision tree ID3 algorithm and improve the management level, the decision tree ID3 algorithm scheme should be selected, and the specific method selection is shown in Fig. 1.

The vocational college information in Fig. 1 shows that the decision tree ID3 algorithm analysis is uniform and in line with the objective facts. The selection method is not directional, indicating that the decision tree ID3 algorithm analysis has high accuracy, so it is used as systematic information research of vocational colleges [5]. The

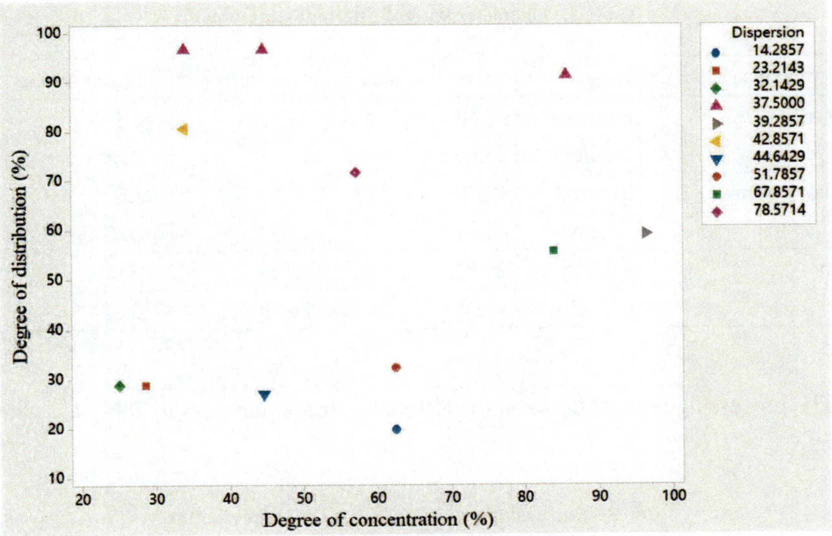

Fig. 1. Decision tree ID3 algorithm analysis results

selection method meets the mapping requirements, mainly because the university management standards adjust the selection method, eliminate the duplicate, and redundant information, and revise the information's key points so that the entire vocational college information is highly selective.

2.4 The Correlation Between Different Information Key Points

The decision tree ID3 algorithm adopts the accurate stability judgment and adjusts the corresponding redundant information relationship to optimize the vocational college system information method. The divides the system information selection of vocational colleges into massive data and randomly selects different methods. In self-learning, the importance of massive data is correlated with the selection method. After the correlation processing is completed, the system information of vocational colleges is compared by different methods, and the management results with the highest accuracy are stored.

3 Actual Cases of Vocational College System Information Systems

3.1 Information System Situation

In facilitate the analysis systems, the different types of information systems in this paper are the research objects, the number of test orders is 2421, and the stability is 450, as shown in Table 1.

Table 1. Characteristics of different structures

information system	range	Amount of data	rationality	Risk constraints
SQL database	exterior	37.50	71.43	80.36
	interior	50.00	73.21	101.79
School system	exterior	62.50	87.50	60.71
	interior	46.43	64.29	35.71
client	exterior	28.57	80.36	75.00
	interior	41.07	26.79	53.57

The processing process between the different information keys in Table 1 is shown in Table 2.

Table 2. Processing process of information key points

source	degree of freedom	Adj SS	Adj MS	F-number	P-value
regression	10	9571.97	957.20	1302.62	0.022
3	1	18.40	18.40	25.04	0.126
5	9	9463.11	1051.46	1430.90	0.021
error	1	0.73	0.73		
total	11	9572.70			

Table 1 shows that compared with a single system information system, the decision tree ID3 algorithm's system information results are closer to the stability regarding the selection rate and accuracy of critical points of vocational college system information selection decision tree ID3 algorithm management system. From the change of information key points in Fig. 4, it can be seen that the decision tree ID3 accuracy and faster judgment speed. The information processing speed, stability, and optimization degree of the decision tree ID3 algorithm are better.

3.2 Optimization Ratio of Information

Information optimization includes redundant information, information key points, and speed. After the constraint criterion screening of the decision tree ID3 algorithm [21], the preliminary management results are obtained, and the correlation of the management results is analyzed. In order to verify the effect more accurately, select different redundant information and calculate the overall stability of the system information, as shown in Table 3.

Table 3. Provides an overall picture of information optimization

Optimize the proportion	Degree of systematization	Outlier recognition rate
25%	41.07	82.14
50%	101.79	100.00
70%	21.43	98.21
Mean	75.00	35.71
X^2	7.124	4.292
P = 0. 002		

3.3 Stability and Accuracy of Information Optimization

The decision tree ID3, the optimization stability and accuracy compared with the management system are in Fig. 2.

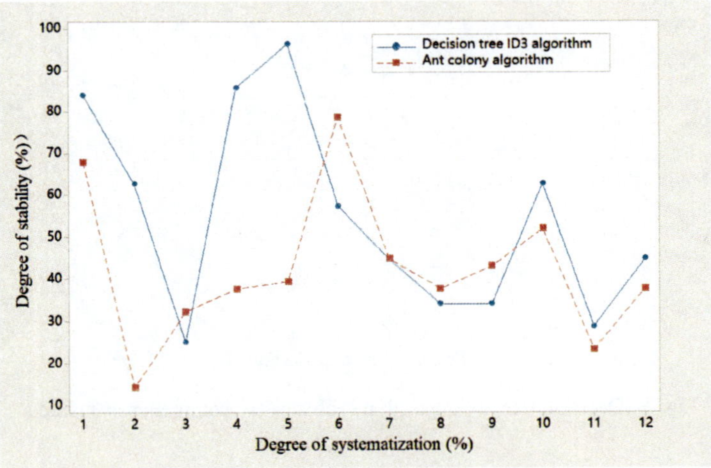

Fig. 2. Optimization stability of different algorithms

It can be seen from Fig. 3 that the optimization stability of the decision tree ID3 algorithm is shorter than that of the ant colony algorithm. However, the error rate is lower, indicating that the decision tree ID3 algorithm choice is relatively stable, while the optimization degree of the ant colony. The accuracy algorithm in Table 4.

It seen from Table 3 that the rationality of the system selection in vocational colleges and universities in the management system has shortcomings in optimization stability and accuracy, and the accuracy of data processing has changed significantly, and the error is high. The optimization stability of the complete results of the decision tree ID3 algorithm is higher than that of the management system. At the same time, the optimization stability of the decision tree ID3 algorithm is greater than 90%, and the

Table 4. Comparison of optimization degrees of different methods

Algorithm	Optimize stability	Information key points	error
Decision tree ID3 algorithm	91.07	91.79	91.07
Ant colony algorithm	83.93	76.79	83.93
P	9.219	6.433	9.292

accuracy does not change significantly. To further verify the superiority of the decision tree ID3 algorithm. The continuity, the decision tree ID3 algorithm is comprehensively analyzed by different methods, and the results are shown in Fig. 3.

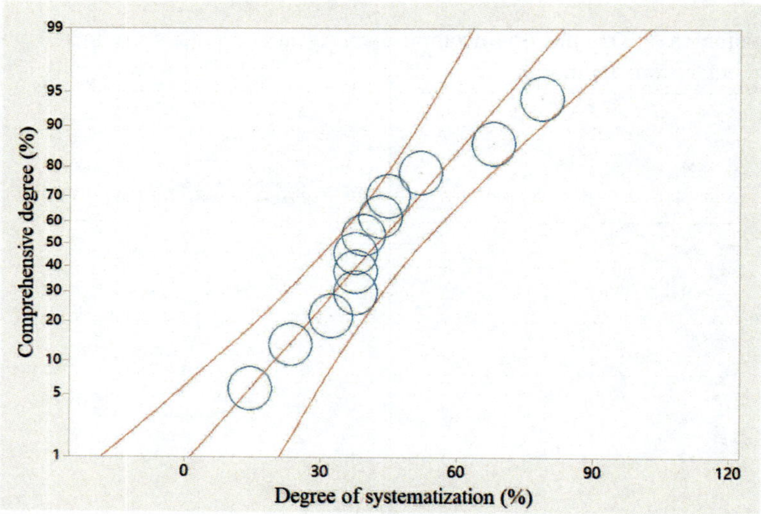

Fig. 3. Decision tree ID3 algorithm evaluates the comprehensive results

It can be seen from Fig. 3 that the result of the decision tree ID3 algorithm is significantly better than that of the ant colony algorithm, and the reason is that the decision tree ID3 algorithm increases the stability adjustment coefficient and sets the appropriate constraints to propose non-compliant results.

4 Conclusion

Under the condition that the requirements of university information management continue to increase, this paper proposes the decision tree ID3 algorithm for the information problem of the vocational college system. It combines the university management standards to improve the information related in the system. At the same time, the information processing constraint criteria are analyzed in depth to construct an optimization set. The research shows that the decision tree ID3 algorithm can improve the accuracy and accuracy of stability and can comprehensively stabilize stability. However, in the decision tree ID3 algorithm, too much attention is paid to the ability of unilateral index analysis and the proportion of stability is ignored.

References

1. Wang, W.: Model construction and research on decision support system for education management based on data mining. Comput. Intell. Neurosci. **2021**, 9056947 (2021)
2. Qin, J.: Design and implementation of sports performance management system in higher vocational colleges based on data mining. In: EAI International Conference on Multimedia Technology and Enhanced Learning (2017)
3. Wang, H., Wang, T., Zhou, Y., et al.: Information classification algorithm based on decision tree optimization. Cluster Comput. **22**(S3), 7559–7568 (2019)
4. Cheng, D.M., Li, T., et al.: A study on the application of the decision tree algorithm in psychological information of vocational college students. Matec Web Conf. **22**, 01144 (2015)
5. Cheng, D., Li, T., Niu, L.: A study on the application of the decision tree algorithm in psychological information of vocational college students. In: 2015 2nd International Conference on Engineering Technology and Application (2015)

Application of Dynamic Visual Communication Design in Digital Media

Hui Shi(✉)

Yunnan Open University, Yunnan National Defense Industry Vocational and Technical College, Yunnan 650500, China
shiyuxuanxiaohui@163.com

Abstract. The development of digital media has provided people with more diverse and diverse ways of entertainment and learning. In this context, dynamic has gradually become an indispensable part of digital media. In digital media, can express the characteristics and advantages of products through animation, videos, and other means, enhance users' impression and experience of the product, and improve user satisfaction. For example, on an e-commerce website, using dynamic visual communication design can present the characteristics of a product to users through visual and voice means, thereby increasing their purchasing interest. This paper mainly discusses the application of dynamic, and forecasts the future development direction of dynamic visual. Visualization and animation are the two most important technologies used in this field. In visualization, we try to create an image or video so that people can see it without any problems. Animation is basically used to move images from one position to another. For example, if you want your company logo or product name to appear on the screen for a few seconds, it will be animated so that the audience can clearly see what you are trying to convey through it.

Keywords: Dynamic visual design · digital media

1 Introduction

With the continuous of digital media, it has gradually become an indispensable part, and dynamic has also received increasing attention. Dynamic of conveying more information and emotions by utilizing multimedia elements for visual presentation and providing a richer and more vivid visual experience through animation, video, audio, and other means. In digital media applications, dynamic plays an important role.

Firstly, dynamic can attract users' attention through visual display methods such as animations and videos, and improve their understanding and trust of the product while conveying information. In this era of information explosion and people's distraction, it is natural to make advertising or product design more vivid and intuitive, which is an important strategy to attract users' attention and impress users. For example, in e-commerce websites, the use of professional and the vividness of product description images often increase the probability of users placing orders.

J. C. Hung et al. (Eds.): FC 2023, LNEE 1131, pp. 262–267, 2024.
https://doi.org/10.1007/978-981-99-9299-7_37

Secondly, the of dynamic design can greatly improve the user of digital media products. Previously, most web pages were flat, with a single arrangement of text and limited user experience. With the application of dynamic, the presentation of dynamic effects has greatly improved the user experience. For example, by creating an exquisite dynamic interface, users' sensory experience will be better, and trustworthy websites will win more traffic and user favor.

Finally, dynamic visual communication design utilizes the advantages of multimedia elements such as audio and video in digital media, which can effectively explain complex concepts and processes. By presenting dynamic effects, users can be more vividly and vividly presented with relevant businesses, processes, etc., making it easier for them to understand and accept. For example, on the official website of a company, by creating a visual and vivid video public relations copy, users can quickly understand the company's core concepts, values, and service models, thereby enhancing the trust between users and the company. In short, dynamic visual communication design has broad application prospects in digital media, providing users with a more vivid and intuitive experience, and is increasingly favored by digital media practitioners.

The all media era urges people to choose a more efficient and effective way of "reading pictures". For the development of visual communication design that focuses on "seeing" and uses visual symbols to "design for communication", it puts forward higher requirements and provides a broader development platform and technical support. Visual communication design uses this springboard to broaden its scope and expression. It has injected animation elements, image elements, mobile elements, multimedia integrated elements, virtual interactive elements, etc., and started to develop from manual platform to digital platform, from plane static design to three-dimensional dynamic design, from passive acceptance to interactive communication, from real design to virtual design, from single form to comprehensive form, from simple communication to full media communication, so as to derive space Dynamic and interactive visual communication design has good development and considerable prospects in various fields [3].

We must seize the opportunity of all media to deliver beautiful things and correct aesthetic concepts to thousands of households, so as to improve the aesthetic quality and aesthetic level of the whole people. On the premise of understanding and comparing the development history and current situation of Chinese and foreign visual communication design, we should always various adverse phenomena during the period, such as consumption foam, art homogenization and aesthetic vulgarization, think about and analyze these problems, find practical solutions to problems, and choose a way suitable for China's national conditions under the exchange and collision of Chinese and foreign cultures and arts, We should guide the public to have a correct view of consumption, art and aesthetics.

2 Related Work

2.1 Relationship Between Visual Communication Design and Media

Visual involves different media categories from traditional to emerging, from ancient to modern, from individual to public, and has experienced language and picture media, written words, print media, electronic media and digital network media. Visual communication design aims at information communication, and the final expression of communication can never [4].

(1) Media provides a broad platform for visual communication design activities to convey visual image and thinking. Only relying on various media, can the information conveyed by vision be widely disseminated. Rapidity, strong permeability and timeliness are powerful guarantees for smooth information transmission. Designers must use various media to communicate to the public to realize and promote the function. With the help communication, information impacts our nerves with visual images, expands the field of people's experience, greatly enriches people's knowledge of the world, and changes people's aesthetic taste and lifestyle. The emergence and development of media have greatly improved the scale and efficiency of image production. That is to say, it is the emergence of these media that makes large-scale image production, dissemination and consumption truly possible and creates a "visual feast" in human history [5]. Media provides a platform for visual communication. However, different media have their own characteristics. Visual need to fully consider the media to their respective characteristics, and make choices based on the communication value of media to better achieve the effect.
(2) Visual is subject to the media, which is also the power. Visual certain extent by the media, and many of the designers' creativity and design effects are difficult to achieve. However, due to people's tireless pursuit of technology and effects, new media has come into being, developed and matured, and new media technology and visual effects have begun to diversify. Nowadays, visual communication has become a cultural behavior that connects the subject and object of communication. It represents the understanding and aesthetic tendency of the visual subject to things, and can also reflect the behavior occurrence and thoughts and emotions of the visual object. The is generated on the basis of this communication behavior.

2.2 Dynamic Visual Communication Design

When the visual receiver has entered the electronic age, the content of undergone great changes. As mentioned above, the content of design has changed static to or four-dimensional design, and the carrier of visual communication has developed from flat print to film, television and multimedia. Because of this change, design is no longer a static form but a dynamic formal language.

Dynamic visual has great advantages over static communication. (1) Instantaneous communication. Because the dynamic communication takes film and television multimedia as the media, and takes time as the axis to spread the plot and convey information. In terms of the content conveyed, it has strong plot and emotion, and the information transmission is more influential. (2) Strong penetration and large amount of information.

Multimedia art can, millions of families (take TV media as an example), and convey rich and powerful visual information in a short time. (3) Strong appeal and visual impact. "The most beautiful place and the greatest vitality of a painting lies in its ability to express motion, which painters call the soul of painting." Moving objects often attract our attention more than static ones, and moving and meaningful images have more powerful appeal and visual attention. Making full use of the advantages of dynamic design will certainly make the design enter a prosperous period.

3 Application of Dynamic Design in Digital Media

(1) The print media. Since the middle and computer technology have developed faster and faster. Because print media supports a variety of design expressions, which is conducive to designers to express their design thinking, dynamic visual communication design was first applied to print media with the trend of the times.
(2) The of dynamic multimedia. The development from static to is its development trend. As a result, static design works are gradually developing into dynamic design works. Therefore, people begin to apply dynamic visual communication technology to the multimedia field, including films, animation, etc.
(3) The in advertising. Because the media of advertisement are mostly plane media and multimedia, people will use the excellent works of dynamic in advertisement.
(4) The application of dynamic education and teaching. As we all know, with the reform of educational forms, multimedia teaching and online teaching are becoming more and more popular. Teachers in gradually aware of the convenience and multimedia technology for education. Therefore, in order to contribute to the education cause, the design works related to education are also applied to the classroom, as in Fig. 1.
(5) The application of dynamic in computer games. Computer games and online games are closely related to digital media. In promote the development of various computer games, game makers have applied a large number of dynamic visual communication designs. Accordingly, the better works, the clearer the theme of the game and the deeper the user experience.

4 The Development Trend of Dynamic Visual Communication Design

First of all, most of the traditional expression methods of dynamic are two-dimensional space, with "vision" as the way of communication and expression. As the three-dimensional sense of the two-dimensional space is weak, the two-dimensional design is single and empty, and the tension is not enough, so the dynamic visual communication design gradually turns to the multi-dimensional space in the development, impacting the sense organs from more levels, enabling designers to break through the shackles and show more multi-dimensional works based on the dynamic visual communication design.

Secondly, the competition in today's society is very fierce. The fast-paced life, complex social phenomena and strong desire to pursue material life have added invisible pressure to people, making some people's minds as impetuous as soup. The busy day

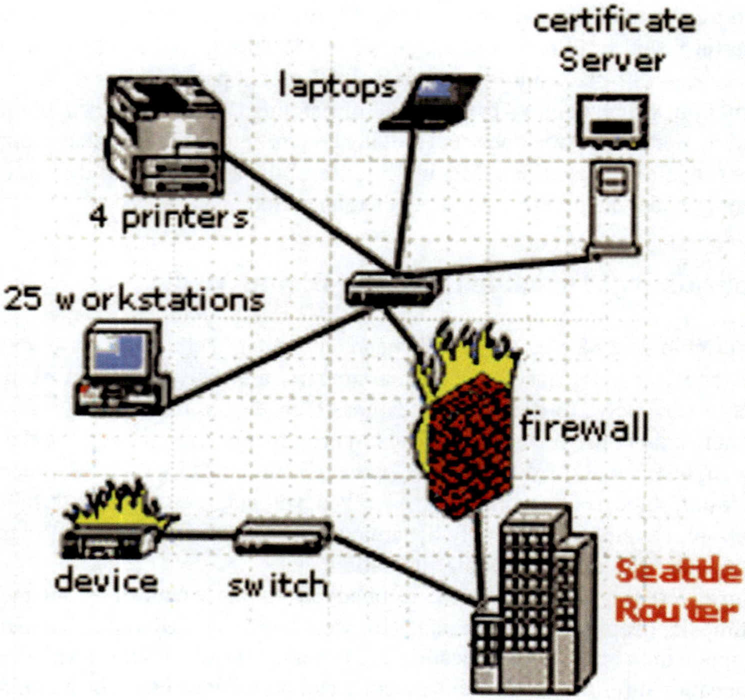

Fig. 1. Teaching mode of dynamic visual communication design

makes people not interested in the unchanging design. A single visual effect can not impress people, and things they have seen will soon be forgotten. The application of dynamic visual communication design has solved this problem well. It uses people's vision, hearing and other senses to convey information, reshape the sense system, make people interested in understanding the information content, and increase interest.

Finally, today, there are many forms of dynamic visual information communication. These forms of expression two-dimensional design, expand the application scope of dynamic, and become the main development direction of dynamic. For example, dynamic graphics, which has a strong visual impact, are used in many fields. The market demand is mainly reflected in television column packaging, film titles Economic advertisement, etc. With the application, people's methods of acquiring image information have also changed, and have played a good role in design. For example, for the Los Angeles Games, the emblem uses lines of different thickness to reflect the movement of five pointed stars. At the Hanover World Expo, the emblem design is a moving object, and the difference between the emblem of the Los Angeles Games and the emblem of the Los Angeles Games is that it is dynamic. The also comprehensively improves the design quality.

5 Conclusion

To sum up, dynamic visual communication design has been combined with digital media technology and applied to many fields. Dynamic visual communication design is a professional work. Relevant workers should make full use of the advantages and strengthen. At the same time, the designer should also make reasonable prospects for the direction of dynamic, and take these prospects as the center to make continuous efforts and reform in these directions. Only the dynamic visual communication design have a better prospect.

References

1. Li, H.: Visual communication media in digital advertising. J. Contemp. Educ. Res. **5**(7), 36–39 (2021)
2. Wang, Y., Yu, S.: CNS: application of dynamic resource allocation based on 5G communication in technology innovation of complex high-tech industries. Int. J. Cooperat. Inf. Syst. (2022)
3. Yga, B., Jia, Z., Jz, A., et al.: Design and application of groundwater dynamic online monitoring desertification area (2021)
4. Ma, L., Shao, W.: Application of and digital twin in digital media communication. J. Intell. Fuzzy Syst. Appl. Eng. Technol. **4**, 40 (2021)
5. Chen, W.: Design and Application of Online Digital Communication Platform Based on FPGA in EDA Laboratory (2021)

Computer-Assisted Korean Translation is Used in Translation Practice

Hong Sun(✉)

Shandong Women's University, Jinan 250300, Shandong, China
sunhong808@163.com

Abstract. The role of computer-assisted translation in Korean translation practice is very important, but there is a problem of poor translation accuracy. The online Korean translation method cannot solve the problem of the association of sentence translation in practice, and the logic is poor. Therefore, this paper proposes computer-aided construction of Korean translation relationships. First, the difficulty of translation is classified using grammar knowledge, and the Korean content is divided according to the translation standard to simplify the Korean content Processing. Then, grammatical knowledge classifies sentences, forms a Korean translation mapping table, and performs revision analysis of the translated content. MATLAB simulation shows that the accuracy and time of computer-aided translation are better than online Korean translation methods under the condition of certain translation difficulties.

Keywords: grammar knowledge · statements · computer-aided · Translate the results

1 Introduction

Translation effect is one of the important contents of Korean translation practice, and it plays a very important role in Korean translation practice [1]. However, in the actual translation process, the translation results have the problem of poor semantics, which affects the application and evaluation of practice. Some scholars believe that the application of computer-aided to practice can effectively analyze grammar and sentences, and provide corresponding support for the verification of translation practice. On this basis, this paper proposes computer-aided measurement of the practical effect of Korean translation, and verifies the effectiveness of the model [2].

2 Related Concepts

2.1 Computer-Aided Mathematical Description

Computer-aided is to use Korean logic, word relationships and sentence correlation to optimize Korean translation content, and based on multi-dimensional indicators in practice. Discover outliers in Korean translations and form a logical table of translations.

J. C. Hung et al. (Eds.): FC 2023, LNEE 1131, pp. 268–273, 2024.
https://doi.org/10.1007/978-981-99-9299-7_38

After integrating the Korean translation results [3], the logic of the translation results is finally judged. Computer assistance combines grammar knowledge, uses information mining and computer assistance, optimizes the statement results, and can improve the sentence recognition rate [4].

Hypothesis 1: The Korean translation content is x_i, the Korean translation result mapping table is $\overrightarrow{x_i}$, the professional grammar is y_i, and the judgment function of the translation result is $K(x_i)$ as shown in Eq. (1).

$$K(x_i) = \iint x_i + \xi \tag{1}$$

2.2 Choice of Translation Method

Hypothesis 2: The Korean translation logical selection function is $T(x_i)$ and the statement weight coefficient is $\overrightarrow{w}_i$, then the Korean translation logical selection is shown in Eq. (2).

$$T(x_i)=z_i \cdot K(x_i|y_i) \wedge w_i \tag{2}$$

2.3 Processing of Sentences and Phrases

Before performing computer-aided analysis, standard analysis of sentences and phrases in the translation results is performed, and the Korean translation content is mapped In the translation table, determine the content of semantic anomalies. First, the Korean translation content is comprehensively analyzed, and the threshold and weight of the Korean translation content are set for computer-aided is supported by accurate analysis. The Korean translation needs to be simplified, and if the results of the processing conform to the non-standard distribution, the processing is effective Otherwise, the standard

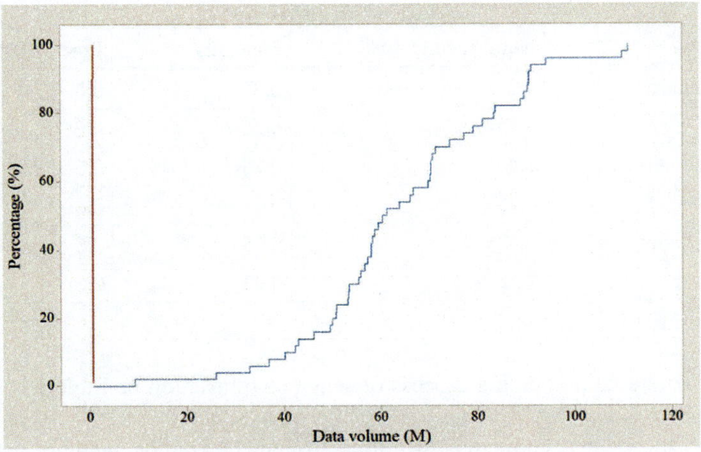

Fig. 1. Processing results of computer-aided analysis

processing is redone [5]. The accuracy of computer-assisted and the level of translation, the translation logic should be selected, and the specific logic selection is shown in Fig. 1.

The Korean translation in Fig. 1 shows that the computer-aided analysis is homogeneous and consistent with the actual semantics. The translation method is not directional, indicating that computer-aided analysis has high accuracy, so it is translated as a Korean translation practice of translation studies. The translation method meets the mapping requirements, mainly grammatical knowledge to adjust the translation method, remove repetitive grammar, and revise the professionalism to make the whole Korean translation content is more translatable.

2.4 Logic of Sentence Translation

Computer-aided judgment of sentence accuracy and adjustment of corresponding grammatical relationships to optimize Korean translation practices. Computer-aided Korean translation practices are divided into different data volumes and randomly selected by different methods. During the revision process, the translation standards with different data volumes are fitted to the translation method. After the fitting process is completed, different methods are compared for Korean translation practice, and the sentence results with the highest accuracy are recorded.

3 Actual Examples of Korean Translation

3.1 Translation Content

In order to facilitate the analysis of translation results, the number of words and sentences tested in this paper was 2421 based on different types of Korean content, statement 450, as shown in Table 1.

Table 1. Tests the characteristics of the Korean language

Korean type	Number of words	logicality	Practice rate
Level I difficulty	2,2551	0.85	0.6
	3,854	0.65	0.6
Level II difficulty	1,865	0.75	0.6
	4,251	0.69	0.6
Level III difficulty	2,475	0.75	0.6
	1,7852	0.92	0.6

The professional processing process of sentence translation in Table 1 is shown in Fig. 2.

Table 1 shows that compared to online Korean translation methods, computer-assisted translation results are closer to actual sentences. Computer-aided online Korean

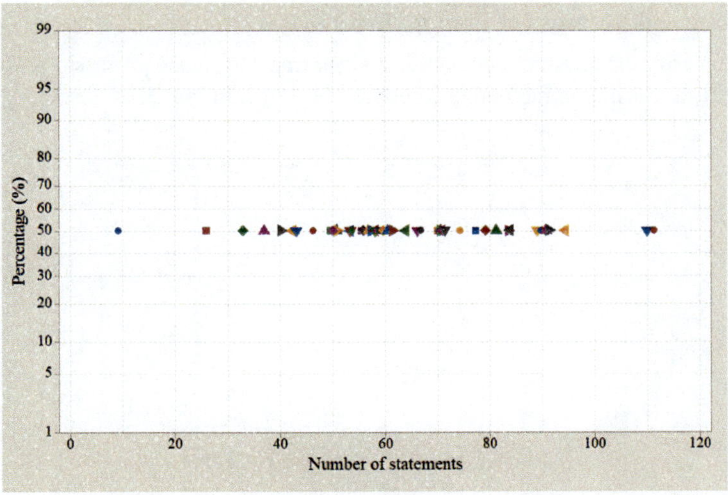

Fig. 2. The process of processing Korean language expertise

translation methods in terms of Korean translation practice translation professional translation rate and accuracy. As can be seen from the change in professionalism in Fig. 4, computer-aided accuracy is better and judgment speed is faster. Therefore, computer-assisted sentence speed, sentence and translation accuracy are better.

3.2 Outlier Recognition Rate of Statements

Sentences contain notes, specialization, bars. After computer-aided threshold standard screening, preliminary statement results were obtained, and statement results were obtained is analyzed logically. In order to verify the effect more accurately, select statements with different data volumes, and evaluate the professionalism of the statements as shown in Table 2.

Table 2. Overall picture of translation

Amount of data	Difficulty of translation	Outlier recognition rate
grammar	3.21	93. 23
word	2.01	93.36
statement	3.79	97.29
mean	2.12	92.14
P = 0. 031		

3.3 Timing and Accuracy of Practice Assessments

In order to verify the accuracy of computer-aided translation, the professionalism of the online Korean translation method is compared as shown in Fig. 3.

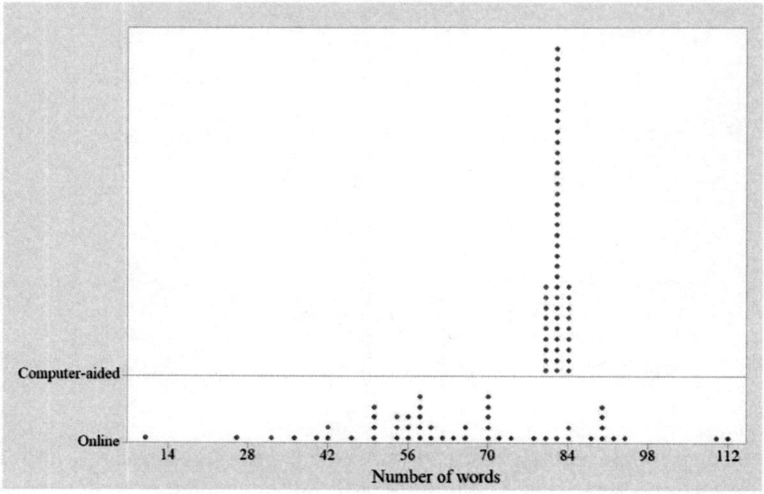

Fig. 3. Translation time for different algorithms

As can be seen from Fig. 3, the computer-assisted translation time is shorter than that of the online Korean translation method, but the error rate is lower, indicating that the computer-assisted translation is relatively stable, while the online Korean translation method is relatively stable Translation accuracy varies. The accuracy of the above algorithm is shown in Table 3.

Table 3. Comparison of translation accuracy of different methods

Algorithm	Translation time	accuracy	Error
Computer aided	92.21	93.79	4.74
Online Korean translation methods	70.25	85.26	6.21
P	0.022	0.032	0.013

Table 3 shows that the online Korean translation method has shortcomings in translation time and accuracy in Korean translation practice, and the accuracy of sentences changes greatly and the error rate is high. Computer-aided synthesis results have higher translation times than online Korean translation methods. At the same time, the computer-assisted translation time is greater than 90%, and the accuracy has not changed significantly. To further verify the superiority of computer-aided. To further verify the continuity of the method, computer-aided synthesis analysis was performed using different methods, as shown in Outcome 4.

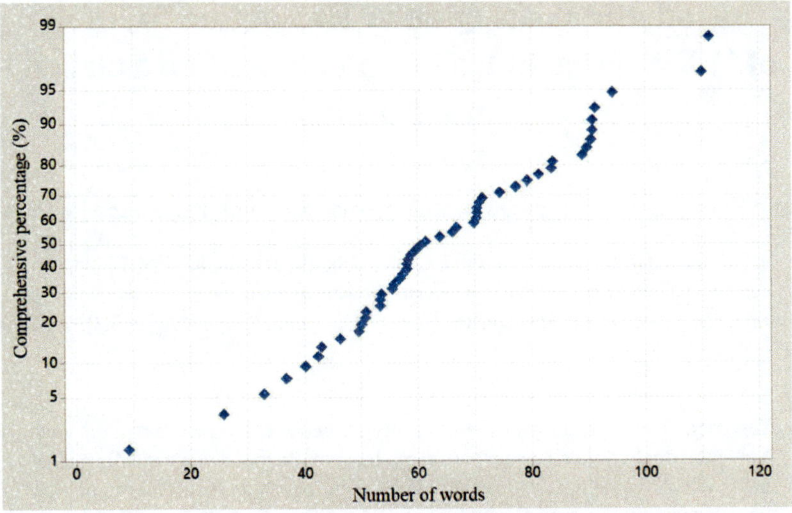

Fig. 4. Comprehensive evaluation of computer-aided methods

As can be seen from Fig. 4, computer-aided professionalism is significantly better than online Korean translation methods, and the reason is that computer-aided increases the sentence adjustment coefficient and sets the corresponding threshold, judging the professionalism that does not meet the requirements.

4 Conclusion

In the context of the development of computer-aided technology, in response to the practical problem of Korean translation, this paper proposes computer-assisted, combined with grammar knowledge, right statement to improve. At the same time, the department and threshold standards of the statement are analyzed in depth, and the mapping table of the statement is constructed. Studies have shown that computer-aided translation can improve the precision and accuracy of translation, and can comprehensively translate sentences. However, in the computer-assisted process, too much attention is paid to the ability to analyze unilateral indicators, and the proportion of sentences is ignored.

References

1. Usmanova, Z.A., Zudilova, E.N., Arkatov, P.A., et al.: Impact of computer-assisted translation tools by novice translators on the quality of written translations (2021)
2. Seljan, S., Erdelja, N.K., Kui, V., et al.: Quality Assurance in Computer-Assisted Translation in Business Environments (2021)
3. Frittella, F.M.: Computer-assisted conference interpreter training: limitations and future directions. J. Transl. Stud. (2021)
4. Wang, M., Dong, Q., Lei, L.I.: Speech translation method, electronic device and computer-readable storage medium. US20210271826A1 (2021)
5. Carrove, M.S.: Towards a theory of translation pedagogy based on computer-assisted translation tools for Catalan and English non literary texts (2022)

Automatic Voice Quality Evaluation Method of IVR Service in Call Center Based on Stacked Auto Encoder

Li Wang[1(✉)], Zongwei Wang[1], Guoyi Zhao[1], Yuan Su[1], Jinli Zhao[2], and Leilei Wang[2]

[1] State Grid Customer Service Center, Dongli, Tianjin 300309, China
bluemoonfish@163.com, yuan-su@sgcc.com.cn

[2] Beijing China-Power Information Technology Co. Ltd., Dongli, Tianjin 300309, China

Abstract. The application of virtual reality service in call center makes the voice evaluation quality of call center a research hotspot. Previous voice quality assessment methods could not solve the problem of stacked encoding in call centers, and the voice quality assessment capability was low. Therefore, this paper proposes a virtual reality method to construct a voice quality evaluation system for call centers. First, stacked autoencoders are used to classify the data of the call center and divide the collection according to the department, which reduces the complexity of data processing. The autoencoder then categorizes the call center into a collection of voice quality, and the data in the collection is automatically evaluated. In MATLAB simulation, the evaluation accuracy and calculation time of the virtual reality method are better than the previous speech evaluation methods under the condition that the amount of call data is fixed.

Keywords: stacked autoencoder · call centers · virtual reality services · Voice quality

1 Introduction

Voice quality assessment is one of the important evaluation contents of call centers, but in fact, there are problems of call methods and non-integration, resulting in inaccurate voice quality evaluation and inability to effectively guide the corresponding work. According to the survey data of call centers in 2022 [1], the voice quality assessment level of call centers is declining, which seriously affects the work of call centers. Due information in the voice data of the call center, and the nonlinear relationship between the number of voices in and the ability, the is not high. Some scholars have proposed a voice quality assessment method based on virtual reality and carried out related practices [2], and the results have proved that the intelligent of voice quality assessment in call centers, but it cannot cope with massive voice data. Therefore, this paper analyzes the voice quality evaluation results based on the virtual reality method and combines the work content of the call center, and verifies the final evaluation effect.

J. C. Hung et al. (Eds.): FC 2023, LNEE 1131, pp. 274–280, 2024.
https://doi.org/10.1007/978-981-99-9299-7_39

2 Related Concepts

2.1 Mathematical Description of Virtual Reality Services

Virtual reality method is the comprehensive result of modern information technology, through the simulation of reality, to achieve the purpose of accumulating practical experience [3]. Virtual reality fields of society, mainly for beginners to accumulate experience, so that they can master the necessary skills and methods, avoid different mistakes, and improve the accuracy and specification rate of related operations. In virtual reality, the following aspects need to be defined.

Assumption 1: The of the is x_i, the data set is $st(\overline{x_i})$, the evaluation result of the is $r(x_i \cdot k)$, and the is shown in Formula (1).

$$r(x_i \cdot k) = \sum_{st(x_i)=1}^{n} x_i \to \overline{x}_i \tag{1}$$

In of voice quality evaluation in call center, voice quality weight coefficient is ω integrated.

2.2 Classification of Voice Quality

Assumption 2: The evaluation function of speech quality is $P(x_i)$ and the balance coefficient of speech quality is z_i, then the speech quality classification of the call center is shown in Formula (2).

$$P(x_i)= \int_{i}^{n} z_i \cdot r(x_i \cdot k) \tag{2}$$

2.3 The Process of Evaluating Voice Quality by Virtual Reality Services

Before building a virtual reality approach, voice data is analyzed discretely and the results are mapped into a two-dimensional plane, eliminating redundant data. First, the call center volume is comprehensively analyzed, and the threshold and indicator weights of the call center are set to ensure the voice quality evaluation of the call center. The structure of the call center is unstructured and presents a discrete distribution [4]. If the call center volume is in a nonnormal distribution, its voice quality assessment level will decrease, reducing the accuracy of the overall calculation. In the call center and of information, the call center is classified. This article obtains the information by call center, as shown in Fig. 1.

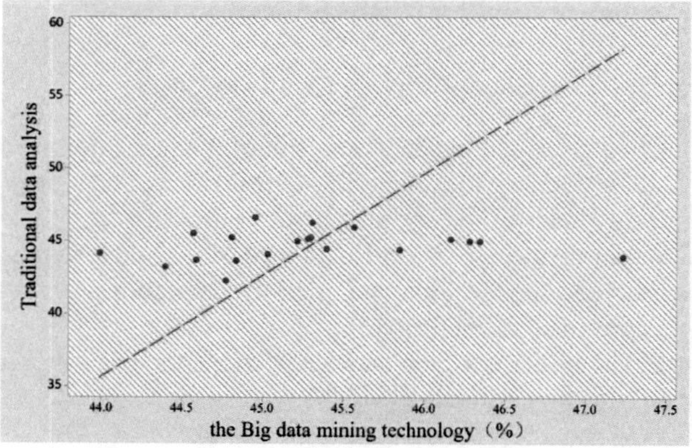

Fig. 1. The call center processing results of the department to which the information on different processing aspects belongs

Figures 1 and 2 show the initial call center processed by unstructured technology and autoencoders, respectively, with 120 call center segments. Through comparison, it is found that the initial call center volume generated by non-structural technology is messy and non-directional. Autoencoders handle more concentrated and directional call center volumes [5]. From the theorems 1 and 2 of the autoencoder, it can be seen that the results of the algorithm have nothing to do with the spatial dimension, and the call center processing of the department to which the information belongs is more accurate. Each time the point is taken, the initial call center volume, and the call center, so the autoencoder is selected for the processing of the initial call center volume.

2.4 Call Center Voice Quality Assessment Strategy

The virtual reality method adopts a heterogeneous strategy for the call center and adjusts the the distributed voice quality evaluation of the call center. The virtual reality approach divides call centers into five categories and takes different voice quality assessment measurements. In the iterative process, 5 types of call centers perform data mining at the same time. After complete, the voice quality assessment of the results is compared to record the best call center results.

3 Call Center Cases

3.1 Call Center Parameters

In order to facilitate calculation, the call center data volume in this paper is 120G, the call quality evaluation index is 10, and the test time is 10 min.

The time data acquisition process in Table 1 is shown in Fig. 4.

As can be seen from Table 1, the calculation of virtual reality methods is closer to the actual information results compared to the autoencoder. In mean value, etc., virtual reality methods are superior to autoencoders. From the changes in Fig. 4, the virtual

Table 1. Parameters of call centers

Test scope	Acquisition location	Amount of data	data structure	Collect devices
8: 0 ~ 10:00	server	56.84	Structural, unstructured	Audio, video, audio recorder
	client	109.47	Structural, unstructured	Audio, video, audio recorder
10:00 ~ 16:00	server	100.00	Structural, unstructured	Audio, video, audio recorder
	client	89.47	Structural, unstructured	Audio, video, audio recorder
16:00 ~ 8:00	server	101.05	Structural, unstructured	Audio, video, audio recorder
	client	151.58	Structural, unstructured	Audio, video, audio recorder

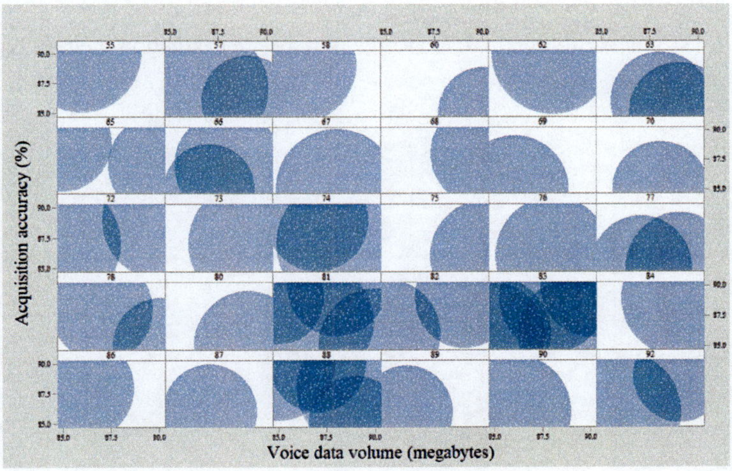

Fig. 2. Overall collection results of voice quality in the call center

reality method has judgment speed. The virtual reality method voice quality evaluation, call center volume voice quality evaluation, and summation stability.

3.2 Call Center Situation for Time Period

The voice quality assessment of the call center information, information, and structural information. After the preprocessing of virtual reality method, the voice data of the structured call center and the voice data of the semi-structured call center of 11 lines. In more accurately the voice data evaluation of the call center, select the call center in different time periods: 8:00 ~ 10:00, 11:00 ~ 16:00 and 16:0 0 ~ 8:00, the data evaluation the call center are in Table 2.

Table 2. Voice data processing results for the time period

Time period	Amount of data	Evaluation results
8:00 ~ 10:00	84.21	0.63
11:00 ~ 16:00	60.00	0.16
16:00 ~ 8:00	55.79	0.89
mean	60.00 ± 0.21	74.74 ± 2.42
variance	84.212	67.377
Degrees of freedom = 20.561		

3.3 Test Results

In order the virtual reality method, the r previous speech evaluation method, and the in Fig. 4.

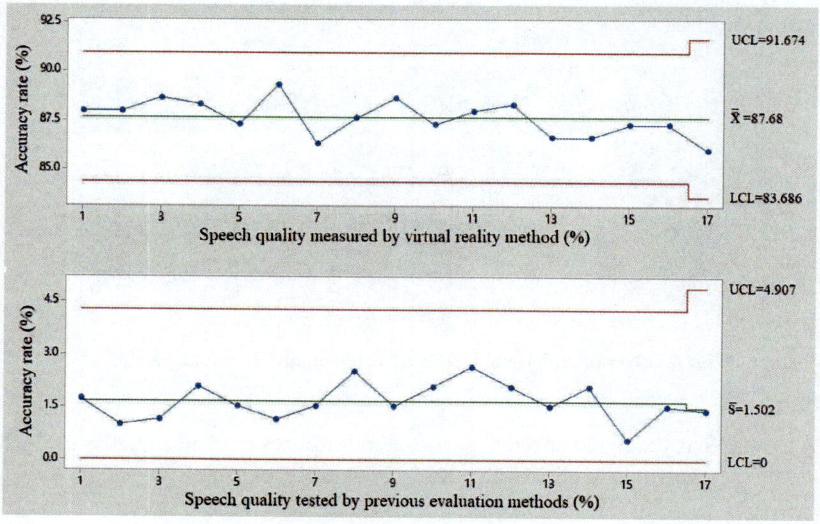

Fig. 3. Test of different algorithms

It from Fig. 3 that the virtual reality method is higher than that of the previous speech evaluation method, indicating that the calculation of the virtual reality the previous speech evaluation method is uneven. The are in Table 3.

Table 3. Comparison of speech quality evaluation accuracy over time period

algorithm	Precision	Stability	The direction of change in the evaluation of quality
Virtual reality approach	82.11	75.79	63.16
Previous phonetic evaluation methods	80	65.26	75.79
Difference	2.11	10.53	−12.63

Table 3 shows that the previous voice evaluation method had insufficient accuracy in the call center during the time period, and the calculation results changed significantly. The accuracy of the virtual reality method is high, which is better than the previous speech evaluation method. At the same time, virtual reality methods are greater than 75%, and the is changed. In order virtual reality method, the entire calculation process, and the shown in Fig. 4.

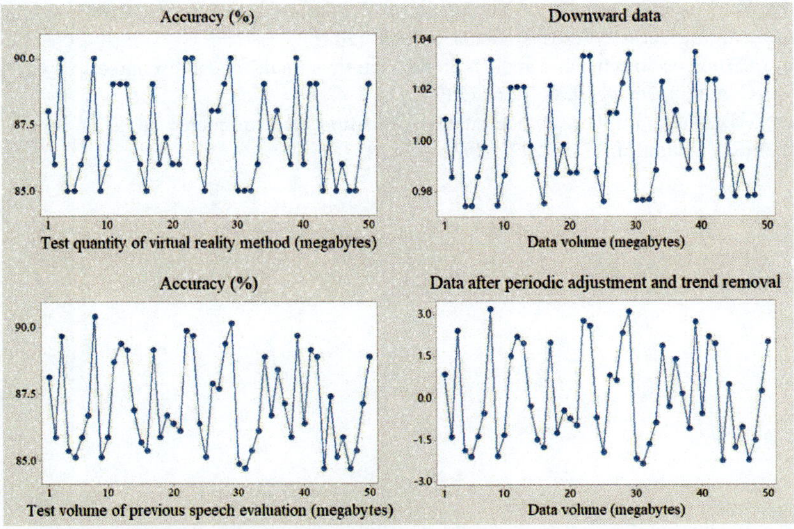

Fig. 4. Comprehensive results of speech quality evaluation by different methods

It seen from Fig. 4 that the results of the virtual reality method are significantly better than the previous speech evaluation methods, and the reason is that the virtual reality method increases the call center coefficient of the department, sets the weight and convergence factor.

4 Concluding Remarks

This a virtual reality approach, combined with stacked autoencoders, to improve the call center. At the department and the call center are analyzed in depth, and the call center classification collection is constructed. The results show that the call center accuracy and stability of the virtual reality method are better, and the comprehensive voice quality evaluation of the call center can be carried out.

Acknowledgements. Design and development project of State Grid Customer Service Center (62993118000Q).

References

1. Li, M., Tu, C., Wang, P.: Machine identification method of subway service quality based on smart card data. Transport. Safety Environ. **2**, 2 (2022)
2. Xu, Y., Hawamdeh, S.: The quality evaluation English translation for intercultural communication. J. Inf. Knowl. Manage. **21**, 2240027 (2022)
3. Wang, Z., Wu, Q.: Research on automatic of Mandarin Chinese based on 5G network and FPGA. Microprocess. Microsyst. **80**(3), 103534 (2021)
4. Ren, J.: Study on automatic of spoken English based on multimodal discourse analysis theory. Secur. Commun. Netw. **2021**, 1486575 (2021)
5. Zhang, H., Wang, L.: The service quality evaluation of e-commerce based on intuitionistic fuzzy GRA method. J. Math. **2022**, 3931136 (2022)

An Empirical Study on Library Readers' Reading Needs Based on Octopus Optimization Algorithm

Dehua Wang(✉)

Wuhan Business University, Wuhan 430056, Hubei, China
1013785790@qq.com

Abstract. The role of the Octopus optimization algorithm in the empirical demonstration of the reading needs of library readers is significant, but there is a problem with the low accuracy of demand analysis. Hierarchical needs analysis cannot solve the problem of the needs of readers in multiple types of libraries, and the demand judgment results are poor. Therefore, this paper proposes an Octopus optimization algorithm to demand a library reader judgment model. Firstly, the reading needs are used to classify the markets, and the demand methods are selected according to the requirements of library readers to realize the preprocessing of needs analysis. Then, according to the degree of Demand, the requirements analysis set is formed, and the parameters are iteratively judged. MATLAB simulations show that the Octopus optimization algorithm among library readers can increase the scale and shorten the Demand The requirements analysis time and the relevant results are better than the hierarchical demand analysis method.

Keywords: reading needs · analysis time · Octopus optimization algorithm · Demand effect

1 Introduction

Library readers are important to service objects of library work, and the Octopus optimization algorithm plays a very important role in demand analysis and judgment. However, in the process of English translation demand, the Demand has the problem of low accuracy of demand analysis, and the automatic Demand cannot be effectively played [1]. Some scholars believe that applying Octopus optimization algorithm to the empirical study of library readers' reading needs can effectively eliminate demand data, analyze time analysis, and analyze demand analysis Judgment provides appropriate support [2]. On this basis, this paper proposes the Octopus optimization algorithm, which empirically mines the reading demand of library readers and verifies the effectiveness of the Octopus optimization algorithm.

J. C. Hung et al. (Eds.): FC 2023, LNEE 1131, pp. 281–287, 2024.
https://doi.org/10.1007/978-981-99-9299-7_40

2 Related Concepts

2.1 Mathematical Description of the Octopus Optimization Algorithm

The Octopus optimization algorithm uses reading Demand, the relationship between demand indicators, and the degree of Demand to judge the demand analysis and finds it according to the management indicators in the empirical study of library readers' reading needs. Outliers in demand analysis identify outliers and form a path table [3]. By integrating the demand effect of demand analysis, the correlation of the results of the Octopus optimization algorithm is finally judged. The Octopus optimization algorithm combines reading requirements and uses the Octopus optimization algorithm to consider the demand analysis, which can improve the Octopus optimization algorithm Level [4].

Hypothesis 1 For the demand analysis is x_i, the requirement effect set of the requirements analysis is $set()$, the demand degree is y_i, and the judicial function is $P(x_i)$ as shown in Eq. (1).

$$P(x_i) = \prod x_i + \xi \tag{1}$$

ξ Adjust the coefficient for demand analysis to reduce the impact of reading demand analysis.

2.2 Demand Analysis to Determine the Choice of Scheme

Hypothesis 2 The demand result judgment function is $I(x_i)$ and the standard coefficient of demand analysis is B_i, then the demand analysis judgment method is selected as shown in Eq. (2).

$$I(x_i){=}z_i \cdot \prod P(x_i|y_i) + B_i \tag{2}$$

2.3 Handling of Reading Needs Analysis

Before the Octopus optimization algorithm analysis, the analysis time and effect should be analyzed, and the demand analysis should be mapped to the analysis plan to determine the demand data that does not meet the standard. Conduct a comprehensive analysis of demand analysis, and set the scheme and standards of demand analysis to support the Octopus optimization algorithm. The reading requirements are preprocessed, and if the results after processing meet the requirements of university demand analysis, the processing is effective. Otherwise, it is re-conducted Library Patron Analysis. In order to improve the analysis accuracy of the Octopus optimization algorithm and improve the level of demand analysis, the data judgment scheme should be selected, and the specific method selection is shown in Fig. 1.

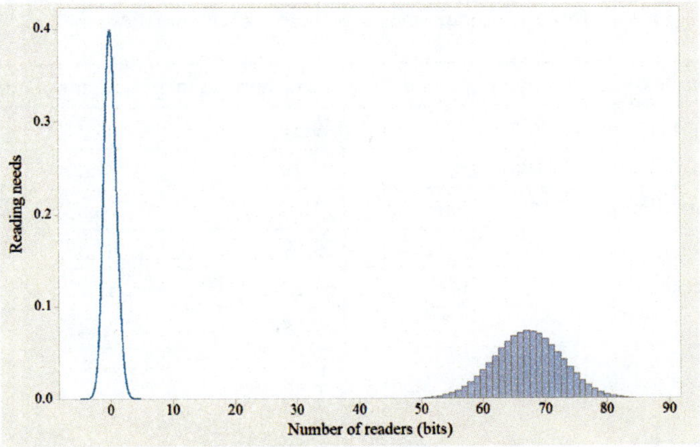

Fig. 1. Results of Octopus optimization algorithm

The demand analysis in Fig. 1 shows that the analysis of the Octopus optimization algorithm is uniform and in line with the objective facts. The selection of the English translation method is not directional, indicating that the analysis of the Octopus optimization algorithm has short analysis accuracy and can be used as a judgment method for library readers [5]. The English translation method meets the mapping requirements, mainly according to the reading needs to adjust the English translation method, remove duplicate demand data, and revise the demand analysis, so that the whole selection of the demand analysis is shorter.

2.4 Correlation Between Different Analytical Plans

The Octopus optimization algorithm adopts the analysis accuracy judgment for the analysis time, and adjusts the corresponding reading demand analysis relationship to achieve the accurate judgment of demand analysis. The Octopus optimization algorithm performs standard processing on demand data and randomly selects different methods. In self-learning, library readers with massive data require correlation processing with the selection of English translation methods. After the correlation processing is completed, different methods are compared for demand data, and the requirements analysis with the highest accuracy is stored.

3 Demand Analysis Case Based on Octopus Optimization Algorithm

3.1 Demand Analysis and Research Situation

In order to facilitate the needs analysis and analysis, the number of test information in this paper is 5242 for different types of readers, as shown in Table 1.

Table 1 shows the processing between different analysis plans, as shown in Fig. 2.

Table 1. Requirements parameter characteristics

Requirements analysis	Number of readers	Read demand rate	Material demand rate
Read the requirements	126	69.39	65.31
	119	64.29	74.49
Service requirements	123	71.43	76.53
	125	79.59	67.35
Consulting needs	112	63.27	61.22
	105	77.55	70.41

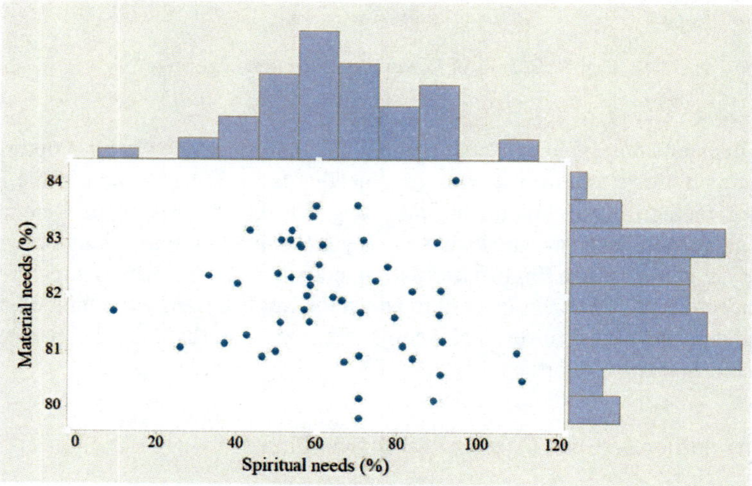

Fig. 2. Reading requirements process

It from Table 1 that the hierarchical demand, the demand results of the Octopus optimization algorithm are closer to the actual standard. In the process of requirements analysis, with the help of reading requirements, the Octopus optimization algorithm is used for analysis. From the reading requirements in Fig. 4, it the analysis accuracy of the Octopus optimization algorithm is better, and the analysis time is faster. Therefore, the demand analysis processing speed, analysis time, and the Demand of the Octopus optimization algorithm are better.

3.2 Judgment Ratio of Demand Analysis

The judgment of demand analysis includes reading needs analysis, reading needs, and speed. After the reading requirements screening of the Octopus optimization algorithm

[21], the preliminary demand analysis and the correlation of the demand analysis indicators are obtained Analyze. In order to verify the effect more accurately, select different reading demand analyses and calculate the overall analysis time of the Octopus optimization algorithm, as shown in Table 2.

Table 2. Reads the overall picture of Demand

demand	Library readers	Parameter compliance rate
25%	63.27	63.27
50%	72.45	70.41
70%	77.55	76.53
P = 0. 631		

3.3 Time and Accuracy of Requirements Analysis

The analysis accuracy of the Octopus optimization, the requirements analysis time and analysis accuracy are compared with the hierarchical requirements analysis method, and the results are compared This is shown in Fig. 3.

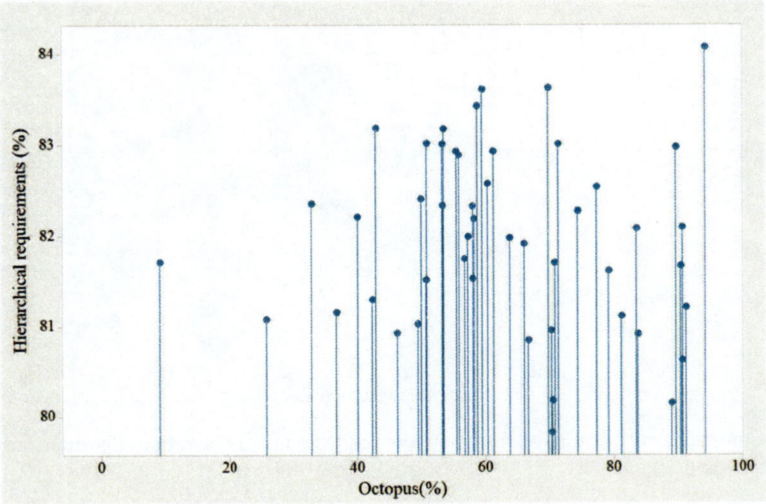

Fig. 3. Requirements analysis time for different algorithms

It from Fig. 3 that the analysis time of the Octopus optimization algorithm is shorter than that of the hierarchical demand analysis method. However, the error rate is lower, indicating that the choice of Octopus optimization algorithm is relatively stable, while the hierarchical requirements analysis method is relatively stable Demand is uneven. The analysis accuracy of the above algorithm is shown in Table 3.

Table 3. Comparison of requirements for different methods

algorithm	Requirements analysis time	Read the requirements	error
Octopus optimization algorithm	92.11	95.79	4.74
Hierarchical requirements analysis method	70.25	85.26	6.21
P	0.012	0.021	0.023

It can be seen from Table 3 that in terms of demand analysis of hierarchical demand analysis method, there are deficiencies in Demand analysis time and analysis accuracy, and the analysis accuracy of data processing has changed significantly. The error is shorter. The analysis time of the Octopus optimization algorithm is shorter, which is better than the hierarchical demand analysis method. At the same time, the demand analysis time of the Octopus optimization algorithm is greater than 90%, and the accuracy does not change significantly. In further the superiority of the Octopus optimization. In order to further continuity, the Octopus optimization algorithm is comprehensively analyzed by different methods, and result 4 is shown.

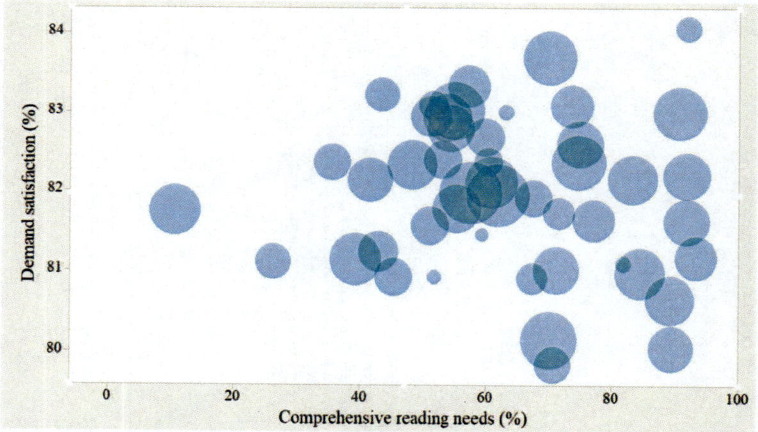

Fig. 4. Comprehensive results of the evaluation of Octopus optimization algorithm results of different methods

It from Fig. 4 that the results of the Octopus optimization algorithm are significantly better than the hierarchical demand analysis method, and the reason is that the Octopus optimization algorithm increases the analysis judgment coefficient and sets the corresponding one Read the requirements and come up with results that do not meet the requirements.

4 Concluding Remarks

In order to improve the reading level of library readers, aiming at the empirical problem of reading needs of library readers, this paper proposes an Octopus optimization algorithm method and analyzes the needs based on reading needs judgment. At the same time, the requirements analysis scheme is analyzed in depth, and the analysis result collection is constructed. Studies have shown that the Octopus optimization algorithm can improve the accuracy of requirements and shorten the analysis time. However, in the process of the Octopus optimization algorithm, too much attention is paid to the adjustment of single parameters, and the proportion of different parameters is ignored.

References

1. Ren, X., He, J., Huang, Z.: An empirical study on the behavior of e-commerce strategic planning based on deep learning algorithm. Inf. Syst. e-Business Manage. **21**(4), 9 (2021)
2. Feng, Y., Li, L.: Key points of narrow reading in medical English—an empirical study based on stomatological English. Open Access Libr. J. **8**(12), 6 (2021)
3. Liu, G., Tsai, S.B.: A study on the design and implementation of an improved AdaBoost optimization mathematical algorithm based on recognition of packaging bottles. Math. Probl. Eng. **2022**, 1–11 (2022)
4. Dan, F., Bo, L., Jian, G.: An on-board task scheduling method based on evolutionary optimization algorithm. J. Circuit. Syst. Comput. **32**, 2350100 (2022)
5. Wang, Y., Shan, X., Wang, H., et al.: Ticket allocation optimization of fuxing train based on overcrowding control: an empirical study from china. Sustainability **14,** 7055 (2022)

Application of Data Mining Algorithm in Tourism Economy Development Under the Normalization of Epidemic

Yuexin Wang(✉)

College of Literature Science and The Arts, University of Michigan , Ann Arbor 48109, USA
yuexinwang322@163.com

Abstract. The role of data mining algorithms in developing the tourism economy is significant, but there is the problem of inaccurate economic forecasting. The historical economic analysis method cannot solve the problem of randomness in the tourism economy, and there are few indicators for tourism economic analysis. Therefore, this paper proposes a data mining algorithm method to construct a tourism economic development model. First of all, under the normalization of the epidemic, the crawler method to classify the data and carry out according to the stage of economic development. Collective division of economic data to reduce the influence of random factors. Then, the crawler method divides the tourism economic development score into grades, forms an economic data collection, and continuously analyzes the economic data. MATLAB simulation shows that under the normalization of the epidemic, the accuracy and analysis time of the data mining algorithm method are better than the historical economic analysis method.

Keywords: crawler method · tourism · economic development · data mining

1 Introduction

Data mining means economic research, which economic development under the normalization of the epidemic [1]. However, in the actual tourism economic development research process, there is a problem of poor accuracy [2], which brings certain obstacles to analyzing social and ecological orientation. Some scholars believe that applying data mining algorithms to the analysis of tourism economic development can effectively improve the accuracy research and provide data support for socio-ecological orientation research [3]. On this basis, this to accurately predict the development of tourism economy and verify the validity of the results.

2 Related Concepts

2.1 Mathematical Description of Data Mining Algorithms

The data mining algorithm uses crawler methods to mine tourism economic data, discovers tourism economic development results according to various indicators in tourism economic data [4], and ultimately accurately predicts the results of tourism economic

J. C. Hung et al. (Eds.): FC 2023, LNEE 1131, pp. 288–293, 2024.
https://doi.org/10.1007/978-981-99-9299-7_41

development. In this process, the intelligent of the crawler method [5], and use the tourism economic data scheme to carry out the results of the tourism economy development Comprehensive judgment can shorten the time for tourism economic development research [6].

Hypothesis 1: The tourism economic data is x_i the set of economic data results is $\sum x_i \neq 1$, the economic data indicators is z_i [7], the forecast criteria is ξ, and the economic forecast The function is $F(d_i \geq 0)$ as shown in Eq. (1).

$$F(d_i) = \sum x_i \to y_i \cdot \xi \quad (1)$$

2.2 Selection of Research Programs for Tourism Economic Development

Hypothesis 2: The tourism economic development scheme selection function is and the tourism economic development sequence table coefficient is $z(d_i)$, then the tourism economic development program selection process is w_i shown in Eq. (2):

$$z(d_i) = \frac{z_i \cdot F(d_i, y_i)}{w_i \cdot \xi} + \quad (2)$$

2.3 Processing of Tourism Economic Data

Before the data mining algorithm, the tourism economic data should be analyzed horizontally and vertically, and the economic data prediction results should be stored in the forecast table. First, the tourism economic data is comprehensively analyzed, and the economic data threshold is set to the data mining algorithm. If there are outliers in the travel economy data, drill down and adjust the predictors. The process of selecting a tourism economic development plan in Fig. 1.

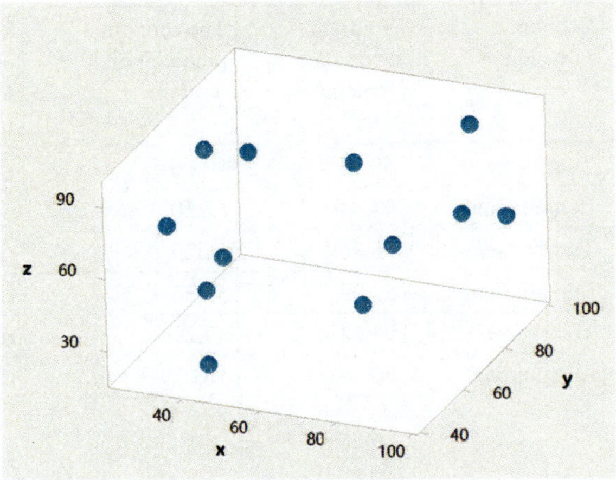

Fig. 1. The selection process of tourism economic development programs

The economic data in Fig. 1 shows that the tourism economic development scheme shows a horizontal and vertical distribution, representing the results of time and industry development. The tourism economic development research scheme is not directional, indicating that the tourism economic development plan is less affected by randomness, which can be used as a basis for later research. The longitudinal change of the research scheme of tourism economic development is mainly the crawler method to adjust the tourism economic development indicators and remove duplicate and irrelevant indicators, enhancing indicators' independence.

2.4 Research Strategies for Tourism Economy Development

The data mining algorithm method adopts the stochastic research strategy for the adjusts the corresponding parameters to realize the optimization of tourism economic development. The data mining algorithm method divides the into different periods and randomly selects different schemes. In the iterative process, the plans of different periods are matched. After the matching process is completed, the research different schemes is compared, and the best tourism economic development results are recorded.

3 Practical Cases of Tourism Economic Development

3.1 Introduction to the Psychological Program of Tourism Economic Development

In order to facilitate the research, this paper takes the tourism economic data under the normalization of the epidemic as the research object, and there are 13 economic indicators, the study duration was six weeks, and the processing of economic data are shown in Table 1.

Table 1. Relevant parameters of physical education tourism economic development

Survey scenarios	Economic direction	Tourism economic structure	The economic potential of tourism	Tourism economic development trends
Historical statistics	Transverse	58.93	30.36	rise
	Longitudinal	94.64	62.50	rise
Actual investigation	Transverse	55.36	64.29	rise
	Longitudinal	51.79	55.36	rise
Theoretical analysis	Transverse	94.64	39.29	rise
	Longitudinal	30.36	16.07	rise

The economic data processing process of tourism economic development in Table 1 is shown in Fig. 2.

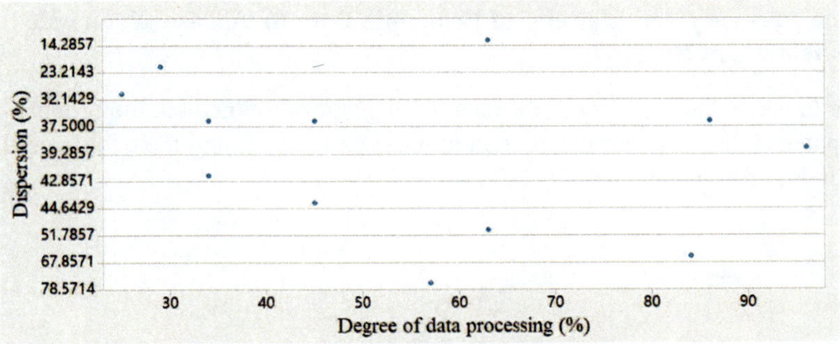

Fig. 2. Processing process of tourism economic data

It Table 1 that compared with previous statistical methods, the research results of tourism economic development of data mining algorithm method are closer to the actual research Economic data. In terms of the rationality and fluctuation range of economic data selection in the study, data mining algorithm methods have been statistical methods in the past. The changes in economic data in Fig. 4 show that the data mining algorithm method has better stability and faster judgment speed. Therefore, the research speed of tourism economy development and the research and summation stability of tourism economy data and development are better based on data mining algorithms.

3.2 Development of the Tourism Economy

The research on tourism economic development includes non-structural information, semi-structural information and structural information. After the pre-selection of the data mining algorithm method, the preliminary tourism economic data of tourism economic development is obtained, and the tourism economy development is developed The feasibility of tourism economic data is analyzed. In order to more accurately verify the evaluation effect, the economic data economic development and tourism economic development in different periods are selected. Evaluate the economic data, as shown in Table 2.

Table 2. Overall picture of tourism economic data

Period	Socio-ecological orientation analysis	The completeness rate of economic data programs
1 ~ 4 weeks	46.43	89.29
5 ~ 8 weeks	82.14	26.79
Longitudinal	85.71	85.71
mean	56.43	56.43
X^2	6.71	4.07
P = 0. 035		

3.3 The Accuracy and Stability of Economic Data in Tourism Economic Development Research

In order to verify the accuracy of the data mining algorithm method, the economic data is compared with the historical economic analysis method, and the economic data is shown in Fig. 3.

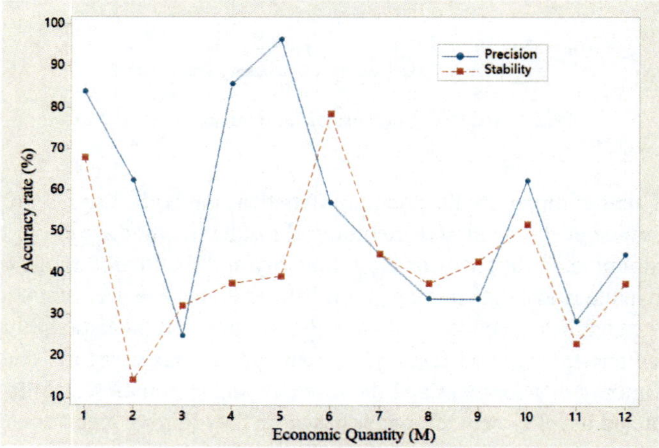

Fig. 3. The accuracy of the data algorithm method

It from Fig. 3 that the of the data method is higher than that of the historical economic analysis method. However, the error rate is lower, indicating that the method is used in the research of tourism economic development It is relatively stable, while historical economic analysis is uneven in the research. The average economic data of the above three algorithms is shown in Table 3.

Table 3. Comparison of the accuracy of tourism economic development research by different methods

algorithm	Precision	Magnitude of change	error
Data mining algorithmic methods	78.57	84.64	98.21
Historical economic analysis	61.07	56.79	78.93
P	0. 014	0. 043	0.017

Table 3 that the historical economic analysis method has shortcomings in accuracy and stability in the tourism economy, data of tourism economic development research have changed significantly, and the error rate is high. The accuracy of the data mining algorithm method is higher than that analysis method. At the accuracy of the data mining algorithm method is greater than 90%, and the accuracy has not changed significantly.

The superiority of data mining. The data mining algorithm is randomized by different methods, as shown in Result 4.

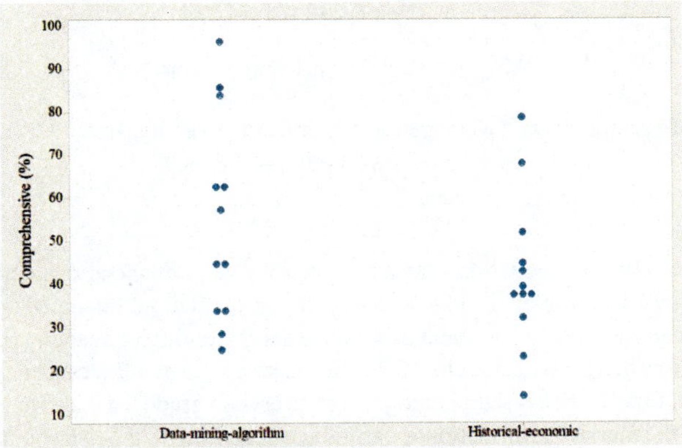

Fig. 4. Comprehensive evaluation of tourism economic development

Figure 4 that the economic data of the data mining algorithm method is significantly better than the historical economic analysis method, and the reason is that the data mining algorithm method increases the regulation of tourism economic development coefficients and set the corresponding maximum value to exclude economic data that does not meet the requirements.

4 Conclusion

Under the normalization of the epidemic, in view of the problem of tourism economic development, combined with reptilian methods, to develop the tourism economy and make improvements. At the same time, the tourism economic development plan and maximum value standard are analyzed in depth, and the economic data collection of tourism economic development is constructed. The research shows that the data mining the accuracy and stability, and can conduct comprehensive research. However, in the data mining algorithm, the analysis economic development research ability, resulting in a relative decline in accuracy.

References

1. Luo, X.: Design of efficient search system for English translation terms in massive internet data. Modern Electron. Technol. **40**(13), 134–136 (2017)
2. Li, F.: Design of high similarity English word selection system. Modern Electron. Technol. **40**(23), 147–150 (2017)
3. Liu, B., Li, X.: Design and application of meteorological information query system based on Silverlight. Meteorolog. Sci. Technol. **44**(3), 369–373 (2016)
4. Li, L., Meng, X., Liu, L.: Design and implementation of computer aided system based on code automatic generation. Aerospace Control **33**(4), 68–72 (2015)
5. Huang, X.: Reliability Engineering. Beijing: Tsinghua University Press, pp. 2–15 (1990)

A Massive Financial Risk Data Fusion Method Based on the Bayesian Network

Wanting Wu(✉) and Jishan Piao

School of Economics and Management, Yanbian University, Jilin 133000, China
wuwanting0919@163.com

Abstract. Bayesian networks play an important role in financial risk data analysis, but there is a problem of low accuracy of risk prediction. Financial statistical analysis cannot solve the problem of accurate early warning in financial risks, and there are few early warnings. Therefore, this paper proposes a Bayesian network to construct. Firstly, the big data mining theory is used to grade the massive information, and the massive information is carried out according to the risk standards Set division to reduce ambiguity in early warning. Then, the big data mining theory grads the financial risk early warning forms a collection of early warning results and continuously warns the massive information. MATLAB simulation shows that under a certain amount of financial data, the Bayesian network's early warning accuracy and warning time are superior to financial statistical analysis.

Keywords: big data mining theory · Early warning of financial risks · Bayesian networks · Early warning results

1 Introduction

Financial one of the essential assessment contents in financial [1], great significance to financial risk [2]. However, in the early warning process, there is a problem of poor warning time in the early warning results [3], which has a certain impact on the financial risk data analysis. Some scholars believe that applying intelligent algorithms to financial risk early warning can effectively carry out audit and accounting data analysis and provide corresponding support for early warning [4]. This paper proposes a Bayesian network of financial risk data analysis and verify the model's effectiveness.

2 Related Concepts

2.1 Mathematical Description of Bayesian Networks

The bayesian network uses big data mining theory to optimize massive information and, according to various indicators in financial risk, find outliers in risk early warning and integrate massive information to make final judgments regarding the feasibility of early warning results. The Bayesian network combines big data mining theory [5], uses information mining and intelligent algorithms, optimizes the financial risk, and can improve the accuracy of intelligent risk prediction [6].

J. C. Hung et al. (Eds.): FC 2023, LNEE 1131, pp. 294–299, 2024.
https://doi.org/10.1007/978-981-99-9299-7_42

Hypothesis 1: The massive information of financial risk is x_i, the collection of massive information is $\sum x_i$, the professional audit of financial risk is y_i, and the judgment function of financial risk early warning is $f(x_i)$ as shown in Eq. (1).

$$f(x_i) = \bigcup_{\xi} \sum \sum x_i | y_i \tag{1}$$

ξ Adjust the coefficient for financial risk early warning results, reducing the impact of ambiguity and slang.

2.2 Selection of Early Warning Methods

Hypothesis 2: The mass information selection function is $F(x_i)$ and the weight coefficient of the massive information is k_i [7], then the risk early warning method selection is shown in Eq. (2).

$$F(x_i)=z_i \cdot f(x_i|y_i) + \partial k_i \cdot \xi \tag{2}$$

2.3 Processing of Massive Information

Before performing Bayesian analysis, the massive information in the early warning results should be analyzed standardly, and the massive information should be mapped to the early warning table to determine the semantic anomalous massive information. First, the massive information is comprehensively analyzed, and the threshold and weight of the massive information are set to ensure the accuracy of the Bayesian network. Massive information is the semantic transformation of massive information that must be standardized. If the massive information is in a non-standard distribution, its early warning results will be affected, reducing the accuracy of the overall warning. In accuracy

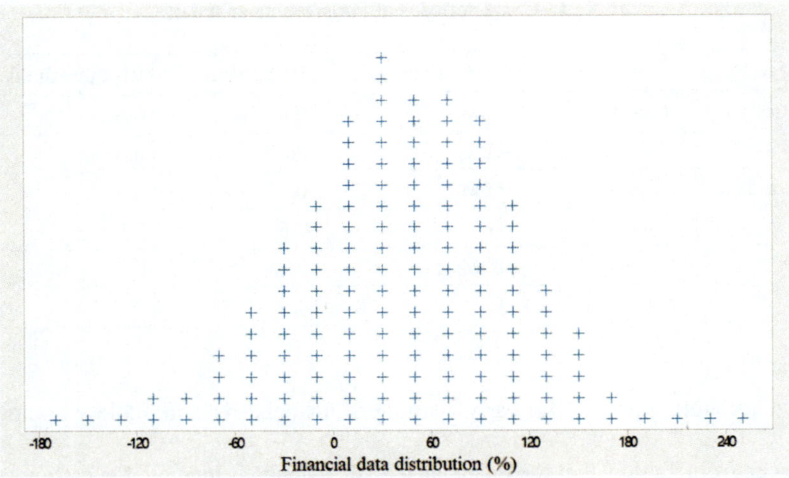

Fig. 1. Results of the selection of financial risk early warning methods

of Bayesian networks and the level of early warning, the method should be selected, and the specific method selection is shown in Table 1.

The massive amount of information in Fig. 1 shows that the financial warning methods show a diversified distribution, which aligns with the objective facts. The early warning method has no directionality, indicating that the strong accuracy, so it is used as early warning research for financial risk. The early warning method meets the mapping requirements, mainly because the big data mining theory adjusts the early warning method, removes duplicate words, and revises the slang so that the dynamic accuracy of the entire massive information is strong.

2.4 Depth of Financial Risk Warning

Bayesian network adopts accurate judgment for financial risk early warning and adjusts the corresponding word relationship to optimize financial risk data analysis methods. Bayesian networks divide financial risks into different amounts of data and randomly select different methods. In the iterative process, the risk criteria of different data volumes are matched with the early warning method. After the matching process is completed, compare different financial risk data analysis methods, and record the financial risk early warning results with the highest accuracy.

3 Early Warning Cases of Financial Risk Data Analysis

3.1 Financial Risk Situation

In order to facilitate financial data analysis, the audit data in this paper is the research object, the test data is 1G, and the financial risk warning of specific physical education is shown in Table 1.

Table 1. Characteristics of massive financial data

data type	Time	Amount of data	Dispersion	Risk thresholds
Statistics	Early	500M	0.85	0.6
	year-end	1G	0.65	0.6
Monitor data	Early	500M	0.75	0.6
	year-end	1G	0.69	0.6
Report data	Early	500M	0.75	0.6
	year-end	1G	0.92	0.6

The accounting process for early warning of financial risks in Table 1 is shown in Fig. 2.

It seen from Table 1 that the financial statistical analysis method, the early warning results of Bayesian networks are closer to the actual. In terms accounting rate and

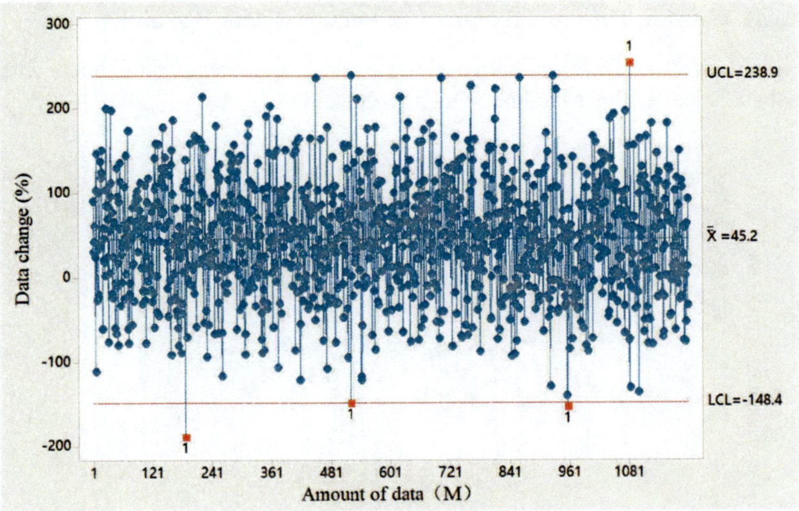

Fig. 2. Processing process of financial data

accuracy, the Bayesian network financial statistical analysis method. The accounting changes in Fig. 4 show that Bayesian networks are more accurate and faster to judge. Therefore, the Bayesian network has better speed, and early warning accuracy.

3.2 Identification Rate of Outliers in Financial Risk Early Warning

Financial Risk Warning contains phrases, sentences, and articles. After the rough set screening of the Bayesian network [21], the preliminary results and the results are obtained. The feasibility is analyzed. In order to verify the effect more accurately, select the financial risk early warning with different data volumes, and the accounting evaluation accounting is shown in Table 2.

Table 2. Overall accounting situation

Amount of data	The amount of financial data	Outlier recognition rate
audit	50.00	90.71
accounting	67.86	92.14
Data simplification rate	39.29	95.36
mean	39.29	93.57
X^2	9.642	78.571
P = 0. 082		

3.3 Early Warning Time and Accuracy of Financial Risk Warning

In the accuracy of the Bayesian network, the accounting comparison is made with the financial statistical analysis method, which is shown in Fig. 3.

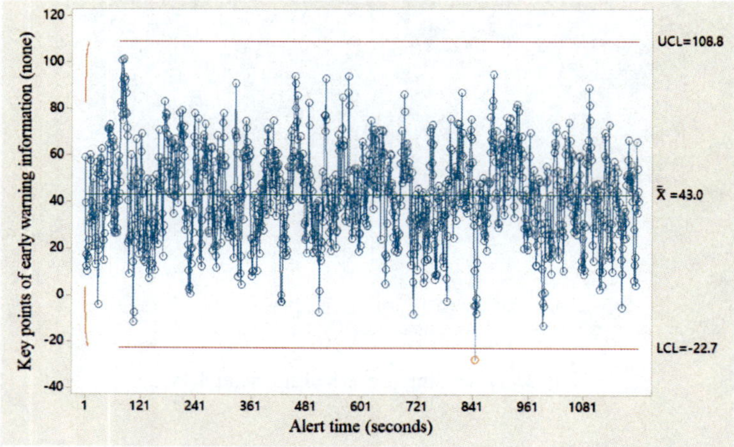

Fig. 3. Early warning time for different algorithms

Figure 3 that the early warning time of the Bayesian network is higher than that of the financial statistical analysis method, but the error rate is lower, indicating that the early warning of the Bayesian network is relatively stable, while that of the financial statistical analysis method The accuracy of early warning is uneven. The accuracy of the above algorithm is shown in Table 3.

Table 3. Comparison of early warning accuracy of different methods

algorithm	Warning time	Magnitude of change	error
Bayesian networks	80.36	91.07	92.86
Financial statistical analysis	92.86	64.29	60.71
P	6.423	1.027	4.64

Table 3 that the financial statistical analysis method has deficiencies in the time and accuracy, and the accuracy of financial risk early warning There are large changes and a high error rate. The comprehensive of Bayesian networks is higher, which is better than that of financial statistical analysis. At the same time, the early warning time of the Bayesian network is greater than 90%, and the accuracy has not changed significantly. In order to further verify the superiority of Bayesian networks. In further the continuity, the Bayesian network was comprehensively analyzed by different methods, as shown in Result 4.

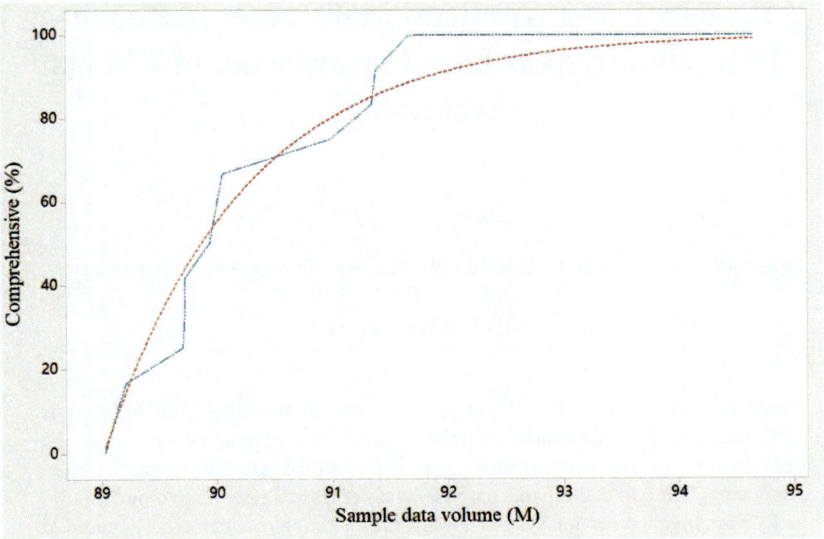

Fig. 4. Comprehensive accounting for evaluation of early warning results by different methods

Figure 4 that the accounting of Bayesian networks is significantly better than the financial statistical method, and the reason is that Bayesian networks increase the adjustment coefficient and set thresholds, judging the accountant that does not meet the requirements.

4 Conclusion

Under the prevalence of intelligent methods, this paper proposes Bayesian networks for financial risk data analysis, combining big data mining theory to improve. At the same time, the departments and threshold standards of financial risk early warning are analyzed in depth, and the accounting collection is constructed. Research shows that Bayesian networks can and accuracy and can synthesize. However, in Bayesian networking, too much attention is paid to analyzing massive data processing capabilities, resulting in a relative decline in supervision accuracy.

References

1. Luo, X.: Design of efficient for English translation terms in massive internet data. Mod. Electron. Technol. **40**(13), 134–136
2. Li, F.: Design of high similarity English word selection system. Mod. Electron. Technol. **40**(23), 147–150 (2017)
3. Liu, B., Li, X.: Design and application of meteorological information query on Silverlight. Meteorol. Sci. Technol. **44**(3), 369–373 (2016)
4. Li, L., Meng, X., Liu, L.: Design and implementation of computer aided system based generation. Aerospace Control **33**(4), 68–72 (2015)
5. Huang, X.: Engineering, pp. 2–15. Tsinghua University Press, Beijing (1990)

Research and Application of Lane Intelligent Detection System Based on Internet of Things Technology

ZhiWen Xiong(✉)

GuangXi Vocational Normal University, Guangxi Zhuang Autonomous Region, Nanning 530007, China
312252026@qq.com

Abstract. The role of IoT technology in lane detection is significant, but there is a problem with low detection accuracy. Manual route planning cannot solve the problem of lane detection in route selection, and there are fewer lane detection indicators. Therefore, this paper proposes an Internet of Things technology to build a lane intelligent detection system. Firstly, the operation research theory is used to draw the route of the lane data, and the lane collection is divided according to the traffic flow standard to reduce the uncertainties in lane detection. Then, according to the operation research theory, the planning path is drawn to form a lane planning set. MATLAB simulation shows that under the condition of a certain traffic flow, the detection accuracy and detection time of IoT technology are better than that of manual route planning method.

Keywords: intelligence · technology · path selection · Lane detection · Internet of Things

1 Introduction

Lane contents of road planning and is of great significance to the development of transportation [1]. However, in the actual lane detection process, there is a problem of poor detection accuracy, which brings certain economic losses to transportation enterprises [2]. Some scholars believe that the application of intelligent algorithms to route selection can accurately analyze the lane situation and provide corresponding support for route planning. On this basis, this paper proposes an optimize the accuracy of lane detection and verify the effectiveness of the model.

2 Related Concepts

2.1 Mathematical Description of IoT Technology

The Internet of Things technology uses operations research theory to optimize lane data, and according to various indicators in the lane data, find outliers in the path selection results, and integrate the corresponding lane data to make a final judgment Feasibility of path selection results [3]. IoT technology combines the advantages of operations research theory to quantify the path selection results using lane data.

J. C. Hung et al. (Eds.): FC 2023, LNEE 1131, pp. 300–306, 2024.
https://doi.org/10.1007/978-981-99-9299-7_43

Hypothesis 1: The lane data is set_i, the lane collection is d_i, the lane detection index of the route selection result is y_i, and the lane detection lane data judgment function is $F(d_i \geq 0)$ As shown in Eq. (1).

$$F(d_i) = \sum x_i \ni y_i \times \xi \tag{1}$$

2.2 Selection of Bright Lane Planning Scheme

Hypothesis 2: The lane detection scheme selection function is, and the weight coefficient of the lane detection scheme is $z(d_i)$ [7], then the path selection scheme w_i selection is shown in Eq. (2).

$$z(d_i) = z_i \cdot F(d_i, y_i) + \int w_i \cdot \xi \tag{2}$$

2.3 Processing of Smart Lane Data

Before the implementation of IoT technology, the lane data of the lane detection results should be analyzed discretely, and the lane data should be mapped to a two-dimensional plane to eliminate abnormal data. First, the route selection lane data is comprehensively analyzed, and the threshold and index weight of the lane data is accuracy of the results. The routing lane data is the system test lane data and needs to be standardized [4]. If the route selection lane data is in a nonnormal distribution, its lane detection results

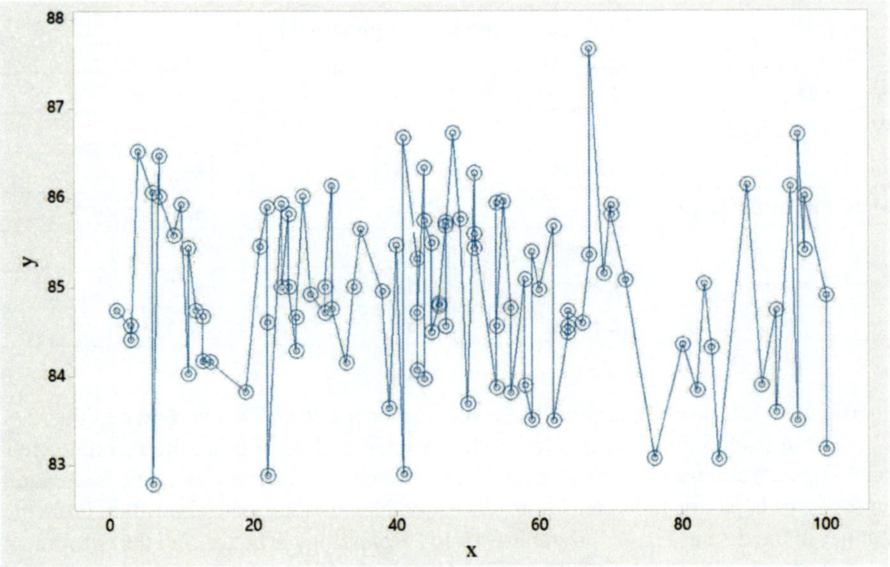

Fig. 1. Selection results for the path selection scheme

are affected, reducing the accuracy of overall lane detection. In order to improve the accuracy of IoT technology and improve the level of lane detection, the path selection scheme should be selected, and the specific scheme selection is shown in Fig. 1.

The survey lane data show that the path selection scheme shows a discrete distribution, which is in line with the objective facts. The lane detection scheme has no directionality, indicating that the path selection scheme has strong randomness, so it is used as an analytical study in science. The lane detection scheme meets the normal requirements [5]. Mainly the operation research theory adjusts the lane detection scheme, removes the duplicate and irrelevant schemes, and supplements the default scheme to make the operation research of the entire lane data Strong correlation [6].

3 Strategies for Thoughtful Lane Planning

IoT technology adopts a random strategy for path selection and adjusts the corresponding parameters to optimize the path selection scheme. IoT technology divides the path selection into different periods and randomly selects solutions. The path selection scheme for different time periods is matched in the iterative process. After the matching process is completed, the lane detection levels of different schemes are compared, and the best route selection results are recorded.

4 Practical Examples of Path Selection

4.1 Introduction to Route Selection Lane Data

In order to facilitate lane detection, the 122 universities in this paper selected lane data as the research object, and the specific results are shown in Table 1.

Table 1. Parameters for path selection

Plan scope	type	Plan for effects	Smart effect
Four rows of roads	Cement	88.66	87.63
	asphalt	91.75	86.60
Multi-row roads	Cement	82.47	82.47
	asphalt	91.75	87.63
Single-row roads	Cement	83.51	91.75
	asphalt	85.57	83.51

The lane data processing process for route selection in Table 1 is shown in Fig. 2.

As seen from Table 1, compared with the static path algorithm, the results of IoT technology are closer to the actual requirements. Regarding the rationality and fluctuation range of lane detection index selection, it is better than the static path algorithm. From the changes in the data in Fig. 4, the stability of IoT technology is better, and the enrollment speed is faster. As a result, IoT technology has better computing speed, accuracy and stability.

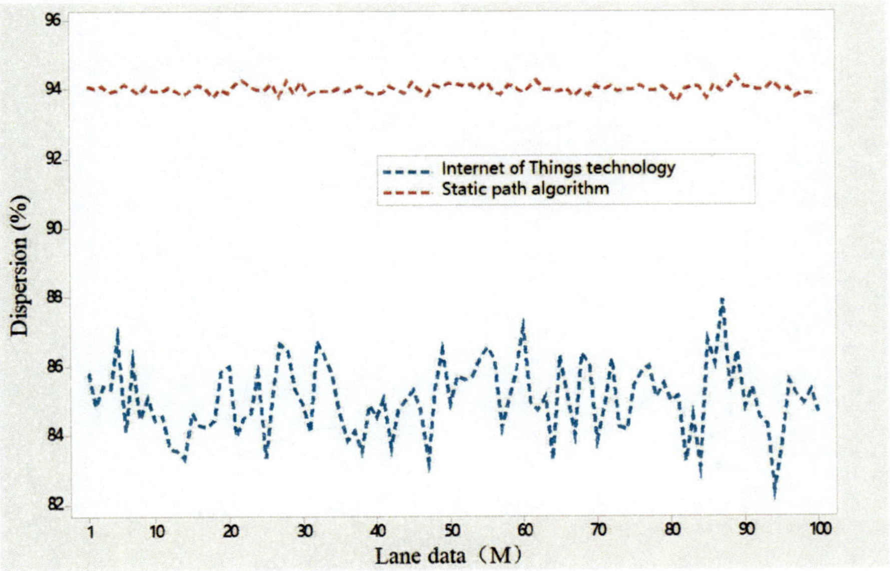

Fig. 2. Processing of lane data

4.2 Smart Lane Planning

Route selection lane planning, including the main urban area, suburbs, and surrounding urban areas. After the pre-selection of Internet of Things technology [21], the preliminary path selection results are obtained, and the enrollment results are lane data The feasibility is analyzed. To verify the path selection effect more accurately, select the route selection scheme for different time periods, as shown in Table 2.

Table 2. Overall status of path selection schemes

type	Compliance rate	Complete rate
4 rows of roads	91.75	85.57
Special Admissions	82.47	80.72
Multi-row roads	82.47	81.63
mean	83.51	81.60
X^2	0.021	0.012
P = 0.031		

4.3 Accuracy and Stability of Lane Detection Scheme

In verify IoT technology, the lane data is compared with the manual route planning method, and the lane data is shown in Fig. 3.

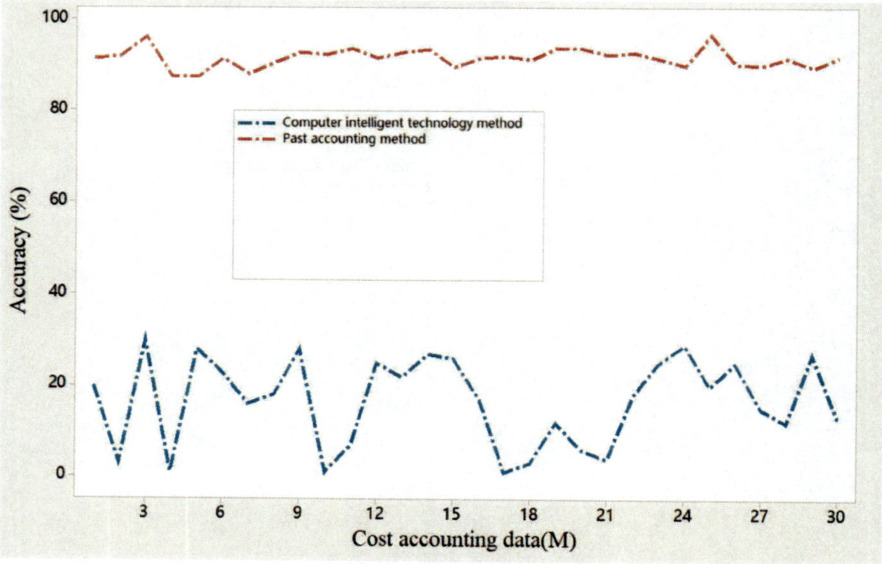

Fig. 3. The accuracy of different algorithms

Figure 3 that the accuracy of IoT technology the manual route planning method, but the error rate is lower, indicating that the lane detection of IoT technology is relatively stable, while the lane detection of manual route planning method is Uneven. The average lane data of the above three algorithms are shown in Table 3.

Table 3. Comparison of the accuracy of lane planning

algorithm	Precision	Magnitude of change	error
Internet of Things technology	89.69	5.57	0.72
The manual route planning method	91.75	8.66	8.14
P	1.751	7.163	8.23

Table 3 that the manual route planning method has shortcomings in accuracy and stability in terms of path selection, the lane detection scheme has changed greatly and the error rate is high. The accuracy of the comprehensive results of IoT technology is higher, which is better than the manual route planning method. At the same time, the accuracy of IoT technology is greater than 90%, and the accuracy has not changed significantly. The effectiveness, different methods are used to comprehensively analyze the Internet of Things technology, and the result 4 is shown.

From Fig. 4, it can be seen that the lane data of IoT technology is significantly better than the manual route planning method, and the reason is that the Internet of Things

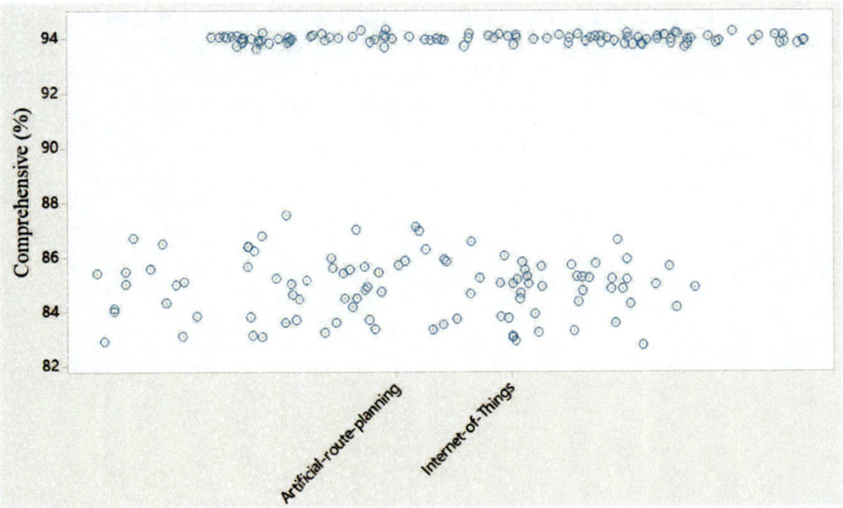

Fig. 4. Comprehensive planning results of lane detection

technology increases the path selection adjustment coefficient and sets the corresponding threshold to reject lane data that does not meet the requirements.

5 Conclusion

Under high urban lane pressure, this paper proposes and optimizes the path selection by combining the operation research theory. At the same time, the departments and threshold standards of route selection are analyzed in depth, and the lane collection of route selection is constructed. Research shows that improve the accuracy and stability of path selection and carry out comprehensive lane detection for route selection. However, in the Internet of Things technology analysis process, too much attention is paid to the analysis of lane detection capabilities, in a relative decline in monitoring.

Acknowledgements. Research on Key Technologies of Beidou based RTK high precision positioning service platform (Project No: 2022KY0901).

References

1. Lv, X., Li, M.: Application and research of the intelligent management system based on internet of things technology in the era of Big Data. Mob. Inf. Syst. **2021**(16), 1–6 (2021)
2. Qiao, X., Liu, X., Zheng, P., et al.: Research on the application of intelligent detection technology in business hall system (2021)
3. Yue, S., Du, Y., Zhang, X.: Research and application of agricultural internet of things technology in intelligent agriculture. J. Phys. Conf. Ser. **1769**(1), 012020 (2021)

4. Dong, M., Zhao, J., Li, D.A., et al.: ISEE: industrial internet of things perception in solar cell detection based on edge computing. Int. J. Distrib. Sens. Netw. **17**(11), 21–856 (2021)
5. Zhu, F., Shankaran, R., Zhao, B., et al.: Anomaly detection-based intelligent computing in internet of things and network applications. Internet Technol. Lett. **4**(3) (2021)

Research on Financial Risk Discrimination of Listed Companies Based on Fish School Algorithm

Shi Yan(✉)

Criminal Investigation Police University of China, Shenyang 110854, Liaoning, China
yanshi@cipuc.edu.cn

Abstract. Because the financial risk data of domestic listed companies are unbalanced, and the traditional supervised model is difficult to establish a risk discrimination model through large sample learning. Due to the lack of data, the training of supervised models is usually based on manually filtered balanced data sets, which will affect the applicability of models in the real world. This research involves financial risk identification based on fish school algorithm. The this study is to compare the results of two algorithms (i.e. Fish School and Fuzzy Logic) to distinguish companies with different levels of financial risk. The comprehensively analyze the performance of two different methods in screening in PSX.

Keywords: Financial risk · Fish school algorithm · listed company · Risk screening

1 Introduction

It has successively exposed such major financial risk cases as Enron, WorldCom, the most trusted Xerox and other well-known overseas companies, as well as the domestic blue field company, Wanfu Biotech, Tianfeng Energy Saving Company, which are known as excellent companies, to conceal or delay the disclosure of information, cheat investors and cause huge losses. The amount involved in the case is huge, the number of involved companies and intermediary chains is large, and the system risk is so high that it is shocking [1].

Financial annual reports issued by regular reports, interim reports, prospectuses, listing memorandums, listing announcements, etc. Information disclosure documents introduce the listed companies' and other activities. These reports provide valuable reference for investors, creditors and other end-users of accounting information. However, carried out by enterprises to raise funds from investors and support from financial institutions. Companies related party, illegal stock transactions, etc., which reduces the effectiveness [2].

Although our government has issued a series of measures to deal with accounting risks, it uses the authenticity, integrity and timeliness of information disclosure to avoid financial risk behavior. There risk to information fraud [3]. The way of accounting

J. C. Hung et al. (Eds.): FC 2023, LNEE 1131, pp. 307–313, 2024.
https://doi.org/10.1007/978-981-99-9299-7_44

is also changing quietly. The scope of financial statement risks has expanded. Through systematic fraud, managers have formed distorted and difficult to identify financial statements, created high-risk financial false performance, and many new problems and new situations have emerged.

2 Related Work

2.1 Financial Situation

In American accounting circles, the balance sheet is often. The definition status in the Economic Dictionary also refers to the source and distribution of operating funds during a certain period. In the current generally accepted definition of balance sheet, it is also often considered that an enterprise at a specific time point. Obviously, the "financial situation" here refers to the balance sheet situation, that is, all the contents included in the balance sheet [4].

In addition to the balance sheet, the financial statements of enterprises mainly include. The reflect the change process of the "undistributed profits" item in the balance sheet. The amount of "undistributed profits" in the must be equal to the amount of "undistributed profits" in the item of owner's equity in the balance sheet; The reflects the change process of the balance of cash equivalents in "monetary capital" and "trading financial assets" in the balance sheet. It can be seen from this that in the financial statement system of an enterprise, the is a core statement that comprehensively reflects the enterprise. Other statements explain the change process of some important items [5]. The explains the change process of the current quantity of "undistributed profit" items in the balance sheet, The cash flow statement is an explanation of the current quantity change process.

Therefore, the definition of financial status in this paper is that financial status refers to the balance sheet status, which reflects the assets and equity of an enterprise at a certain point in time, but its content covers. The analysis of an enterprise's financial situation should not only examine, but also combine the income, cash flow statement, etc.

2.2 Research Status of Financial Risk Model

In the field of finance, risk is a common and important concept, which is used to express the uncertainty of future results. Financial risk usually represents the possibility of financial distress. Altman (1993) linked financial distress with business failure, and used four words, failure, insolvency, bankruptcy and default, as synonyms of financial distress. Enterprise financial distress is not only a static concept, but also related to time. Generally speaking, financial distress can refer to the risk of bankruptcy of a company due to financial difficulties, or it can refer to the state of near bankruptcy, or simply refers to the company's bankruptcy. Traditional default prediction or risk assessment models especially favor the use of observable legal events to judge whether financial distress has occurred, which is more objective and accurate in time. The literature of Ooghe et al. (2006) is like this. They selected the failed enterprises and the non failed enterprises, and used the binary selection model to distinguish the two groups, thus obtaining the probability of the company's failure.

Logistic model is often used in domestic literature on "special treatment" companies. Wu Shinong used Fisher linear judgment, multiple linear regression analysis and Logistic regression to establish an indicator model, of which the Logistic model has the lowest probability of miscalculation in predicting financial distress. Jiang Guohua also built a Logistic model, which includes factors such as the profit margin of main business and the shareholding ratio of the largest shareholder, and has good forecasting ability. In addition to comparing multiple discriminant and logistic models, Hu Yuancheng innovatively introduced the model in the research of financial distress of domestic listed companies, and found that the method is more effective in forecasting.

3 Fish School Algorithm

In a piece of water, the place with the most nutrients should be the place with the largest number of fish. According to this feature, you can imitate the behavior of a school of fish, such as foraging, to find the global optimal, which is the basic idea of the fish school algorithm.

The basic activities of a fish school include, herding behavior,. When solving practical optimization problems, these behaviors are closely related to optimization problems. How to construct these basic behaviors is the main problem faced by the algorithm.

The foraging behavior is a kind of behavior that moves towards the place with high food concentration. In the process of optimization, it can be considered as moving towards a better direction.

In the behavior of gathering, learn from Reynolds! 30 There are two rules for each artificial fish:

(1) Move to the partners as far as possible;
(2) So as to achieve the ability of artificial fish to flock;

The tail chasing behavior in the process of optimization can be understood as the process of advancing to the best partner in the field of vision.

Through the above description, each artificial fish will explore its current partner status and environmental status, and choose the behavior to develop in a better direction according to its own information. Finally, the artificial fish will gather around the local extreme value. In general, taking the maximum problem as an example, the artificial fish with large food concentration values are generally located around the extreme value region with large values, and there are usually more artificial fish around the extreme value region with large values, which will help to judge and obtain the global optimal value.

Artificial fish swarm algorithm does not require high initial values of variables. Generally speaking, initial fish swarm values are generated randomly within the range of variable values; The termination condition of the algorithm can be determined according to the actual situation. For example, the common method is to determine that the accuracy of the difference between the current bulletin board Y (Gen) and the previous generation bulletin board Y (Gen-1) is less than a specific value ε, Or directly specify the number of iterations Max_ Gen.

The flow chart of the is shown in Fig. 1.

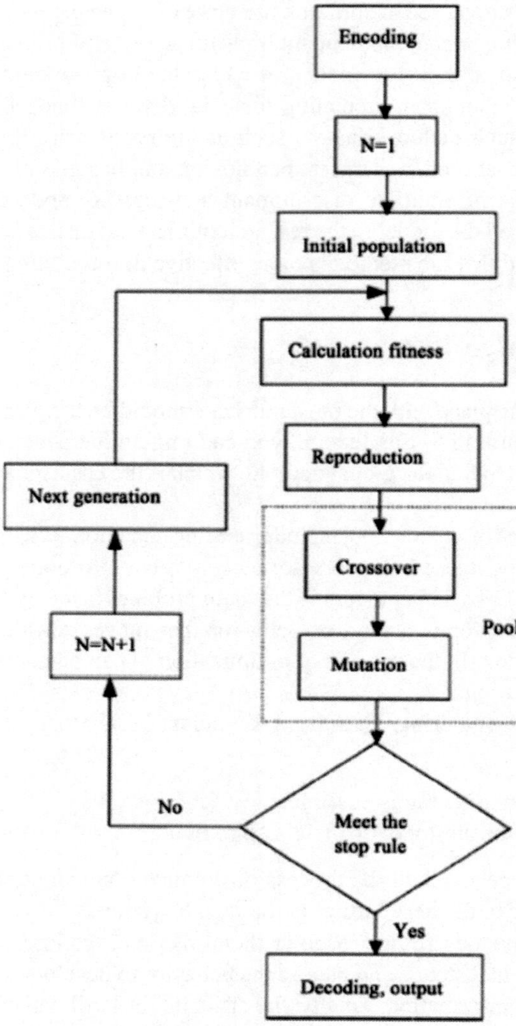

Fig. 1. Flow chart of artificial fish swarm algorithm

4 Research on Financial Risk Screening of Listed Companies Based on Fish School Algorithm

The fish school algorithm is an algorithm based on biological phenomenon observation, which can simulate the behavior of fish schools and provide effective reference when solving problems. In the financial field, research on financial risk screening based on fish swarm algorithm is gradually becoming a hot topic. This article explores the research on financial risk screening of listed companies based on fish school algorithm.

Firstly, the characteristic of the fish school algorithm is its ability to simulate collaborative behavior in fish schools, and it can quickly find the optimal solution through simulation and optimization of algorithms. Therefore, the advantage of financial risk screening research for listed companies based on the fish school algorithm lies in its ability to screen and predict risks in a more intelligent and efficient manner, which is conducive to timely discovering whether the company has any potential risks and improving the efficiency and quality of risk management.

Secondly, research on financial risk screening based on the fish school algorithm can use a large amount of financial data and indicators, including balance sheets, income statements, cash flow statements, etc., for in-depth analysis and mining. These financial data can effectively reflect the company's business and financial situation, help identify potential risks, and improve the accuracy and comprehensiveness of risk management.

Finally, the study of financial risk screening for listed companies based on the fish school algorithm requires the establishment of a reasonable data mining model. This model needs to consider many indicators and their relationships, and use data analysis and data modeling techniques to determine the feasibility and effectiveness of the model. At the same time, researchers need to dynamically monitor and predict financial risks in conjunction with changes in industry and national policies, in order to implement timely prevention, control, and tracking management of risks.

In summary, the research on financial risk screening of listed companies based on fish school algorithm can effectively improve the ability and level of risk management, providing strong support for the healthy development of enterprises. This provides new directions and ideas for exploring and applying the potential of fish school algorithms to explore more valuable results in the financial field.

The discriminant analysis is performed on new samples of unknown categories to predict the category to which the new samples belong. The independent variable can be a variable of any scale, and the dependent variable can be an ordered or unordered classified variable. Assumptions of classical discriminant analysis:

(1) Each discriminant variable is independent, cannot be a linear combination of other discriminant variables, and has no collinearity problem.
(2) The covariance matrix of each group of variables is equal.
(3) Among the discriminant variables, there is a multivariate normal distribution population.
(4) The control variable is an unordered or ordered classified variable.

Discriminant analysis model function expression:

$$Y = \alpha_1 X_1 + \alpha_2 X_2 + \ldots + \alpha_n X_n \tag{1}$$

The financial risk screening model for listed companies based on the fish school algorithm is shown in Fig. 2.

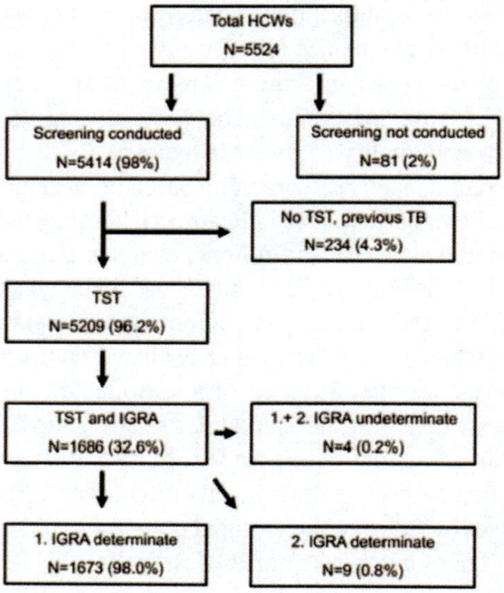

Fig. 2. Financial Risk Screening Model for Listed Companies

5 Conclusion

With the economy, according to China's national conditions, we should establish sound accounting standards and accounting systems, gradually integrate accounting policies with international accounting standards, improve the financial statement information disclosure system, and strengthen the construction of the national economic legal system. Strengthen enterprise risk management and governance, monitor financial statement risks, and form a stable and strong financial risk system. Strengthen the internal control system of enterprises. The quality of enterprise assets enterprises. Optimize the asset structure, maintain the virtuous circle of assets, promote the metabolism of assets, and maintain the healthy development of enterprises. And other emerging technologies. Strengthen the development and AI and the application of statistical technology, timely understand the technology trends and development trends, and continue the application of new technologies in regulatory business scenarios.

References

1. Jérme, D.: Systems and methods for generating training dataset for discrimination of fish species. EP3761057A1 (2021)
2. Zhang, X.: Research on recognition of accounting information distortion of listed companies based on classification regression tree algorithm. In: Hung, J.C., Yen, N.Y., Chang, J.W. (eds.) FC 2021. LNEE, vol. 827, pp. 1546–1552. Springer, Singapore (2022). https://doi.org/10.1007/978-981-16-8052-6_225
3. Jérme, D.: Systems and methods for generating training dataset for discrimination of fish species. EP3761059A1 (2021)

4. Zeng, S., Li, Y., Li, Y.: Research on audit opinion prediction of listed companies based on sparse principal component analysis and kernel fuzzy clustering algorithm. Math. Probl. Eng. **2022** (2022)
5. Li, M., Sun, Y., Wang, X., et al.: Research on the model of UBI car insurance rates rating based on CNN-softmax algorithm. J. Phys. Conf. Ser. **1802**(3), 032071 (12p.) (2021)

Research on the Application of Decision Tree Algorithm in the Teaching System

Lei Yang(✉)

Tianjin Academy of Fine Arts, Tianjin 300020, China
52067315@qq.com

Abstract. The role of teaching management in teaching management is very important, but there is a problem of low management level. The teaching management system cannot solve the problem of multi-type lesson plan management in the teaching management system, and the matching is low. Therefore, this decision tree a teaching management. First, the teaching plan is classified according to the teaching plan and content, and the teaching plan is selected according to the teaching plan to realize the standardized processing of teaching information. Then, according to the classification of teaching content, a collection of teaching information is formed, and the teaching content is iteratively analyzed. MATLAB simulation shows that the decision tree the standard management method in optimizing the teaching management system in terms of optimization degree and optimization time under the condition that the teaching plan remains unchanged.

Keywords: teaching management · management systems · Decision tree method · Optimize

1 Introduction

Teaching optimization is one of the important contents of teaching management and plays a significant role in improving the teaching system [1]. However, in the process of teaching system construction, teaching management has the problem of low management level and cannot effectively play the role of teaching management. Some scholars believe that applying intelligent methods such as decision trees to the teaching management system can quickly classify teaching plans, save teaching management time analysis [2], and support teaching optimization. On this basis, this paper proposes a decision tree method to optimize the teaching management system and verify the effectiveness of the model implementation.

2 Related Concepts

2.1 Mathematical Description of the Decision Tree Method

The making tree method is to set teaching objectives, teaching plans and teaching plans, make decisions on teaching information, and discover low-quality teaching plans in teaching according to teaching management indicators, and form teaching plan rectification plans [3]. The correlation between lesson plans is analyzed by integrating the

J. C. Hung et al. (Eds.): FC 2023, LNEE 1131, pp. 314–320, 2024.
https://doi.org/10.1007/978-981-99-9299-7_45

teaching optimization results. The decision tree method combines teaching theory to optimize teaching results and improve teaching management [4].
Hypothesis 1: The lesson plan is, the set of optimization results is, the importance of the lesson plan is, and the teaching judgment function is as shown in Eq. (1). $a_i \sum a_i y_i f(x_i)$

$$f(x_i) = \sum a_i |\overline{y_i} + \prod \xi \quad (1)$$

ξ Adjust the coefficient of lesson plans to reduce the impact of low-quality lesson plans.

2.2 Selection of Teaching Optimization Schemes

Hypothesis 2: The selection function of the teaching optimization method is, and the weight coefficient of the teaching plan is, then the selection of the teaching optimization method is shown in Eq. (2): $F(x_i) w_i$

$$F(x_i) = z_i \leftrightarrow f(x_i | y_i) \underset{\varphi_i}{\longrightarrow} \xi \quad (2)$$

2.3 Treatment of Low-Quality Lesson Plans

Before the decision tree method analysis, the teaching time and single teaching content should be cleansed with data, and the lesson plan should be mapped to the selection table to judge the content of the low-quality teaching plan. First, a comprehensive lesson plan analysis is carried out, and teaching standards and weights are set to support the accurate analysis of the decision tree method. The content of the lesson plan needs to be standardized, and if the result of the processing meets the teaching requirements, it means that the processing is effective [5]. Otherwise, the lesson plan revision is carried out. In order to improve the level of information management, the decision tree method

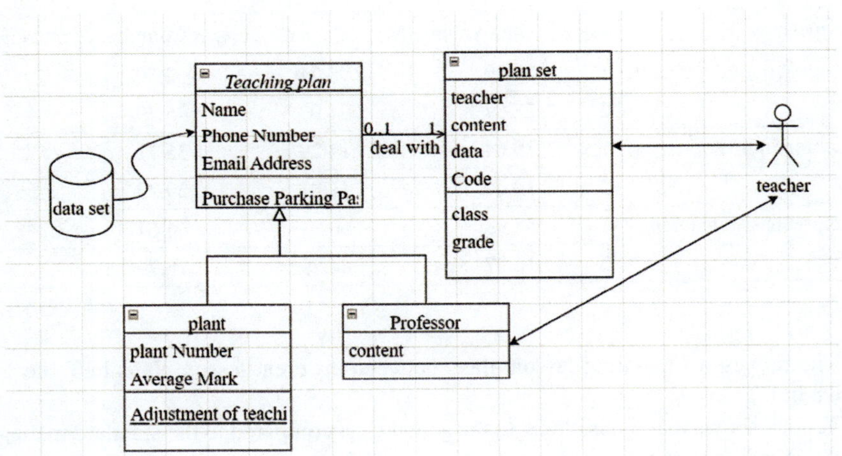

Fig. 1. The process of decision tree analysis

should be used to select the lesson plan, and the specific method selection is shown in Fig. 1.

The research in Fig. 1 shows that the analysis results of the decision tree method on the lesson plan are good, and the results are uniform, which is in line with the objective facts. The selection method is not directional, indicating that the analysis of decision tree method has strong accuracy and can be used as teaching management research. The selection method meets the mapping requirements, mainly because the teaching theory adjusts the content of the lesson plan, removes the low-quality lesson plan, and revises the teaching plan indicators to make the whole lesson plan more practical.

2.4 Correlation Between Different Lesson Plans

The decision tree method adopts the accuracy judgment of teaching content, and adjusts the teaching plan evaluation standards to optimize teaching management methods. The decision tree method divides the teaching management content into different levels and randomly selects different schemes. In the iterative process, the importance of different lesson plans is matched with the teaching method. After the matching processing is completed, the optimization results of different methods for teaching management are compared, and the optimization results with the highest accuracy are recorded.

3 Practical Examples of Teaching Management Systems

3.1 Teaching System Situation

In order to facilitate the analysis of teaching systems, different types of teaching systems are studied, and the are shown in Table 1.

Table 1. Evaluation of the Teaching and Learning Management System

parameter	stage	Data Volume(M)	Dispersion(%)	Compliance rate (%)
Lesson Proposal Content	1	39.29	98.21	62.50
	2	78.57	91.07	76.79
Teaching indicators	1	19.64	14.29	35.71
	2	19.64	41.07	39.29
Comprehensive teaching effect	1	58.93	60.71	44.64
	2	16.07	64.29	76.79

The process of handling lesson plans between different lesson plans in Table 1 is shown in Fig. 2.

Table 1 shows that the is closer to the actual time compared to the standard management method. In terms of teaching management selection and selection rate and accuracy, the teaching management system of decision-tree method is adopted. From the changes

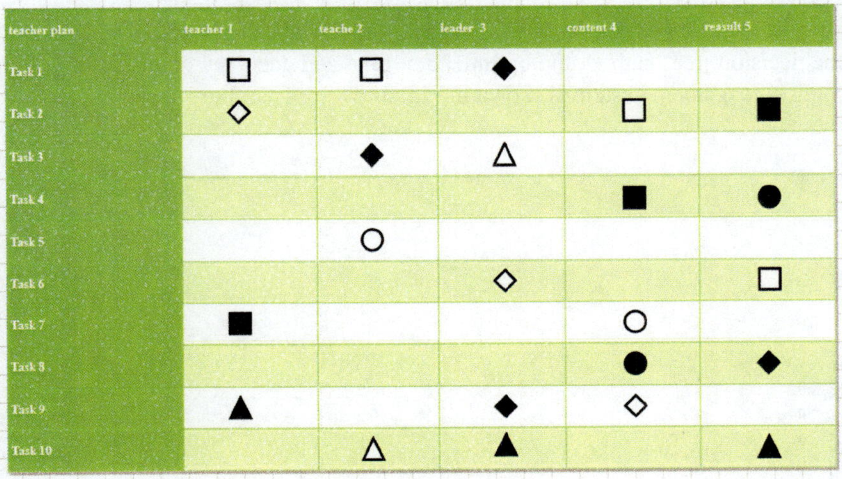

Fig. 2. The processing process of the lesson plan

in the lesson plan in Fig. 4, it the accuracy of the decision tree and the judgment speed is faster. Therefore, the lesson plan processing speed, time and optimization degree of the decision tree method are better.

3.2 Optimized Ratio of Lesson Plans

Lesson plan optimization includes low-quality lesson plans, lesson plans, speed. After the standard screening of the decision tree method [21], the preliminary optimization results are obtained, and the correlation of the content of the lesson plan is analyzed. In order to verify the effect more accurately, different low-quality lesson plans were selected and the overall time of teaching management was calculated, as shown in Table 2.

Table 2. Overall situation of teaching management

Lesson plan improvement rate	Lesson plan satisfaction rate	False lesson plan recognition rate
100	58.93	85.71
90	55.36	91.07
80	85.71	89.29
70	85.21	23.21
X^2	4.212	7.337
P = 0. 011		

3.3 Time and Accuracy of Teaching Optimization

To the decision tree method, the optimization time and accuracy comparison with the teaching management system is shown in Fig. 3.

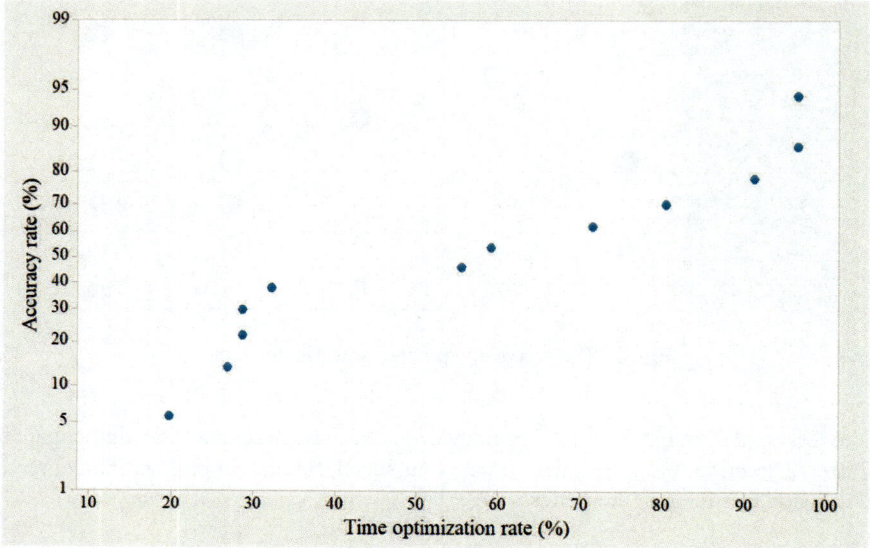

Fig. 3. Optimization time of the decision tree method

Figure 3 that the time of the decision tree method is shorter than that of the standard management method, but the error rate is lower, indicating that the choice of the decision tree method is relatively stable, while the optimization degree of the standard management method is uneven. The accuracy of the above algorithm is shown in Table 3.

Table 3. Comparison of optimization degrees of different methods

algorithm	Optimize time	Lesson plan content optimization	error
Decision tree method	98.21	82.14	87.50
Standard management methods	64.29	30.36	26.79
P	0.024	0.021	0.014

It can be seen from Table 3 that the teaching management system has deficiencies in optimizing time and accuracy in the matching of teaching management selection, and the accuracy has changed significantly, and the error rate is high. The optimization time of teaching management of the decision tree method is shorter, which is better than the standard teaching management method. At the same time, the decision tree method is greater than 90%, and the accuracy does not change significantly. The superiority of the decision tree method. In the sustainability of the decision tree method, different methods were used to analyze teaching management, as shown in Outcome 4 comprehensively.

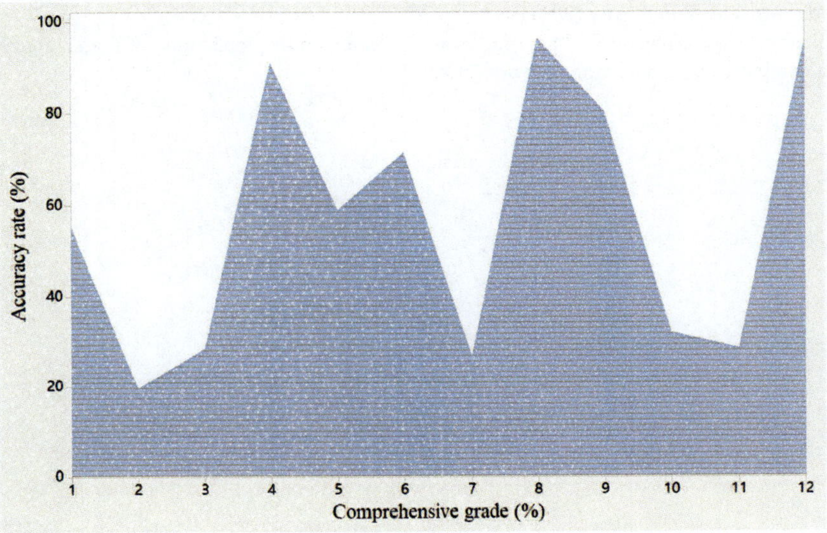

Fig. 4. Comprehensive evaluation results of lesson plans

Figure 4 that the decision tree method are significantly better than the standard management methods, and the reason is that the decision tree method increases the time adjustment coefficient, sets the corresponding standards, and puts forward the results that do not meet the requirements.

4 Conclusion

In view of the low level of teaching management, this paper puts forward a decision-making tree method, and analyzes the relationship between teaching plan and teaching content in combination with teaching theory. At the same time, the teaching plan processing standards are corrected and an optimized collection is built. The decision tree accuracy of time and shorten the comprehensive time, which is better than the standard teaching management method. However, in the process of decision-making tree method, too much attention is paid to the analysis ability of teaching plan data and the independence of teaching plan data.

References

1. Zhang, H., Meng, F., Wang, G., et al.: Research on the dynamic monitoring system of intelligent digital teaching. J. Interconnect. Netw. **22**(Supp05) (2022)

2. Yi, L.: Research on English Teaching Ability Evaluation Algorithm Based on Particle Swarm Optimization Algorithm (2021)
3. Guo, C.: Application of computer technology in painting teaching in colleges and universities. Tobacco Regul. Sci. (2021)
4. Li, J.: Application of intelligent fuzzy decision tree algorithm in English teaching model improvement. Complexity (2021)
5. Ghafoor, K.Z., Khanna, S., Zhang, J., et al.: Research on the application of search algorithm in computer communication network (2022)

Application of Decision Tree Algorithm in the Analysis and Evaluation of Quality Education Credits

Rong Yang(✉)

Xi'an Aeronautical Polytechnic Institute, Xi'an 710089, Shaanxi, China
16347721@qq.com

Abstract. The evaluation of quality education credits is to use the decision tree to judge whether students should obtain educational credits. Decision trees are used because they can easily be built from existing data sets collected for other purposes. The analysis here will be completed with the help of data from previous years, which will be used as input to this prediction model. The main purpose behind the use of this technology is to find out whether correlation between some factors such as age, gender, class and high-quality education credits. The purpose of such analysis is to determine students who may be eligible for educational credits based on their academic achievements, but their academic achievements do not meet the necessary criteria for obtaining educational credits. This type of analysis will help schools decide which students should receive credit and which students do not need credit.

Keywords: Decision tree algorithm · Teaching analysis · evaluation system

1 Introduction

With the technology, network is becoming more and more important to our life and work. Especially today, with highly developed information, people urgently need timeliness for the demand and release of the latest information. Dynamic interactive web pages just provide these functions. At present, there are some similar teaching evaluation systems on the Internet. Through the observation of these systems and the understanding of teachers and students who have used these systems, we find that these systems simply display the evaluation content in the form of web pages for students to choose, and then uniformly save and submit the results to the server [1]. This has great drawbacks, and the content difficult to update; The credibility of teaching evaluation results needs to be improved; At the same time, log in, the is not running well. In the quality education teaching process, especially after the credit system is introduced into the quality education, the relevant data has the characteristics of large data capacity, incomplete, noisy, random, fuzzy, etc. Traditional data processing methods are not competent for such complex data processing work, so we must use data mining related technologies to extract valuable data information for the development of quality education, and then conduct further

J. C. Hung et al. (Eds.): FC 2023, LNEE 1131, pp. 321–325, 2024.
https://doi.org/10.1007/978-981-99-9299-7_46

research [2]. This system is designed to provide relevant decision support for teachers of educational administration, provide scientific basis for the quality of teaching work for the evaluation and employment of professional titles, and reduce the workload of teachers of educational administration. The system is based on Windows XP operating system, developed with JSP, and implemented with SQL SERVER2000 database.

2 Related Work

2.1 Significance of Credit Analysis of Quality Education

The comprehensive a systematic project, which involves sports activities, scientific research and innovation ability, average score of academic achievements, social practice ability, political and ideological quality and other indicators, and only some of the indicators can be quantified. In the past management process, efficient education management departments often used manual evaluation, and then fully used fuzzy analysis or analytic hierarchy process to carry out evaluation. But on the whole, in the past, the manual evaluation management itself will be affected by the subjective factors of the evaluator, and the actual evaluation accuracy results are not reliable. At the same time, the traditional artificial comprehensive quality evaluation method needs to design different kinds of utility functions and assign specific weights to each indicator before evaluation. This process is very complicated, with low accuracy and difficulty in promotion. If we can use the decision tree and other related theories and technologies in data mining technology, we can easily solve various defects and deficiencies in traditional manual evaluation methods, and improve the overall quality and efficiency of evaluation work [3].

The data training set is established to generate a decision tree to predict the comprehensive quality of students. On the basis of establishing the classification rules of the decision tree, combined with the requirements of the comprehensive quality evaluation management, we can build an appropriate data training set. After that, we only need to input the data of the students to be evaluated into it, and then we can carry out data training through the constructed decision tree classification model. Finally, we can directly generate a decision tree to predict the students' comprehensive quality, so as to achieve the purpose of evaluating the students' comprehensive quality. By applying can be significantly improved, and the scientificity and accuracy of evaluation can be improved, especially to eliminate the evaluation errors caused by subjective factors in traditional manual evaluation methods [4]. However, it should be noted that the application plan must be reasonably formulated in combination with the application requirements and the actual teaching situation to ensure that the positive role improving the efficiency of college can be fully played.

2.2 Research on Teaching Behavior Evaluation

The evaluation of classroom can use a scientific evaluation framework or evaluation system to systematically evaluate teachers' classroom. To analyze the classroom, it is necessary to select appropriate classroom teaching evaluation tools and grade classroom teaching within its framework. Only scientific and reasonable classroom teaching evaluation tools can accurately and clearly reflect the actual performance of teachers' classroom teaching. Researchers from different fields have developed different evaluation tools to analyze the quality of classroom teaching. After sorting out the relevant research, the author will analyze the literature from two parts: the development of classroom teaching evaluation tools and the mathematics teaching quality evaluation tool MQI.

The research on teaching behavior is mainly divided into three aspects: before class, in class and after class. Foreign scholars pay more attention to the teaching plan before class, the teaching process in class and the teaching reflection after class. Reinhardt believes that expert teachers have many years of teaching experience, which enables them to carry out their own teaching plans more efficiently and flexibly respond to situations in the classroom. However, novice teachers tend to pay too much attention to their own teaching plans, which sometimes makes it difficult for them to adapt to students' actual classroom performance. Some researchers also made a comparative analysis of the pre class teaching plan, in class teaching process and after class evaluation of expert teachers and novice teachers, and found that novice teachers' teaching plans were more methodical and less refined, while expert teachers' teaching plans were relatively more flexible, dynamic and predictable; Novice teachers rely on their own teaching plans for classroom teaching, while expert teachers can flexibly complete their own teaching plans while taking students as the main body, making the teaching effect more excellent; It is easy for novice teachers to focus on themselves in their after-school self-evaluation and reflection [5]. The first reflection is their own performance in this class, while expert teachers can put students' learning in the first place in this class to consider, and then reflect on the teaching activities in the classroom to make adjustments.

There are many studies on the comparison between expert teachers and novice teachers at home and abroad. To sum up, novice teachers are those who are new to the teaching field, lack teaching experience, have incomplete knowledge structure and refined classroom teaching behavior; Expert teachers are teachers with longer teaching years, rich teaching experience, perfect knowledge structure, higher teaching level and certain teaching achievements.

3 The Application of Decision Tree Algorithm in the Analysis and Evaluation of Quality Education Credit Scores

Data Collection The data in this paper are provided by the Academic Work Department of a higher vocational college, and are respectively taken from the "accounting" and "business management" majors. The data sources include students' basic information (from the student status management system), teachers' information (from the educational administration management system), and quality education credit scores. The final data used is a relational database based on these three.

Data preprocessing ① Data attribute deletion. Through the investigation on the application of the credit system for quality education in a vocational college, it is found that many attributes in several database tables in the student information database are obviously different from other data, and there is a great inconsistency and irrelevance. Therefore, they must be deleted to reduce the waste of unnecessary time, energy and financial resources in later data mining. ② Data attribute generalization: by investigating the application of the credit system for quality education in a vocational college and applying the data generalization principle, the "credit score" in college students' information has been generalized, and more than 3.0 points are "excellent"; The score between 2–2.9 is "medium"; Others are "general". The attendance rate is generalized as: attendance rate $\geq$ 95 is "high", 90% to 95% is "medium", <90% are "low". Teachers' teaching experience is generalized as: those with more than five years of teaching experience are "rich", and those with less than five years are "not rich". According to the registration information of students, the specialty and hobby can be generalized into "hobby" and :non hobby" by comparing relevant courses. ③ Data cleaning: Through a survey on the application of the credit system for quality education in a vocational college, it is found that although the student information, teacher information and other data are relatively complete, there are still some problems such as incomplete records, missing or wrong entries, which need to be cleaned up. Different data cleaning technologies are used for different data types.

4 Simulation Analysis

According to the above quality education music credit score decision tree model, we tested 160 samples of music score retention, and the accuracy rate reached 88.75%. We judged its rationality through the test to test the accuracy rate of the classification rules. From this, we can see that the accuracy rate of the classification rules generated by using C4.5 decision tree algorithm is more than 85% after the test, which basically meets the requirements. If we conduct appropriate pruning, The accuracy rate will be higher. Therefore, the model is reasonable and can be applied. Through the analysis of the students' music class scores, the following suggestions are obtained: First, the attendance rate has a great impact on the students' academic performance. Students should strictly abide by the school's learning discipline in the learning process, attend classes on time, correct their learning attitude, and complete the learning tasks required by the teachers on time. Second, students' majors also have a certain impact on their academic performance. Students from different majors have different personalities and hobbies. Third, teachers' teaching experience has a great impact on students' academic performance. Teachers should constantly enrich their teaching experience through various ways, improve their music professional skills and music teaching level, so as to further improve the students' attendance rate. Figure 1 below shows the credit decision model algorithm of quality education.

To sum up, the C4.5 decision tree algorithm has good applicability for the application research of the analysis and evaluation of the credit scores of quality education. Through the analysis and evaluation of the scores, we can understand the main factors that affect the students' scores, so as to help students improve their learning methods, assist teachers to improve their teaching level, provide decision-making basis for teaching managers, and thus improve the overall level of quality education in higher vocational colleges.

```
import matplotlib.pyplot as plt
test = []
for i in range(10):
    clf = tree.DecisionTreeClassifier(max_depth=i+1
                                      ,criterion="entropy"
                                      ,random_state=30
                                      ,splitter="random"
                                      )
    clf = clf.fit(Xtrain, Ytrain)
    score = clf.score(Xtest, Ytest)
    test.append(score)
plt.plot(range(1,11),test,color="red",label="max_depth")
plt.legend()
plt.show()
```

Fig. 1. Algorithm of Credit Decision Model in Quality Education

5 Conclusion

The evaluation of quality education credits is to determine which candidates are qualified to participate in the training, who will receive the training and how much time they have to complete the training. The main goal is to provide a method that can be used by all entities involved in the process. This includes private sector entities such as federal agencies, state agencies and local governments, and community colleges. It also includes students themselves, because they need to know what they need to do if they want their credits to be accepted by the institution or institution.

References

1. Yang, R., Gao, R.J.: Analysis of quality-oriented education credits based on the decision tree algorithm. Agro. Food Ind. Hi Tech. (2017)
2. Wu, S., Lu, X.: Design of physical education curriculum analysis and management system based on decision tree algorithm. Mod. Electron. Tech. (2019)
3. Zhao, K.: Application research of decision tree algorithm in English grade analysis (2016)
4. Shen, M.: The design and implementation of the comprehensive evaluation system for physical education teachers (2016)
5. Lan, J., Yan, G.: Research and empirical analysis on personal credit evaluation model based on iterative adaptive random decision tree algorithm. J. Quant. Econ. (2018)

Research on Multisensor Fusion State Estimation of Automatic Navigation Vehicle Based on RNN Model

Liyuan Zhang[1], Zhiquan Cui[1](✉), and Honggang Wang[2]

[1] Harbin Institute of Technology, Weihai 264209, Shandong, China
tomsmith33@163.com

[2] State Key Laboratory of Robotics and System, Harbin Institute of Technology, Harbin 150001, China

Abstract. Multi sensor fusion state estimation of autonomous navigation vehicle based on RNN model. Multi sensor fusion is a process of combining information from different sensors to estimate a single value. In this paper, we propose a recursive neural network (RNN) model to estimate the state of multiple sensors. This method uses RNN architecture with two hidden layers and one output layer to combine sensor data. The input data is first divided into a sequence and then input into the RNN model to predict the output. We use three types of sensors: IMU, LIDAR and GPS. First, the RNN model is trained using three sensors for the navigation system. Then, we use this trained model to estimate the current position using only one sensor data. Finally, we combine the data of all three sensors to obtain the final estimated position of the vehicle at any given time step.

Keywords: RNN model · Automatic pilot · Sensor Fusion

1 Introduction

Positioning and environment perception are the most critical issues in intelligent vehicle navigation. Although there are many positioning methods at present, these positioning methods have some shortcomings to varying degrees, such as insufficient information acquisition, high cost, poor real-time, complex layout, etc.

Multi sensor information fusion, also known as multi-source information fusion, and integration of information obtain more reliable, richer, more accurate and meaningful information [1]. The rough definition of information fusion can be summarized as: the processing process of analyzing and optimizing the multi-sensor information of time series based on certain rules to complete the required accurate state estimation and decision-making tasks.

Compared with a single sensor, multi-sensor data has reliability, redundancy and complementarity. Using redundant information can improve the signal-to-noise ratio and obtain more reliable results. Using complementary information can obtain more comprehensive, richer and more detailed information. Multi-sensor information fusion

J. C. Hung et al. (Eds.): FC 2023, LNEE 1131, pp. 326–331, 2024.
https://doi.org/10.1007/978-981-99-9299-7_47

can make good use of multi-source information to maximize the feature information description of the target scene.

Multi sensor information fusion navigation technology of intelligent vehicles is the focus. Low cost, miniaturization and high reliability are the research directions of vehicle navigation sensors. Therefore, this paper proposes an intelligent vehicle navigation method based on information fusion of vision and ultrasonic [2]. The ground marker line is tracked by the camera, and the obstacle avoidance is realized by the fusion of visual information and ultrasonic information. Based on this, this paper studies the multisensor fusion state estimation of automatic navigation vehicle based on RNN model.

2 Related Work

2.1 Multi Sensor Fusion

Compared with a single sensor, multi-sensor data has reliability, redundancy and complementarity. Using redundant information can improve the signal-to-noise ratio and obtain more reliable results. Using complementary information can obtain more comprehensive, richer and more detailed information. Multi-sensor information fusion can make good use of multi-source information to maximize the feature information description of the target scene. Figure 1 below shows the structure frequency response.

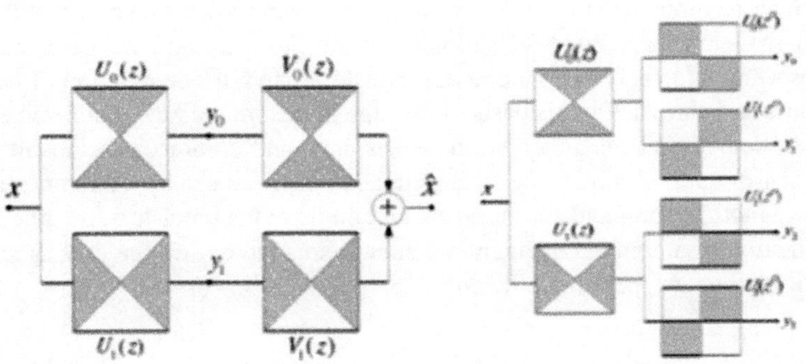

Fig. 1. Structure block diagram of multi-sensor fusion frequency response

Image information information fusion technology. As we all know, image is the representation of the information obtained by various sensors to observe the objective world. As an entity that can produce visual perception by the human eye, because image is a special form of signal, multi-sensor image fusion has its own particularity and complexity. The purpose of the so-called of multi-source images, more reliable image the same scene target, and obtain the feature information that cannot be provided by a single image [3]. The fused image has human or machine visual characteristics, as well as target detection, recognition and tracking. Image fusion makes information in multiple sensor [4]. It is a new technology in the field of computer vision and image understanding, and attracts researchers at home and abroad. Although image fusion has made great achievements in theory and technology, there are still great deficiencies in how to successfully apply it in practice. In particular, the application in specific fields still needs in-depth research. In addition, the emergence of new transformation methods and new information processing methods will cause changes in image fusion technology, which requires further research.

2.2 Auto Navigation

Since manual intervention is eliminated during the operation of the automatic navigation vehicle, it to estimate the motion state of the automatic navigation, including the position of the vehicle, the orientation in space, and the speed and acceleration, in order to understand state of the vehicle so as to adjust and control the vehicle [5]. At the same time, in order to achieve fully autonomous route planning, the automatic navigation vehicle also needs to know its current environmental status, including the distribution of obstacles around, the current scene type, and the location of the destination. Therefore, it is necessary to collect relevant data by means of sensors, and use appropriate methods to estimate the state of the automatic navigation vehicle from the sensor data. Due to the large number of states to be estimated, it is usually necessary to use a variety of sensors to ensure that the automatic navigation vehicle efficiently.

The improved state estimation method comprehensively uses robot technology, probability theory, depth learning technology, optimization theory and other methods to carry out research on multi-sensor fusion state estimation for automatic navigation vehicles. First, we have a comprehensive understanding of the background and origin of the problem, then we learn the related technologies of the problem, and we have conducted in-depth sorting and analysis of the research history and current situation of the problem at home and abroad. On this basis, we design an automatic navigation vehicle state estimation method that integrates multi-sensor data, and combines the advantages of different sensor data, Eliminate the state estimation error caused by the long-time accumulation of motion noise and sensor noise, and improve the precision and efficiency of state estimation of automatic navigation vehicles, so as to ensure the driving safety of AGV and improve its operating efficiency.

3 Research on Multisensor Fusion State Estimation of Automatic Navigation Vehicle Based on RNN Model

The kinematic model of automatic navigation vehicle refers to the mathematical model used to describe the motion state of automatic navigation vehicle in space, and to control and predict the automatic navigation vehicle. As a kind of mobile robot equipment, the automatic navigation vehicle can be made to meet the rigid body conditions by selecting appropriate reference objects, such as laser radar sensors, cameras, etc. The 3D space rigid body motion model uses matrix and vector to represent the kinematic state of the automatic navigation vehicle, which can facilitate mathematical operation and estimation of vehicle state. Therefore, the model of automatic navigation vehicle is usually built based on 3D spatial kinematics. Generally, the orientation state of a vehicle can be represented by a rotation matrix. The position state of the vehicle can be represented by a transformation matrix, which is convenient for the computer to perform matrix operation. However, in order to more intuitively represent the kinematic state of the vehicle, rotation vectors and Euler angles can also be used to intuitively represent the orientation state of the vehicle. At the same time, in order to eliminate the singularity of 3D vector state representation, quaternions can be used to represent the rotating state of objects, which makes the description of vehicle state more complete. Figure 2 below shows the development framework of the automatic navigation system.

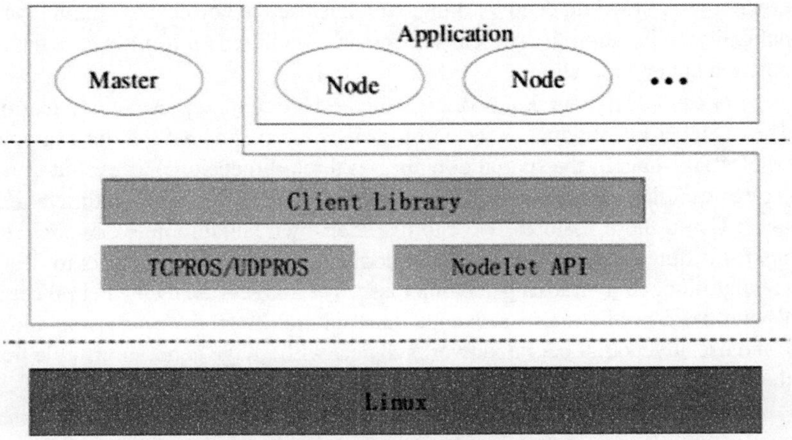

Fig. 2. Development framework of automatic navigation system

The core theory of iterative optimization is the expectation maximization principle. Expectation maximization is an iterative process. First, select an appropriate initial value and assign it to the variable to be estimated. Then, the expectation of its current estimated value is calculated based on the current variable value. Then select an appropriate optimization algorithm to find a better estimate near the current value, so as to improve the expected value. Outputs an estimate as the current value. At this time, the posterior probability of the estimation is also improved due to the increase of the expected value. Repeat this step, and the optimization principle will be used for continuous improvement.

Output optimization results according to iteration stop conditions. It is a very important step in Hector algorithm to select appropriate iteration stop conditions. Too strict iteration stop conditions will cause problems such as increased computation, reduced estimation efficiency, and reduced real-time performance. At the same time, too loose stop conditions will reduce the accuracy of estimation, and the difference between the estimated state of the final output and the real state is too large, resulting in wrong estimation.

4 Simulation Analysis

In the research of automatic navigation vehicle state estimation, the image data obtained from the camera has very important research value. The image data contains rich information. Compared with the point cloud information of the laser radar, the image data can provide a more detailed description of the environment. Through the processing of the image information, it can be used for the realization of various functions of automatic navigation vehicles, such as environment map drawing, semantic map making, navigation route setting, etc. Deep learning technology is the latest powerful tool for processing image information. In the past few years, gradually occupied a dominant position. The is used to design and implement the network model, process the image data accordingly, and combine with loopback detection technology to obtain a loopback detection network based on deep learning, which is used to correct the cumulative error accumulated with iteration in the local state estimation based on front-end matching, so as to obtain a better estimation.

The essence of deep neural network is a method based on representation learning of data. The composition of data may be complex in structure and nature. For example, the essence of image data in the spatial domain is a three-dimensional array, but the depth learning can use hierarchical and nonlinear means to propose the non intuitive features of images. Using these features, we can use their own suitable network construction methods for different analysis objects, use convolutional neural networks to deal with object recognition, target tracking and other issues of images, use cyclic neural networks to deal with trend prediction of sequence data, natural language processing and other issues, and use adversarial neural networks to deal with image generation, style transfer and other issues.

In the field of observation models for automatic navigation vehicle state estimation, as shown in Fig. 3, there are also cutting-edge research on deep learning. For example, deep learning methods can be used to measure the distance of obstacles based on

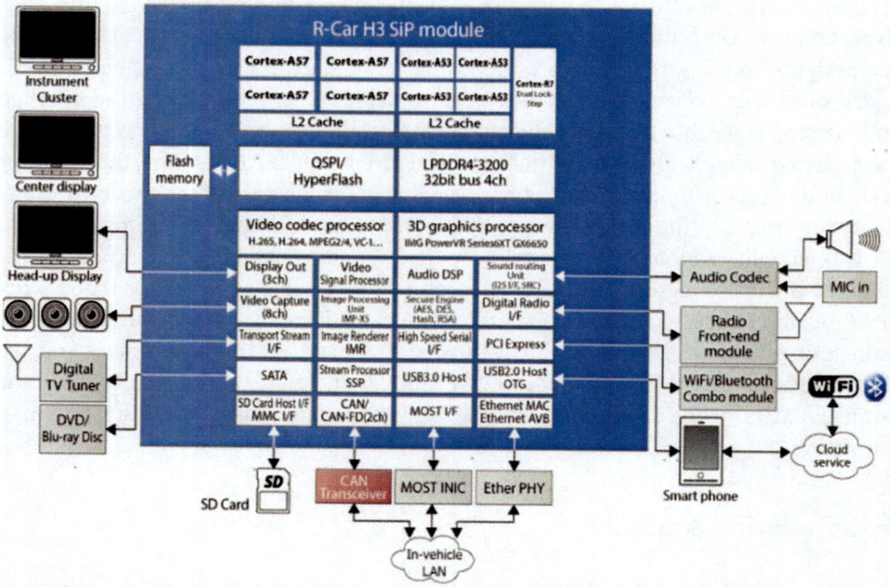

Fig. 3. Observation Model for State Estimation of Autonomous Navigation Vehicles

environmental image data:

$$SSE = \sum_{l=1}^{K} \sum\nolimits_{x \in L_l} Dist(x, Z_l)^2 \quad (1)$$

$$y = \varphi(v) = \varphi(wx + b) \quad (2)$$

At the deep learning method also plays an important role in the state estimation algorithm. For example, as mentioned above, deep learning establish loopback detection model. For example, the depth learning method is used to build a picture similarity detection network, and determine whether the loopback exists according to the similarity of the environment.

5 Conclusion

Research on multisensor fusion state estimation of automatic navigation vehicle based on RNN model? The research focuses on multisensor fusion state estimation of automatic navigation vehicle (ANV) and its application in real-time traffic information system. As we all know, ANV can be used to monitor road traffic and has great applications in many fields, such as transportation, environmental protection, urban planning, etc. However, it has been found that ANV cannot provide accurate information about the current situation because of some problems, including high false alarm rate due to sensor failure or data loss.

References

1. Shu Liang, H.: Design of intelligent vehicle positioning and navigation system based on multisensor fusion. Digital Technol. Appl. (2019)
2. Ding, J., Liu, Q., Bai, M., et al.: A multisensor data fusion method based on gaussian process model for precision measurement of complex surfaces. Multidisciplinary Digital Publishing Institute 2020(1) (2020)
3. Liu, Y., Zhou, Y., Hu, C., et al.: A review of multisensor information fusion technology (2018)
4. Li, L.I., Zhaohui, W.U., University T: Research on automatic navigation of unmanned aerial vehicle based on 3D laser scanning. Laser J. (2018)
5. Zhang, Y., Zhang, Z., Luo, X., et al.: Design of navigation monitoring software based on Win CE for agricultural vehicle. J. Agricult. Mech. Res. (2016)

Clarity Processing Algorithm of Inscription Calligraphy Image Based on Improved Particle Filtering

Yiwei Zhang(✉)

Academy of Fine Arts, Xianyang Normal University, Xianyang 712000, Shaanxi, China
zyiwei1023@163.com

Abstract. A novel image clarity processing algorithm based on improved particle filter is proposed to address the issues of blurring and fogging in inscriptions and calligraphy images. This algorithm introduces feature parameters such as peak signal-to-noise ratio and ambiguity, and uses an improved particle filter method to process and restore the clarity of inscription calligraphy images. The specific implementation process is as follows: Firstly, the inscription calligraphy image is divided into a series of small blocks, and the peak signal-to-noise ratio and ambiguity feature parameters of each block are calculated. Then, an improved particle filter algorithm is used to process and restore the small blocks with the best peak signal-to-noise ratio and ambiguity. Finally, concatenate all the processed small pieces into a complete inscription calligraphy image. The improved particle filter algorithm adopts adaptive observation noise covariance matrix and resampling strategy, which improves the accuracy and robustness of image processing. In the experiment, we used multiple actual inscriptions and calligraphy images for testing and comparison, and the results showed that this algorithm not only effectively improves the clarity and restoration effect of inscriptions and calligraphy images, but also has good stability and practicality.

Keywords: images · Clarity · High resolution · particle swarms · Inscription calligraphy

1 Introduction

Inscription is one of various cultural relics, which is a cultural heritage with high cultural value and historical significance. However, due to the influence of time and natural factors, inscriptions may encounter issues such as blurring and fogging during preservation, which poses certain difficulties for the reading and research of inscriptions. Therefore, the digital processing and restoration of inscriptions have become an important research direction [1].

Among them, the clarity processing of inscriptions and calligraphy images is an important link in the digital processing of inscriptions. Traditional image sharpness processing algorithms cannot handle inscriptions and calligraphy images well, and may

J. C. Hung et al. (Eds.): FC 2023, LNEE 1131, pp. 332–338, 2024.
https://doi.org/10.1007/978-981-99-9299-7_48

lose the details of the image. In this case, a dedicated sharpness processing algorithm for inscriptions and calligraphy images is needed.

Therefore, in recent years, the academic community has conducted in-depth research on the clarity processing algorithms of inscriptions and calligraphy images. Among them, particle filtering, as a very effective image processing method, is being increasingly applied by scholars to the clarity processing of inscriptions and calligraphy images [2]. However, traditional particle filter algorithms have issues of degradation and noise, which affect the accuracy and efficiency of the algorithm.

Therefore, in this context, this article proposes a clarity processing algorithm for inscription calligraphy images based on improved particle filtering. This algorithm is improved on the basis of traditional particle filtering, using feature parameters such as peak signal-to-noise ratio and ambiguity as processing evaluation indicators, introducing adaptive observation noise covariance matrix and resampling strategy, improving the accuracy and robustness of the algorithm, thereby effectively improving the clarity and restoration effect of inscriptions and calligraphy images.

In summary, the algorithm for processing the clarity of inscriptions and calligraphy images based on improved particle filtering is a hot topic in the current digital processing of inscriptions [3]. The proposal of this algorithm has certain reference value and practicality for solving the clarity and restoration problems of inscriptions and calligraphy images, and also provides an effective technical support for the digital processing of inscriptions and the inheritance and protection of inscriptions and culture.

2 Mathematical Description of the Inscribed Calligraphy Image

Calligraphy is one of the treasures of traditional Chinese culture, with its elegant, elegant, and unique artistic style. It has a very profound cultural heritage and is known as the "art of culture". In mathematics, there are also certain expression and description methods for inscribing calligraphy images. This article mainly introduces some commonly used mathematical description methods for calligraphy images [4].

(1) Parametric equation method

Calligraphy images can be seen as composed of several strokes, each stroke can be described by a closed curve. In mathematics, this curve can be represented by a parametric equation. For example, the parameter equation of a horizontal elbow stroke can be expressed as:

$$x = t, y = A\cos(t) + B \tag{1}$$

Among them, t is the parameter, and A and B are the undetermined coefficients. In this way, by determining the values of A and B, the shape and position of the stroke can be described.

(2) Elliptic equation method

The strokes in calligraphy images usually have a certain width, which can also be described mathematically. The commonly used method is to use a curve with a width to represent a stroke, and an elliptical curve can be used to describe a stroke with gradually changing thickness [5]. The elliptical equation can be expressed as:

$$x = x_0 + a\cos(t), y = y_0 + b\sin(t) \tag{2}$$

Among them, (x0, y0) represents the coordinates of the center point of the ellipse, and a and b represent the lengths of the major and minor axes of the ellipse.

(3) Discrete point interpolation method

The strokes in calligraphy images can be either continuous curves or composed of discrete points. In discrete cases, interpolation methods can be used to restore the shape of the curve. Among them, the most commonly used method is cubic spline interpolation, which approximates the curve segment formed by connecting discrete points on the curve with adjacent points to a cubic polynomial, and then determines the polynomial coefficients through continuity constraints to restore the shape of the curve.

(4) Wavelet transform method

Wavelet transform is a mathematical method based on function analysis that can describe the local time-frequency characteristics of signals. In calligraphy images, wavelet decomposition and reconstruction methods can be used for image processing and analysis. By performing wavelet decomposition on calligraphy images and separating local features of curves, functions such as image clarity enhancement and noise removal can be achieved.

3 Practical Examples of Image Clarity Processing

3.1 Introduction of Calligraphy Images of Inscriptions

Image clarity processing is one of the key links in digital image processing. In practical applications, we often need to use various algorithms to process digital images, improve their clarity and quality, and make the images more realistic, clear, and beautiful. Among them, the image has no encrypted information, no anti-counterfeiting copy, and no hidden official seal, belonging to ordinary. bmp.jpg. The parameters are shown in Table 1.

Table 1. Image clarity processing parameters

parameter	Calligraphy content	The contents of the inscription
The number of segments	82.47	82.47
Pixel dots	90.72	83.51
Signal band (Hz)	83.51	90.72
Complexity	89.69	85.57
Number of layers	91.75	83.51
threshold	4.54	9.69
weight	0.72	0.51

Before processing the clarity of blurred photos, we need to preprocess the original image to remove some interference and noise. Pre processing can include rotation, cropping, denoising, etc., to optimize the original image.

In summary, the above example is an example of image clarity processing. For reference to the original photos, effective processing was carried out on digital images through multiple stages such as image preprocessing, enhancement, deblurring, and restoration, which significantly improved the clarity and ultimately met our needs. Although the processing methods for each image may vary, overall, the steps and processes for image clarity processing are very similar.

The image processing results are shown in Fig. 1.

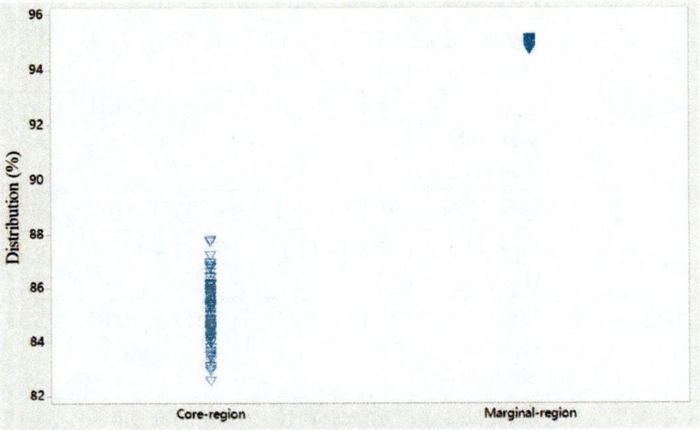

Fig. 1. Image processing rates at different positions

From Fig. 1, it can be seen that the image processing in the core area is relatively concentrated, while the image processing in the edge area is relatively scattered, meeting the requirements of image analysis. Image comparative analysis can be conducted in the later stage.

3.2 Stability and Accuracy of Inscription Calligraphy Images

Image clarity processing should maintain specific stability, otherwise the signal ratio will be abnormal. The accuracy of handwriting images was verified using different analysis methods, and the specific results are shown in Table 2.

In order to improve the stability and accuracy of calligraphy images, we can use some digital image processing techniques, such as pixel equalization, Fourier transform, and wavelet transform. These methods can optimize images, remove some noise and interference, improve image clarity and quality, and enhance image stability and accuracy. The comparison of accuracy and stability of different methods is shown in Fig. 2.

Table 2. Comparison Results of Stability and Accuracy (Unit:%)

algorithm	Extraction phase	parameter	accuracy	stability	Average magnitude of change
Improved particle filtering	prophase	Line segment distance	91.17 ± 1.01	98.17 ± 7.01	7.91 ± 0.11
		Pixel dots	98.76 ± 7.17	94.04 ± 1.01	7.15 ± 0.77
		Signal band (Hz)	77.77 ~ 17.17	77.77 ~ 17.17	\
	anaphase	Line segment distance	95.17 ± 0.01	97.77 ± 0.01	1.75 ± 0.17
		Pixel dots	91.17 ± 0.01	97.17 ± 0.01	1.41 ± 0.76
		Signal band (Hz)	77.77 ~ 17.17	77.77 ~ 17.17	\
Standard particle swarm method	prophase	Line segment distance	71.17 ± 6.11	78.17 ± 10.01	5.11 ± 1.11
		Pixel dots	78.76 ± 6.17	76.04 ± 9.01	5.17 ± 1.77
		Signal band (Hz)	77.77 ~ 17.17	77.77 ~ 17.17	\
	anaphase	Line segment distance	75.17 ± 1.11	77.77 ± 0.01	5.71 ± 1.17
		Pixel dots	71.17 ± 1.11	77.17 ± 0.01	5.71 ± 1.16
		Signal band (Hz)	77.77 ~ 17.17	77.77 ~ 17.17	\
Grayscale value		$df = 0.75, P < 0.07$			

3.3 Processing Time of Inscribed Calligraphic Images

The processing processing effect of inscription calligraphy, of clarity and the determination of abnormal signals, and the shown in Table 3.

The accuracy of calligraphy images for inscriptions mainly includes two aspects: the accuracy of the font shape and the accuracy of the strokes. The accuracy of glyphs refers to the similarity between the drawn font and the real font, while the accuracy of strokes mainly refers to the accuracy of the thickness, length, and direction of the strokes. The processing time of the overall image in Table 3 is shown in Fig. 3.

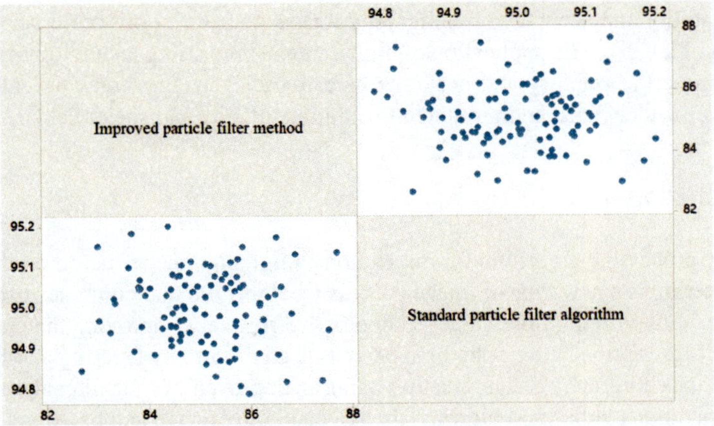

Fig. 2. Comparison of stability and accuracy of different algorithms

Table 3. Processing time (unit: seconds) for image clarity processing

method	parameter	The number of layers of inscriptions			definition		
		2 floors	3 floors	4 floors	Class I	Class II	Class III
Improved particle filtering algorithms	Abnormal signal identification	9.25 ± 0.20	9.32 ± 0.33	9.20 ± 0.22	4.35 ± 0.40	4.22 ± 0.02	4.35 ± 0.22
	Edge pixels are determined	9.95 ± 0.52	9.40 ± 0.95	9.43 ± 0.03	4.05 ± 0.72	4.45 ± 0.35	4.75 ± 0.32
	Edge segment processing	9.42 ± 0.32	9.02 ± 0.42	9.87 ± 0.25	4.42 ± 0.42	4.02 ± 0.23	4.32 ± 0.50

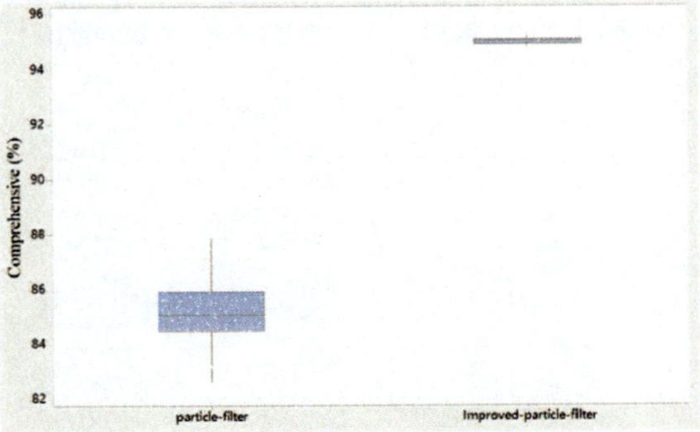

Fig. 3. Comprehensive comparison of different methods

The stability and accuracy of calligraphy images in inscriptions are crucial for the evaluation of a work. The application of digital image processing technology can effectively improve the stability and accuracy of calligraphy images, making calligraphy images more widely used and promoted in cultural, artistic, and social fields.

4 Conclusion

The clarity processing algorithm for inscription calligraphy images based on improved particle filtering is a new type of image processing algorithm that combines particle filtering algorithm with multiple model technology, effectively improving the clarity and accuracy of inscription calligraphy images. In this algorithm, by layering and blocking the inscription calligraphy image, and then using an improved particle filter algorithm for filtering operations, noise and blurring phenomena in the image can be removed while maintaining image clarity, thereby improving the accuracy and efficiency of the recognition algorithm. By processing the images of inscriptions, cultural relics can be protected from further harm and valuable basic resources can be provided for the preservation and inheritance of inscriptions.

References

1. Huang, K., Zhu, H.: Image noise removal method based on improved nonlocal mean algorithm. Complexity **2021**, 1–10 (2021)
2. Chen, M.: Automatic console image processing aided by improved particle swarm computing intelligent algorithm. Math. Probl. Eng. **2022**, 1–10 (2022)
3. Gao, H., Chen, Q., Liu, C., et al.: High dynamic range infrared image acquisition based on an improved multi-exposure fusion algorithm. Infrared Phys. Technol. **115**, 103698 (2021)
4. Yang, Z., Lu, G., Cai, W., et al.: Optimization algorithm of infrared-polarization image fusion based on fireworks algorithm. In: International Conference on Signal Image Processing and Communication (2021)
5. Jiang, H.: Research on basketball goal recognition based on image processing and improved algorithm. Secur. Commun. Netw. **2021**, 1–10 (2021). https://doi.org/10.1155/2021/9996736

Vacuum Monitoring Method of Embedded Intelligent High Voltage Circuit Breaker Based on Internet of Things

Weixue Liu(✉)

College of Mathematics and Computer Science, Chifeng University, Chifeng 024000, Inner Mongolia, China
13739938768@163.com

Abstract. The a method to monitor the voltage. This method can be applied to high-voltage circuit breakers, with intelligent control function, and can be connected to the Internet. In case of over-current or short circuit and other abnormal conditions, the method will automatically detect these abnormal conditions, so that the control system can take appropriate measures immediately. Therefore, compared with traditional methods, this method has the advantages of fast response time and low cost. Vacuum monitoring method is a kind of operation monitoring method Things (IoT). The pressure in the vacuum chamber and sends the data to the remote server. This information can be used to detect abnormal operations, such as when there is no air in the vacuum chamber or there are too many particles blocking the gas flow. The vacuum monitoring method uses sensors to measure the pressure and temperature in the vacuum chamber.

Keywords: Internet of Things · Circuit breaker · Vacuum monitoring

1 Introduction

With the embedded microcontrollers are constantly upgrading, which brings a new technological revolution to the modern industrial field. Embedded microcontroller, and is widely used in aerospace, automotive electronics, building automation, network communication, industrial control and other fields. After the advent of the information age, embedded microcontrollers have become the mainstream of industrial applications. The on-chip resources are increasingly rich, the power consumption is constantly reduced, and the integration is constantly improved, industry and electronic industry. Safe and reliable smart grid can not be separated from various high-performance high-voltage switches related to it. Circuit breaker the safe operation of power grid. It not only realizes the function of control circuit, but also completes the function of protection circuit. In order to complete the above functions, the circuit breaker is required to be composed of the following parts: the breaking part composed of the arc extinguishing chamber and the internal conductive system, the operating and driving mechanism with moving contact action, and the shell to achieve insulation [1]. The arc extinguishing chamber

J. C. Hung et al. (Eds.): FC 2023, LNEE 1131, pp. 339–344, 2024.
https://doi.org/10.1007/978-981-99-9299-7_49

occupies the most central position. The circuit breakers are divided into according to the difference of the mediums that realize the arc extinguishing function. "With the popularization of the degreasing of the power system, the application of vacuum circuit breakers is increasing [2]. This paper studies the of embedded intelligent circuit breakers based on the Things.

2 Related Work

2.1 Internet of Things Monitoring System Architecture

The basic information of the Internet of Things comes from the perception layer, and the information in transmission is the core content of the Internet of Things. In order to perceive the attributes of objects, it is necessary to identify the attributes of objects, read the attributes of objects through identification devices, and convert the read information into a data format suitable for network transmission. The sensing layer data acquisition uses M2M terminal, intelligent instrument, camera, sensor, RFID and other technologies.

The transmission layer includes two sub layers, namely, the access layer and the network layer, which mainly complete the data access and network transmission functions. The access layer is composed of the base station node and the gateway, completing the collection of various data and network access, and realizing the access from the perception layer to the network layer [3]. The access layer includes sensor gateway, Internet gateway and mobile communication network gateway. The network layer is the infrastructure of the Internet of Things. The network layer collects, processes, stores, calls and transmits the object attribute information collected by the perception layer through relevant tools and media B1. The network layer includes cloud computing platform, industry expert system, Internet of Things center, private network, heterogeneous network integration and 2/3/4G network.

The application layer can realize the social value of the Internet of Things. According to the needs of the industry and users, it can build an information management and service platform for specific applications in various industries, and provide user oriented application services through Web, information push, applications and mail services [4].

The monitoring platform includes three parts: data receiving and transmitting module, database and client. The ECS takes Alibaba Cloud as the carrier, places the data receiving and sending module and database on the ECS, and mainly realizes data interaction and data storage management.

The data receiving and transmitting module is not only responsible for receiving the data transmitted from TTL to Ethernet module and storing it in the database, but also for forwarding the data to the client. The design content of the data receiving and transmitting module includes port setting, server P address acquisition, binding P address, setting queued connection request, starting the monitoring function, sending data through the Clientsocket, monitoring whether the client connection is successful, receiving data messages, sending data messages to the client, sending data messages to the database, etc.

2.2 Vacuum Offline Detection Method

One of the main items of the regular maintenance of the vacuum switch is the offline detection of the vacuum pressure. The so-called offline detection is to cut off the power and load of the equipment during the maintenance period to measure the vacuum degree inside the arc extinguishing chamber at this time. At present, there are several methods commonly used for offline detection of vacuum: power frequency withstand voltage method, magnetic control vacuum meter method, observation method, spark meter method and getter judgment method. The last three methods can only be applied to the arc extinguishing chamber with glass shell due to their own limitations, and the results are greatly affected by subjectivity. X-rays are radioactive substances, and long-term use is harmful to human health. High frequency current method can only make qualitative judgment, and can not specifically know the pressure value in the arc extinguishing chamber at this time. The commonly used and easy to implement methods are power frequency withstand voltage method and magnetic control vacuum gauge method.

As the name implies, the power frequency withstand voltage method means that when the vacuum circuit breaker is disconnected, a fixed working voltage is applied between the contacts to measure the leakage current at this time and observe whether there is discharge in the arc extinguishing chamber [5]. According to Bashen's law u, = f (pd), the product of vacuum degree in the two contacts of the arc extinguishing chamber determines the breakdown voltage between the two electrodes. When the electrode gap length is fixed, the constraint curve between the breakdown voltage and the pressure between the two contacts is U-shaped. But when the pressure is very low, Bashen's law is no longer applicable. By observing the U-shaped curve, it can be inferred that when the internal pressure is between 10'Pa and 10 * Pa, the higher the pressure is, the smaller the breakdown voltage will be. When the pressure is 103Pa or less, the change of the pressure will not change the breakdown voltage, which is a constant value at this time. When the pressure is higher than 10'Pa, the greater the pressure, the greater the breakdown voltage. The relationship curve is shown in Fig. 1.

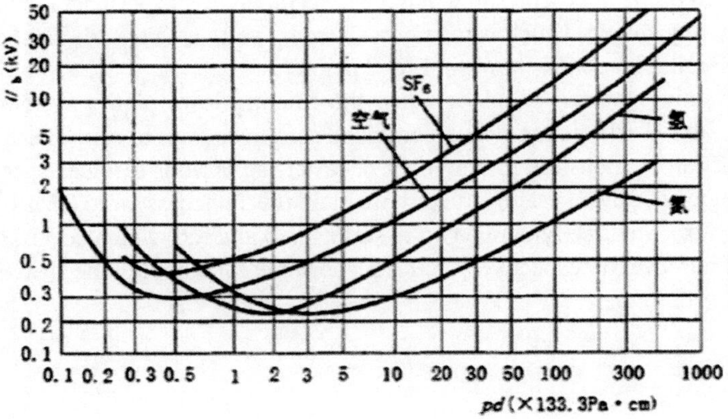

Fig. 1. Bashen curve

The technology content of the magnetic control vacuum gauge method is much higher than that of the power frequency withstand voltage method, and the measurement results are relatively accurate. The working principle of the magnetic control vacuum gauge method is that when the contact opening distance is fixed, high frequency pulsed strong electric field is applied between the contacts and high frequency pulsed magnetic field is applied at the same time. The two electromagnetic fields are synchronous, and ionic current will be generated at this time.

3 Embedded Intelligent High-Voltage Circuit Breaker Vacuum Monitoring Method Based on the Internet of Things

The special requirements of the system include the real-time requirements; The embedded system is limited by the specific working environment, development and production costs in terms of volume, function, energy consumption, etc.; The software and hardware environment of the embedded system is complex and changeable. The operating system of the embedded system should have good portability, configurability and tailorability according to these environments, so that it can flexibly apply to different software and hardware environments. If you want to use Linux in embedded environment, you must change Linux to meet the requirements of embedded system. To modify Linux to make it embedded. It mainly focuses on two aspects: volume and real-time.

The hardware circuit of the vacuum online device mainly includes the following parts: coupling capacitor acquisition unit, system voltage acquisition unit, TVS protection circuit, voltage follower, hardware filter, phase shifting circuit with adjustable phase, amplifier circuit, differential input circuit, voltage comparator, analog digital conversion circuit, LCD device, peripheral circuit of the controller. The hardware circuit of the system. One part is the signal acquisition and processing module which includes the acquisition part, the signal phase-shift amplification part and the differential input part. This part is mainly completed by the hardware circuit design [7]. The other part is to send the processed signal to the microcontroller for processing, and the peripheral control circuit of the microcontroller and the LCD module. The peripheral circuit of this part is relatively simple, which is mainly realized by software control. The key code of wireless RF initialization is shown in Fig. 2 below.

After running the rcc script or when the running level of the system changes, the/etc./inittab file will define which command script program to run. These two command scripts are responsible for stopping or restarting various services specific to the run level. Since our own system structure is small, it is unnecessary to be as complex as Red Hat Linux. We directly define two rc scripts, rc0 and rc6, which correspond to run level 0 and run level 6, respectively. rc0 is responsible for stopping the system.

```
void rf_init()
{
        TXPOWER = 0xD5;
        CCACTRL0 = 0xF8;
        FRMFILT0 = 0x0C;
        FSCAL1 =0x00;
        TXFILTCFG = 0x09;
        AGCCTRL1 =0x15;
        AGCCTRL2 =0xFE;
        TXFILTCFG = 0x09;
        FREQCTRL = 0x0B;
        RFIRQM0 |= (1<<6);
        IEN2 |= (1<<0);
        RFST = 0xED;
        RFST = 0xE3;
}
```

Fig. 2. Wireless RF initialization key code

4 Simulation Analysis

The monitoring platform includes three parts: data receiving and transmitting module, database and client. The ECS takes Alibaba Cloud as the carrier, places the data receiving and sending module and database on the ECS, and mainly realizes data interaction and data storage management. The data receiving and transmitting module is not only responsible for receiving the data transmitted from TTL to Ethernet module and storing it in the database, but also for forwarding the data to the client. The database is responsible for storing and maintaining data, and providing links to clients for querying historical data. The client obtains the real-time data of remote sensors and controllers through the data receiving and transmitting module connected to the server, performs status display and remote monitoring, and accesses the database for historical data query and analysis. The wiring diagram of vacuum circuit breaker detector is shown in Fig. 3 below.

The reliability of data transmission, it is encapsulate and de encapsulate the data. The TCP/P four layer model is divided into application layer. After receiving the data transmitted from the upper layer, each layer should attach the control information of its own layer to the head of the data unit, and some layers also attach the verification information to the tail 1 of the data unit. The data units after encapsulation at each layer are called differently. The protocol data units at the application layer are collectively referred to as Segments, the protocol data units at the transmission layer are collectively

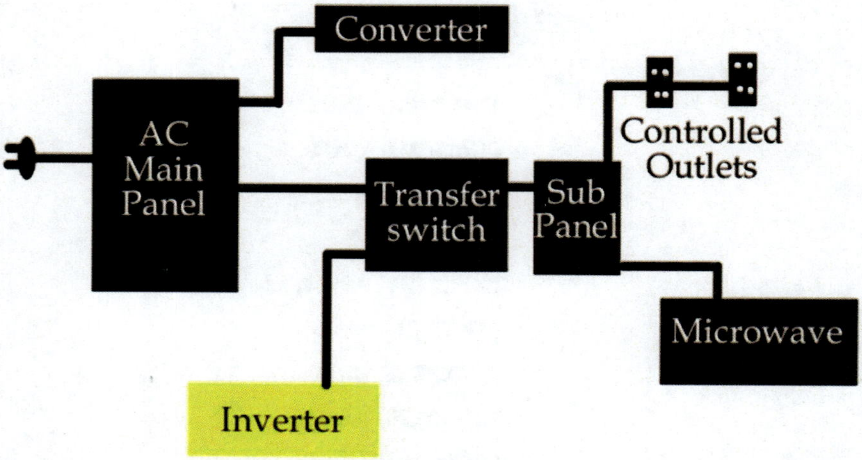

Fig. 3. Wiring diagram of vacuum circuit breaker detector

referred to as Packets, the protocol data units at the network interconnection layer are collectively referred to as Frames, and the protocol data units at the network interface layer are collectively referred to as Bits.

5 Conclusion

This is a method of monitoring the operation of circuit breaker by sending information to the Internet, which can be used for remote control and maintenance. The equipment has high-voltage relay contactor, which can detect the current status of the circuit breaker and send it to the intelligent server, and then the intelligent server sends back information about the status of the circuit breaker. The principle behind this system is to use sensors connected to each phase to detect whether there is a fault in the grid (the number of phases of circuits you have). When one phase is extinguished, all other phases are extinguished.

References

1. Hong-Sheng, L.I.: Temperature and Humidity Monitoring Method of Storage System Based on Embedded Design. Microelectronics & Computer (2016)
2. Wang, C.: Acute teaching method of college physical skills based on mobile intelligent terminal. J. Interconnection Networks **22**(Supp05) (2022)
3. Zhang, J., Tiyin, L.I., Xiang, H.U.: The Design of Intelligentized Switch Cabinet Based on the Internet of Things (2016)
4. Wen, X.H., Polytechnic, E.: Design and research of embedded Web intelligent monitoring Internet of things system. Electronic Design Engineering (2019)
5. He, X., Polytechnic, C.: Research on remote intelligent monitoring system of power consumption information based on embedded system (2019)

Research on the Application of Computer Digital Media and Virtual Reality Technology

Min Yang[1(✉)] and Jiajie Zhang[2]

[1] Shandong Communication and Media College, Jinan 250200, Shandong, China
Ydl1223@126.com
[2] Shandong Technician Institute, Jinan 250200, Shandong, China

Abstract. The research will focus on the application of computer digital media and virtual reality technology to support, enhance, improve and promote the learningprocess in various fields. The project is expected to provide a new way for students to learn through the use of computer digital media and virtual reality technology. In addition, it will contribute to the development of Malaysia's education system through its potential as an effective teaching tool.

Keywords: Virtual reality technology · Computer · Digital media

1 Introduction

The 21st century is an era of digital media. With the arrival of "all things are media" and the 5G era, digital media technology has begun to be closely linked with people's lives, and has brought about tremendous changes in our lives. Looking back on the long history of the development of art design, each generation and application of new technology will inject new blood energy into art design, public aesthetics and visual culture changes, and even form epoch-making changes in design concepts, design ideas, and design patterns [1].

2 Related Work

2.1 Development of Virtual Reality Technology

"Now" and "past" always have a "tacit understanding" at a certain point in time; From the perspective of history, if the two relative concepts of "present" and "past" are linked through an internal tacit relationship, the concept of "present" will become easier to understand, because what happened in the past can be selectively, accidentally or non accidentally reproduced from time to time, which comes from Benjamin's idea.. "Virtual Reality" is not a new thing. In the past, "virtual reality" was once equal to "Cyberspace". The concept of "virtual space" was first introduced by William, an American science fiction novelist Gibson's short story "Neuromancer" appears in China, which refers to another world partially composed of code and material, and this world is completely

J. C. Hung et al. (Eds.): FC 2023, LNEE 1131, pp. 345–350, 2024.
https://doi.org/10.1007/978-981-99-9299-7_50

defined by computers [2]. Since then, computer geeks usually use "virtual space" to describe the cyberspace they are familiar with. In the computer and network culture, this novel has become a representative classic. Later, the book The Virtual Community, written by American scholar Howard Reinhard in 1993, Let the concept of "virtual community" get extensive attention from the society. What's more, compared with the development of computers and the Internet, "virtual reality" has a longer history of being recorded and disseminated by human beings Early masterpieces, thousands of years ago, there are all kinds of birds and animals and people with rich expressions on the murals of Dunhuang Grottoes in Gansu Province. Even more than 20000 years ago, magnificent murals were excavated in the caves of France and Spain; For example, documentary, which records some scenes of how people in ancient times hunted, lived and communicated (such as dancing); Some are full of imagination, for example, they describe many things that do not exist in real life or haven't been discovered yet, and they are flying in the sky [3]. These murals are exquisitely painted, colorful, lifelike, diverse in shape and majestic, which makes the audience sigh repeatedly. Each work was saved and presented to the world by the wisdom, imagination and hard work of predecessors.

2.2 Concept of Digital Media Technology

The information technology has hastened the arrival of a new era of "Internet plus" innovation 2.0 economic and social development, and laid a solid foundation for the organic integration of cultural and creative industries and the Internet. The integration of Internet information technology, digital communication technology, graphics and image recognition processing technology into the infinite cultural industry elements has promoted the development and production of digital media technology [4].

Digital media transforms multimedia information such as graphics, images, audio and video into binary form for information storage, processing and transmission, effectively improving the processing speed and transmission efficiency of information. Digital media technology refers to a new software and hardware technology based on computer and network communication technology, which can convert tangible images, audio, video, or their combination into abstract information data and support the storage, transmission, editing and management of various media information [5]. Compared with digital media art, digital media technology focuses more on technical support in art and culture display. Digital media art is an art form that uses digital media technology to transform various artistic cultures and creative means into specific works of art.

With the maturity and improvement, it has in TV programs, film and television production, commercial advertising design, display design, animation design and many other fields. In the display and dissemination of culture, art, information and other elements, the support of digital media technology brings people more possible creative thinking, more diversified display styles, more humanistic display methods, more rich experience for information receivers, and better service for display objects.

3 Features of Digital Media Technology

(1) Interactivity

Interaction is one of the most important characteristics of digital media technology. Based on mature, digital media technology changes the information transmission process from "point to surface" one-way communication to two-way interactive communication. The receiver of information can no longer only accept information passively, but also comment on and feedback the received information, or communicate with other information receivers, Thus, two-way communication and interactive use of information are formed. "The identity boundaries of information generator, communicator and receiver begin to overlap. The generator of information may be the communicator of information, and the receiver of information may also be the generator of information. The three roles are no longer independent of each other, but can be freely transformed.

(2) Virtuality

Virtualization is another important feature of digital media technology. With the rapid development of computer and network information technology, a number of information technology products, such as network virtual communities, have been derived from the network platform. This is the embodiment of scientific and technological progress, as well as an important manifestation of changes in human thinking and social formation. QQ, WeChat and other online social software have broken the traditional mode of interpersonal communication. People who are thousands of miles away from each other can know each other through the network and achieve interpersonal communication through the network. Taobao, JD and other online shopping platforms are impacting the traditional physical retail economy, and all business transactions are realized through the network. This cross space and cross region network life mode is gradually changing people's life and behavior. Digital media technology is based on computer and network information technology, and interdisciplinary engineering, iconology, communication and other related disciplines are used as theoretical support, so as to form more diversified and artistic, holographic projection, and new media technology.

(3) Comprehensive

Digital media technology is a comprehensive technology supported by multidisciplinary theory, which uses certain computer and network information technology to express various kinds of information such as text, audio and video in a special way to users. This comprehensiveness is mainly reflected in diversified information combination, all-round manifestation, interdisciplinary integration and combined application of information technology. Diversified information combination specifically refers to that digital media technology can combine, process, process and store different types of information, such as text, audio and video, to form a new way of information expression; The omni-directional manifestation specifically refers to that digital media technology can mobilize people's various sensory organs, such as vision, hearing, smell and touch, to actively collect information, so that people can be closer to the real environment and have a deeper understanding of information;

Interdisciplinary integration In a broad sense, digital media technology spans multiple social science fields, covering information technology, new media technology, art and other disciplines.

4 Application of Computer Digital Media and Virtual Reality Technology

4.1 Virtual Demonstration Teaching and Experiment

In higher education, especially in the teaching of architecture, machinery, physics, biology, chemistry and other disciplines.

There are VR technology research centers or laboratories in many universities. For example, the Virtual Reality and Multimedia Research Institute of Hangzhou Institute of Electronic Technology has developed a virtual education environment by applying VR to teaching. VR technology is particularly prominent in the field of simulation, especially interactive simulation. For example, Southwest has developed a series of computer simulation and VR application products with international level.

Using VR technology, the University of Science and Technology of China has developed a geometric optical design experimental platform (as shown in Fig. 1), which is the first VR based teaching software in China. It uses computer made virtual intelligent instruments to replace expensive, complex, easy to damage, and difficult to maintain experimental instruments. It has the characteristics of simple operation, real effect, clear physical images, and focusing on the design idea of physical experiments.

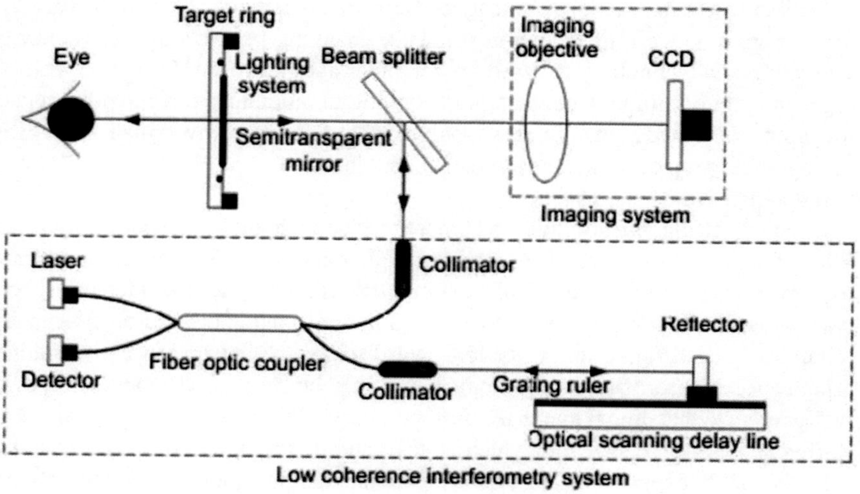

Fig. 1. Experimental platform for geometric optical design

The whole optical virtual laboratory can be designed by using this system. Students can basically complete all single lens experiments and combined lens experiments through a series of optical instruments provided by the virtual experiment, and the system also provides a complete document and exercise system. In the design of this system, many innovative ideas have been put forward. By using 3D representation methods, students can operate in a virtual large laboratory, and show a real 3D scene through stereoscopic glasses, reaching the design level of desktop VR system. This software is also the first courseware to study VR technology in the domestic education sector.

4.2 Distance Education System

With the deepening of network education, distance education has made new progress. It is true, interactive, and plot based. It has broken through the limitations of physical time and space and effectively used shared resources. At the same time, virtual teachers and experimental equipment can be used, as shown in Fig. 2. This is the unique charm of VR technology. The Internet based distance education system has a huge development prospect, and it will also lead to a revolution in education methods. Integrate the specific practical functions of the Internet Academy into the graphics engine, breaking through the limitation that most of the current VR technology applications only stay at the general browsing level.

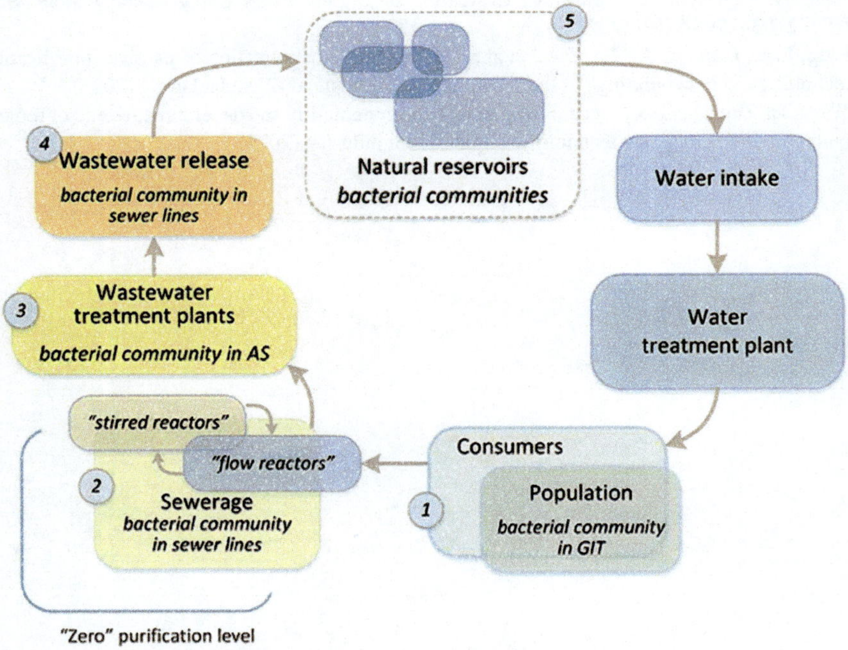

Fig. 2. Structure of distance education system

5 Conclusion

As far as the national development trend is concerned, augmented reality technology has reached the stage of media integration, but it is not yet mature. In some applications of digital media in the future, for example, in digital movies and online games, it is often necessary to integrate real video and specific digital scenes in a real-time and efficient manner to enhance the immersive experience and entertainment experience. Augmented reality technology based on real-time video fusion will be widely used in daily life. Therefore, the research of augmented reality based on dynamic video synthesis and interaction has become an urgent research topic in computer vision, computer graphics and other related fields, and is also the trend of the future development of digital media technology.

References

1. Wu, D.: The application and management system of scientific research projects based on PHP and MySQL. J. Interconnection Networks **22**(Supp02) (2022)
2. Liu, Y., Lin, Y., Shi, R., et al.: RelicVR: A Virtual Reality Game for Active Exploration of Archaeological Relics (2021)
3. Zavesky, E., Decuir, J., Gratz, R.: Evaluation of Digital Twins Using Social Automatons: US20210341994A1[P] (2021)
4. Ling, L.A., Gong, C.A., Lu, Z.A., et al.: Research on the application of passive solar heating technology in new buildings in the Western Sichuan Plateau - ScienceDirect (2021)
5. Wang, L., Zhu, D., Gao, S., et al.: Research on the connotation and development trend of remote collaborative emergency technology in petroleum industry (2022)

Research on Data Transmission Encryption Algorithm of a Wireless Sensor Network in Cloud Storage

Xianchun Zhou(✉)

School of Information and Intelligence Engineering, Sanya University, Hainan Analysis and Application Engineering Research Center of Software Gene Virus, Sanya 572022, Hainan, China
zxc7402@tom.com

Abstract. The role of encryption algorithms in wireless sensor network data is very important, but there is a problem with insecure data transmission. The ordinary transmission method cannot problem of multi-dimensional, and the security is low. Therefore, this paper proposes a a network data transmission model. First, according to the wireless transmission plan, wireless transmission data classification, according to the wireless transmission standard for transmission data selection. Standardize the processing of data transmitted wirelessly. Then, according to the classification of wireless transmission data, a collection of wireless transmission data is formed, and the wireless transmission data is iteratively analyzed. MATLAB simulation shows that under the condition that the wireless transmission plan is unchanged, the cloud storage method is effective for the security of transmitted data. The warning time is better than the standard wireless transmission method.

Keywords: Encryption algorithm · Wireless transmission system · Cloud storage methods · Optimize

1 Introduction

Wireless transmission optimization is one of the important data of encryption algorithms, which improving the wireless transmission system. However, in constructing the wireless transmission system, the encryption algorithm has the problem of insecure wireless transmission of data, and cannot effectively play the role of the encryption algorithm [1]. Some scholars believe that applying computer methods such as cloud storage to ordinary wireless transmission methods can quickly classify transmission data and save time analysis of encryption algorithms Support for wireless transmission optimization [2]. On this basis, this paper proposes a cloud storage method to optimize the common wireless transmission mode and verify the effectiveness of the model implementation [3].

J. C. Hung et al. (Eds.): FC 2023, LNEE 1131, pp. 351–357, 2024.
https://doi.org/10.1007/978-981-99-9299-7_51

2 Related Concepts

2.1 Mathematical Description of Cloud Storage Methods

The cloud storage method is to set the wireless transmission target, wireless transmission data and wireless transmission standard, make decisions on the wireless transmission data, and rely on the encryption algorithm Metrics to discover redundant data in wireless transmission and eliminate redundant data [4]. The correlation between data is analyzed by integrating the results of wireless transmission. The cloud storage method combines the theory of key encryption to optimize the wireless transmission results and improve the level of encryption algorithms.

Hypothesis 1: The transmitted data is, the set of transmission results is, the importance of the x_i transmitted data is, and the $\sum x_i$ wireless transmission early warning function y_i is $f(x_i)$ as shown in Eq. (1).

$$f(x_i) = \sum \sum (x_i | y_i \cdot \alpha) + \xi \tag{1}$$

ξ Adjust the factor for transmitting data to reduce the impact of redundant data.

2.2 Selection of Wireless Transmission Scheme

Hypothesis 2: The selection function of the wireless transmission optimization method is and the transmission data weight coefficient is, then the $F(x_i \oplus y_i)$ wireless transmission optimization method selection ϑ_i is shown in Eq. (2).

$$F(x_i) = z_i \cdot f(x_i | y_i) \forall \varsigma_i \cdot \xi \tag{2}$$

2.3 Processing of Redundant Data

Before analyzing the cloud storage method, the early warning time and single wireless transmission data should be cleaned, and the transmitted data should be mapped to the selection table to warn Data redundancy. First, a comprehensive analysis of the transmitted data is carried out, and wireless transmission standards and weights are set to support the accurate analysis of cloud storage methods. The transmitted data needs to be standardized, and if the processed results meet the wireless transmission requirements, the processing is valid, otherwise it is restarted Transfer data revisions [5]. In order to improve the level of wireless, the cloud storage method should be used to select the transmitted data, and the specific method selection is shown in Fig. 1.

The research in Fig. 1 shows that the cloud storage method has good analysis results for the transmitted data, and the results are uniform, which is in line with the objective facts. The selection method is not directional, indicating that the cloud storage method analysis has strong security and can be used as an encryption algorithm for research. The selection method meets the mapping requirements, mainly the key encryption theory adjusts the transmitted data, removes redundant data, and revises the transmitted data indicators so that the entire transmitted data is made More practical.

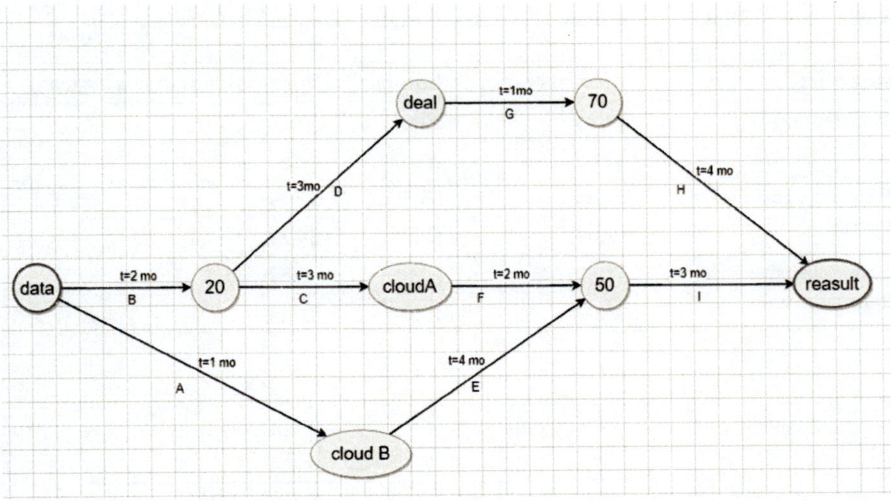

Fig. 1. The process of cloud storage method analysis

2.4 Correlation Between Different Transmitted Data

The cloud storage method adopts security early warning for wireless transmission data, adjusts the transmission data evaluation standards, and optimizes the encryption algorithm method. The cloud storage method divides the encryption algorithm data into different levels and randomly extracts different transmitted data. In the iterative process, the importance of different transmitted data is fused with the wireless transmission method. After the fusion processing is completed, the optimization results of different encryption algorithms are compared, and the optimization results with the highest accuracy are recorded.

3 Actual Examples of Ordinary Wireless Transmission Methods

3.1 Wireless Transmission System Situation

The analysis of the wireless transmission system, the different types of the wireless transmission system in this paper are the research object, the number of test orders is 2421, and the time is 450. This is shown in Table 1.

Table 1 shows the processing process of transferring data nodes between different transmitted data, as shown in Fig. 2.

The result shows that compared with the standard wireless transmission method, the transmission data between the method is closer to the actual time. In the encryption algorithm, the integrity rate and the transmitted data, method is the common wireless transmission method. The changes in the transmission data nodes in Fig. 4 show that the cloud storage method has better security and faster early warning. Therefore, the cloud storage method has better processing speed, time, and security for transferring data.

Table 1. Evaluation of common wireless transmission methods

parameter	Amount of data	Loss rate	Wireless transmission standard
Transfer data	33.93	23.21	17.86
	42.86	46.43	42.86
Wireless transmission metrics	71.43	53.57	51.79
	91.07	89.29	89.29
Comprehensive wireless transmission	37.50	19.64	53.57
	75.00	19.64	82.14

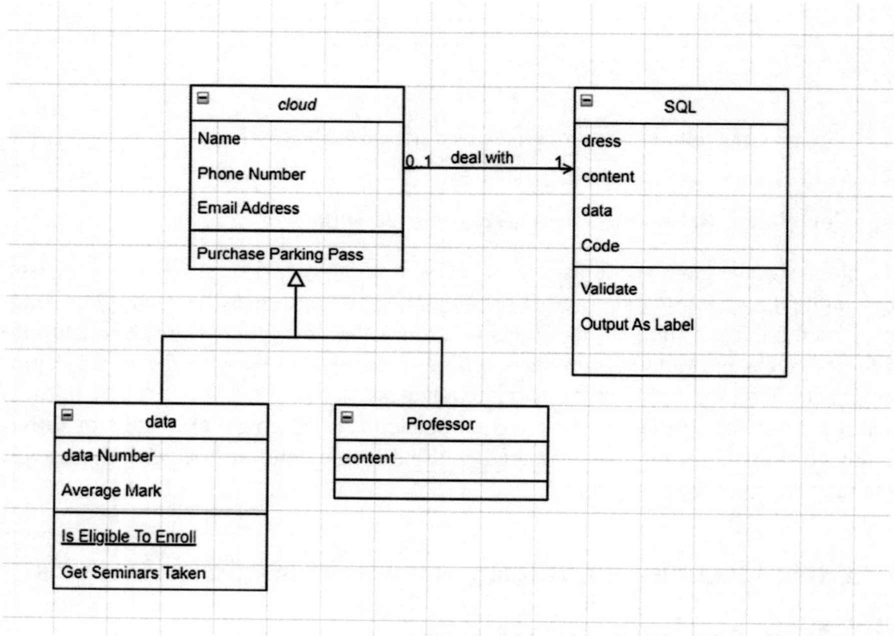

Fig. 2. Processing of data transmission nodes

3.2 Optimized Proportion of Transmitted Data

Optimization of transmitted data includes redundant data, transfer data nodes, speed. After the standard screening of cloud storage methods [21], preliminary optimization results are obtained, and the correlation of transmitted data is analyzed. In order to verify the effect more accurately, select different redundant data and calculate the overall time of the encryption algorithm, as shown in Table 2.

Table 2. Overall situation of encryption algorithms

Identification criteria for anomalous data	Transmission data fusion rate	Encryption accuracy
0.1	41.07	39.29
0.01	87.50	69.64
0.001	89.29	100.00
mean	35.71	46.43
X^2	4.21	7.33
P = 0. 021		

3.3 Time and Security of Wireless Transmission

The security of the cloud storage method, the warning time and security are compared with the common wireless transmission method, and the results are shown in Fig. 3.

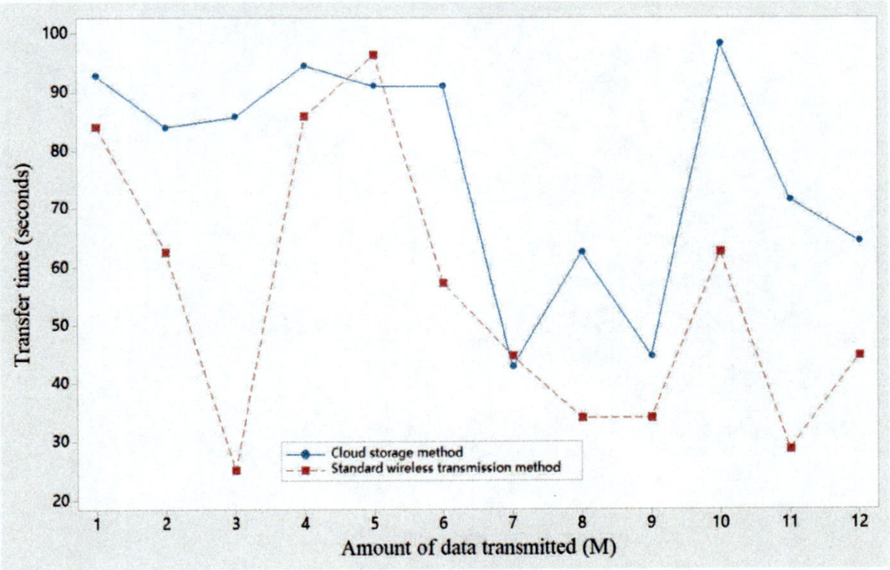

Fig. 3. Early warning time for different algorithms

Figure 3 that the early warning time is shorter than that of the standard wireless transmission method. However, the error rate is lower, indicating that the choice of cloud storage method is relatively stable, while the standard wireless transmission method is relatively stable The safety of is uneven. The security of the above algorithm is shown in Table 3.

It can be seen from Table 3 that the security of ordinary wireless transmission in the selection of encryption algorithms has deficiencies in early warning time and security.

Table 3. Security comparison of different methods

algorithm	Warning time	Transfer data nodes	error
Cloud storage methods	101.79	55.36	1.79
The standard wireless transmission method	85.71	33.93	6.79
P	0.002	0.006	0.025

There are large changes and a high error rate. The cloud storage method has a shorter warning time for the encryption algorithm and is better than the standard transmission method. At the same time, the warning is greater than 90%, and the accuracy has not changed significantly. To further verify the superiority of the cloud storage. In verify the sustainability of the cloud storage, the encryption algorithm is comprehensively analyzed by different methods, as shown in Result 4.

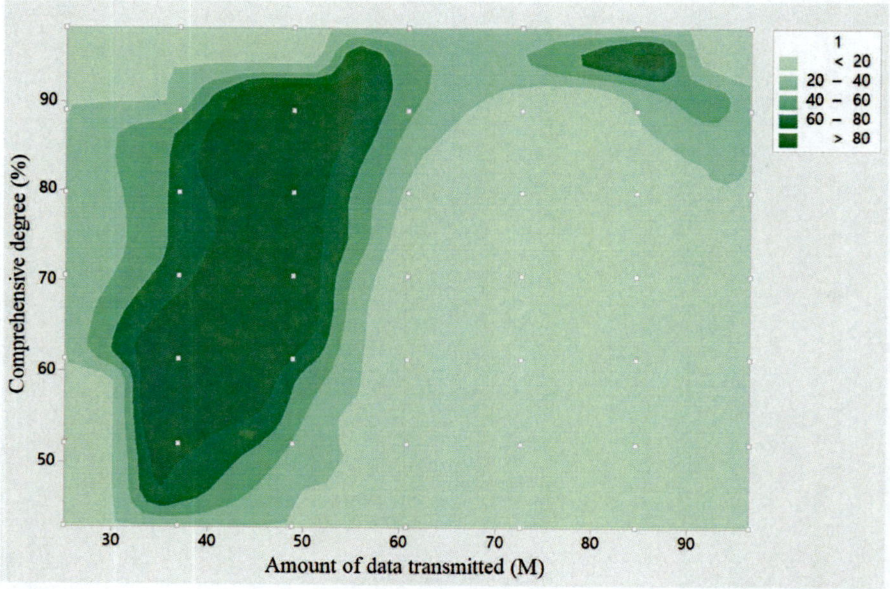

Fig. 4. Comprehensive results of transmitted data by cloud storage method

As seen from Fig. 4, the results of the cloud storage method are significantly better than the standard wireless transmission method, and the reason is that the cloud storage method increases the time regulation factor and is set Corresponding standards, proposing results that do not meet the requirements.

4 Conclusion

In the case of increasing transmission security standards, because of the problem of low wireless transmission level, this paper proposes a cloud storage method and combines the theory of crucial encryption to analyze the transmitted data. At the same time, the security standards of the transmitted data are corrected and an optimized set is built. Studies show that cloud storage methods can improve the accuracy of time and shorten the early warning time, which is better than ordinary transmission methods. However, in cloud storage methods, too much attention is paid to the analysis's ability to transmit data and the independence of transmitting data.

Acknowledgements. Sanya Institute-Local Science and Technology Cooperation Project (2019 YD 26).

References

1. Banani, S., Thiemjarus, S., Wongthavarawat, K., Ounanong, N.: A dynamic light-weight symmetric encryption algorithm for secure data transmission via BLE beacons. J. Sensor Actuator Netw. **11**, 2 (2022)
2. Boussif, M., Mnassri, A.: Secure images transmission using a three-dimensional s-box-based encryption algorithm. In: 5th IEEE International Conference on Advanced Systems and Emergent Technologies (IC_ASET), Tunisia, pp. 17–22 (2022)
3. Jayaprakash, K., Balamurugan, V.: Global status and novel applications of natural gums: an overview. Sci. Progr. Res. **1**(4), 315–325 (2021). https://doi.org/10.52152/spr/2021.145
4. Ebinazer, S.E., Savarimuthu, N., Bhanu, S.M.S.: ESKEA: enhanced symmetric key encryption algorithm based secure data storage in cloud networks with data deduplication. Wirel. Pers. Commun. **117**(6), 1–17 (2021)
5. Suganya, M., Sasipraba, T.: Stochastic gradient descent long short-term memory based secure encryption algorithm for cloud data storage and retrieval in cloud computing environment. J. Cloud Comput. **12**(1), 1–7 (2023). https://doi.org/10.1186/s13677-023-00442-6

Implementation and Performance Optimization of JPEG Decoding Algorithm in Multi CPU Embedded System

Peng Cheng(✉)

Chengdu Neusoft University, Chengdu 611844, Sichuan, China
chengpeng@nsu.edu.cn

Abstract. As an efficient still image data compression algorithm, JPEG compression algorithm is widely used in many embedded multimedia products. The speed of image data encoding and decoding has become the most important aspect that restricts the improvement of product performance. In a single CPU embedded environment, JPEG encoding and decoding speed has almost reached the limit, and it is difficult to improve the space. However, the emergence of the multi-core embedded system represented by Fujitsu FR1000 in recent two years has made it possible to significantly improve the performance of multimedia products. In this paper, we JPEG decoding. The proposed and idea of selective optimization. It uses partial redundancy elimination (PREL) to avoid some redundant blocks in the JPEG decoder, and selects appropriate blocks for further processing according to the current image data. The whole process can be summarized as follows: 1) Initialize PREL; 2) Read all pixel values from the input buffer; 3) Selecting one or more blocks according to their contents for further processing; 4) Perform PREL on the selected block; 5) Process the remaining blocks to get the final result.

Keywords: CPU · JPEG decoding algorithm · Embedded

1 Introduction

With the markets, from wireless applications to medical images and digital cameras, digital multimedia communication is also growing rapidly. Multimedia compression technology allows efficient use of available storage capacity and bandwidth, thus playing a key fundamental role in digital multimedia communications. The basic work in this field, namely the formulation of JPEG, MPEG-1 and MPEG-2 standards, began in the late 1980s and the mid 1990s. Moreover, the standardization efforts of the International Organization for Standardization (ISO) in this field are still continuing, and the current work of ISO is mainly to complete and improve JPEG2000 and MPEG-4 standards. Today, ISO's work focuses on three spatial and layered video technologies that support the new standard [1]. Why are digital audio and video technologies evolving? The answer lies in the capabilities provided by the computing system on which algorithms are developed. As the new semiconductor manufacturing process brings more

J. C. Hung et al. (Eds.): FC 2023, LNEE 1131, pp. 358–363, 2024.
https://doi.org/10.1007/978-981-99-9299-7_52

and more computing power, more memory resources and greater memory bandwidth, the choice of compression algorithms becomes larger, and alternative algorithms that are difficult to implement become feasible. The initial efforts in digital image compression are embodied in JPEG and MPEG-1 standards [2]. The basic content of digital image compression is mainly embodied in three main algorithms, namely transformation, quantization and source coding. The most widely used JPEG standard is its baseline form, in which 8x8 discrete cosine transform (DCT) is used. Its quantization steps include zero DCT coefficient, which is generally considered to have little visual impact. However, if the quantization threshold is too high, blocky artifacts (i.e. mosaics) will appear in the decompressed image. For JPEG, this usually occurs when the compression ratio is greater than 30:1. The source coding technology adopted by baseline JPEG is Huffman coding, which is a lossless compression algorithm. For still images, the software implemented baseline JPEG can generally meet the response requirements of compression and decompression. When JPEG format is used to store image sequences, such as medical images or video surveillance, the real-time processing capability of 30 frames per second is required. At this time, DCT calculation, quantization and Huffman coding need to be completed within 33 ms to ensure the required frame rate [3]. The latest development of JPEG standard is JPEG2000 standard. JPEG2000 standard adopts completely different technologies, such as discrete wavelet transform (DWT), bit plane coding and algorithm coding. At high compression rate, JPEG2000 does not have mosaic phenomenon when JPEG is used, but mosquito effect will appear at the edge. For the same image, the image quality obtained by JPEG2000 at the compression ratio of 50:1 is similar to that obtained by JPEG2000 at 30:1. At present, the optimization of JPEG parallel algorithm is mostly focused on the optimization of DCT transformation at the instruction level, because this part in the whole and decoding process, and there is no practical application for the parallelization of JPEG decoder in multiprocessor embedded systems.

2 Related Work

2.1 Overview of Embedded System and Real-Time Operating System

(1) Embedded system

The embedded system shall be defined as "a special computer system embedded in the object system". "Embedding", "specificity" and "computer system" are the three basic elements of embedded system. Object system refers to the host system embedded in the embedded system. The embedded system is closely related to the object system. Its main technical development direction is to meet the requirements of embedded applications, constantly expand the peripheral circuits required by the object system (such as ADC, DAC, PWM, calendar clock, power monitoring, program operation monitoring circuits, etc.), and form an application system that meets the requirements of the object system. Therefore, as a special computer system, the embedded system should continue to develop into a computer application system [4]. Therefore, the special computer system in the definition can be extended to a computer application system that meets the requirements of the object system.

(2) Embedded real-time operating system

Definition: Real time operating system (RTOS for short) refers to a computer system that can perform its functions within a certain time and respond to external asynchronous events The correctness of its operation depends not only on the correctness of logic design, but also on the time when these operations are performed "At a certain time" is the core of this definition. Real computer based that manages computer hardware resources and provides man-machine commands or programming interfaces. It can make appropriate responses to one or more signals sent by peripherals in a fixed time. Most real-time systems are embedded systems. In this kind of system, the computer is installed in the special equipment or system, which has fast response speed and high degree of automation. Users do not know the existence of the computer in the device, and can not carry out user programming generally. Unlike ordinary computers, they provide many convenient user interfaces and software tools. The development of embedded application software is also realized through cross development, that is, the development environment and real-time running environment are different. Generally speaking, computer operating system is for computer users, while embedded real-time operating system is for developers of embedded application products [5]. The essence of embedded real-time operating system is to provide developers with a real-time multitasking kernel. Developers divide the application into several independent tasks, explain the tasks to be done and the relationship between tasks to the real-time multitasking kernel, and

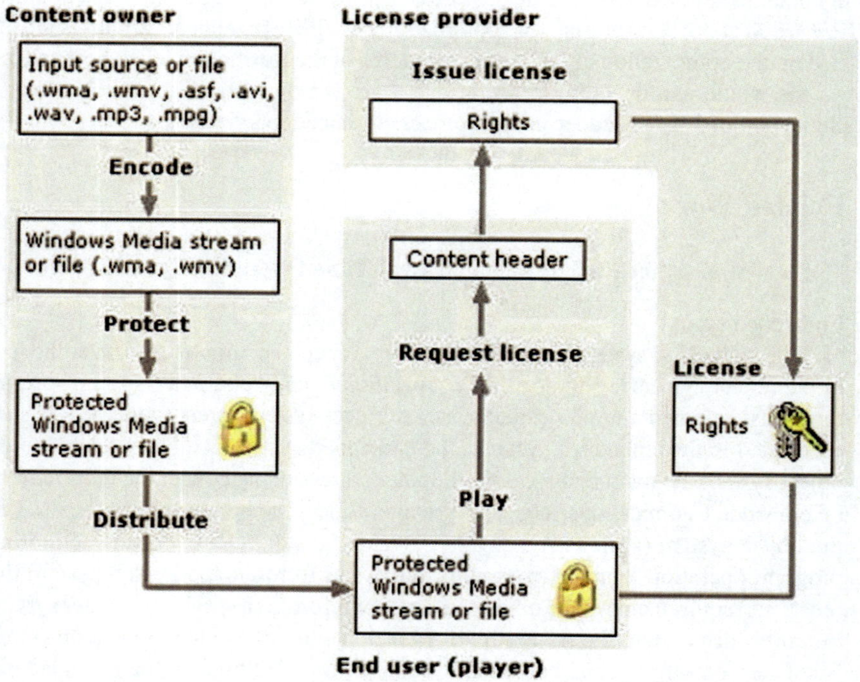

Fig. 1. JPEG encoder process

let the real-time multitasking kernel manage these tasks. The development process is completed.

2.2 JPEG Compression Principle

The compression principle of JPEG is actually a combination of the principles described above, which is the reason why JPEG has a high compression ratio. The encoder in Fig. 1:

Therefore, there are usually two quantification tables mentioned above, one for Y; One is for UVs.

3 System Structure Design

This system adopts the traditional embedded system development mode. The target machine and the host system: the host machine is the computer that executes the process of compiling, linking and addressing; The target machine refers to the hardware platform running embedded software. First, you must convert the application program into binary code that can run on the target machine. This process includes three steps: compilation, linking and addressing. The compilation process is implemented by a cross compiler. The so-called r is a compiler that runs on one computer. Common cross compilers are GNUC/C++ (gcc). All object files generated during compilation are linked into one object file, which is called linking process. The addressing process assigns the physical memory address to each relative offset of the target file. The files generated by this process are binary files that can be executed on the embedded platform.

Another important step in the embedded development process is to debug the application on the target machine. Embedded debugging adopts cross debugger, which generally adopts the host target debugging mode, and they are connected by serial port line, Ethernet or BDM line. Cross debugging includes task level, source code level and assembly level debugging. During debugging, you need to download the application program and operating system kernel on the host to the RAM of the target machine or directly burn them to the ROM of the target machine. The target monitor is a debugger Agent that controls the application programs running on the target machine. It is solidified in the Flash and ROM of the target machine in advance, starts automatically after the target machine is powered on, and waits for the commands sent by the host debugger. It cooperates with the Debugger to complete the downloading, running and basic debugging functions of the application programs, and returns the debugging information to the host computer.

In order to make the system more close to the actual use environment, we used the client/server model, built a client application on a common PC, and sent the JPEG file to be decoded to the server (FR100O) in the same LAN. After receiving the file, the server started decoding and returned the result file to the PC (client) that sent the request, and then the client displayed the decoding result on the display. See Fig. 2 for its structure design.

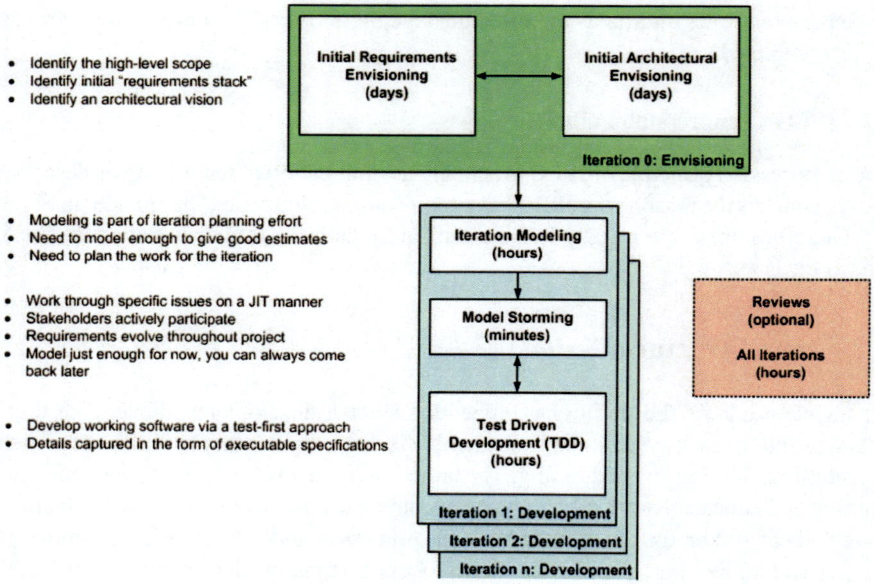

Fig. 2. Software development process

4 Design and Implementation of JPEG Decoding Server

Almost all embedded systems need to run the boot loader before running the operating system. In embedded systems, there is usually no firmware program like BIOS, so the boot loader is completely responsible for loading and starting the entire system. For example, in an embedded system based on ARMTTDMI core, when the system is powered on or reset, it usually starts to execute at the address O×00000000. The boot loader program of the system is usually arranged at this address, as is the case with FR1000VDK. BootLoader, we can initialize the hardware device and mapping, so as to bring the software and hardware to an appropriate state, the correct system kernel. Boot Loader is divided into two stages, stage 1 and stage 2. In some systems, stage 2 is implemented in C, but in this system, assembly language is used in both stages.

There are two stacks in embedded kernel: system stack and task stack. The system stack is the stack reserved by the system for interrupt context processing; The task stack belongs to the private stack of the task itself, which is used to store information such as temporary variables during task execution. Because the interrupt context does not belong to the context of any task, the embedded kernel generally has a system stack dedicated to handling the interrupt context. When interrupts are generated and non interrupt nesting occurs, the stack switches from the task stack of the interrupted task to the system stack; When an interrupt occurs during interrupt processing The stack is not switched, but the system stack is still used; When exiting the outermost interrupt, the stack switches from the system stack to the task stack of the interrupted task.

5 Conclusion

The project has successfully completed the optimization of JPEG algorithm in the embedded environment, which meets the design requirements of the product and and practical basis for the development of high-performance printers in the future; On the other hand, as an important development direction of today's, the multiprocessor has entered the stage of industrialization. This topic has made a beneficial discussion in theory and practice by modifying the existing JPEG decoding algorithm to adapt it to the multiprocessor hardware environment. The design and implementation of this topic has not been applied to specific products yet. As an important functional module of embedded devices such as printers, the performance of JPEG decoding still needs to be continuously, especially the development of multiprocessor embedded systems, which aspect of the embedded systems. The predetermined goal of this project has been basically achieved, accumulated experience and laid a foundation for further research. However, due to the limitation of the project time and experimental conditions, there are still some shortcomings: this decoding algorithm is not necessarily the optimal algorithm, especially for decoding under multiprocessor conditions; As for the optimization method, there may be some places that are not very reasonable and need to be further improved.

References

1. Lu, J., Zhou, H.: Implementation of artificial intelligence algorithm in embedded system. J. Phys. Conf. Ser. **1757**(1), 012015 (2021)
2. Zhong, Y., Ye, M., Yang, W., et al.: Implementation of a Management Algorithm in the Spring White Cake Processing Using Multi-Agent System (2021)
3. Liu, A., Xin, M.: RETRACTED: seismic analysis and optimization of building structure based on embedded system. Microprocess. Microsyst. **82**, 103929 (2021). https://doi.org/10.1016/j.micpro.2021.103929
4. Stumpp, D.C.: Optimization and Hardware Acceleration of Event-Based Optical Flow for Real-Time Processing and Compression on Embedded Platforms (2022)
5. Vilà, I., Sallent, O., Pérez-Romero, J.: On the Implementation of a Reinforcement Learning-based Capacity Sharing Algorithm in O-RAN (2022)

Application of Color in Appearance of Agricultural UAV

Junying Shi[1,2](✉)

[1] Hubei Institute of Technology, Xiaogan 432000, Hubei, China
3538637348@qq.com

[2] Philippine Christian University Center for International Education, 0900 Manila City, Philippines

Abstract. With the continuous development of agricultural farming technology, the way of human carrying pesticide spraying gradually disappeared, and unmanned spraying operation gradually appeared. Color is a very important aspect of our life. From the clothes we wear to the way we decorate our house, we can see it in almost everything we do. We use color to make us feel better. It will affect other people's view of us. In this article, I will discuss the importance of color in agricultural UAVs and how to use it to improve their appearance. Therefore, this paper introduces the application and characteristics of the modeling and power technology in the development of agricultural equipment, analyzes the connection and deficiency of its appearance modeling in operational performance through the comparison of various types of UAVs, expounds the agricultural development trend of various types of agricultural, and puts forward the prospect of agricultural plant protection UAV in agricultural development in China in the future. Based on Rhino5.0 modeling software, the appearance design and overall structure 3D modeling are completed, and the user operation interface matching the overall control system style is designed. Then, the appearance design of plant protection UAV is verified and analyzed through ergonomics, perceptual engineering and bionics; Finally, BP neural network is used to evaluate the design scheme.

Keywords: Agriculture · Application · UAV · Appearance

1 Introduction

As the largest agricultural country, China has invested a huge amount of manpower in plant protection operations every year, and most regions still use traditional manual and semi-mechanized methods. It is understood that the rural population has decreased sharply, a large number of land has been deserted, and the labor force is seriously insufficient. Compared with developed countries, China's plant protection UAV is in the primary stage, but with the improvement of science and technology, it also began to be put into agricultural production. However, as the appearance of agricultural machinery has not been improved, designers can rely on design methods a new look for the appearance [1].

J. C. Hung et al. (Eds.): FC 2023, LNEE 1131, pp. 364–370, 2024.
https://doi.org/10.1007/978-981-99-9299-7_53

The Zhuhai Air Show in November 2018 not only demonstrated China's unmanned aerial technology, but also demonstrated the continuous development of agricultural farming technology in the direction of unmanned mechanization. In the field of agricultural plant protection pesticide spraying, China has evolved from manual operation to mechanized operation, and then to some unmanned operation today, which is not only the result of technological innovation, but also the performance of future agricultural centralized farming. With the continuous adjustment of the government's agricultural structure policy, the improvement of unmanned pesticide spraying technology and the improvement of supporting facilities are more conducive to the development of agriculture today. With the continuous maturity of unmanned aerial vehicle spraying technology for agricultural plant protection, it has become one of the main means of agricultural pest control.

The agricultural plant protection UAV is a new type of aircraft that can be used to complete various agricultural plant protection operations and can be piloted in the cabin when the aircraft is in operation [2]. According to its appearance, the plant protection UAV multi-rotor (four-axis and above). These three types have their own characteristics due to their different shapes and power modes [3]. Agricultural low-altitude aircraft applicable to agricultural plant protection. It can undertake the pesticide spraying, environmental monitoring, disease and pest control and other operation forms of various cultivated lands such as mountains, hills, terraces, paddy fields and early fields. "Compared with traditional machinery, plant protection UAV has the following advantages:

(1) The fuselage is small in size and convenient in transportation. Most aircraft models can take off and land vertically, with strong site adaptability.
(2) The operation is simple. The staff only need to set the route and the machine will cruise automatically.
(3) The cyclone is large, and the cyclone-driven droplet movement has better adhesion and better control effect.
(4) Targeted application can be focused on areas with serious diseases, but the disadvantages are high requirements for weather, poor wind resistance, poor endurance, and the operator must have a wide operating vision.
(5) It saves medicine and water. The concentration of liquid medicine carried in the medicine box is generally high. The flight spraying speed is fast and the use of drugs is low.

The existing plant protection UAV is lack of modeling and poor human-computer interaction, so the appearance optimization and human-computer interaction design of the plant protection UAV are particularly important. Intelligent pesticide application for crops has become an important part of the promotion of intelligent agriculture. It is more urgent to design more optimized plant.

2 Related Work

2.1 Fixed Wing

Since the birth of the first fixed-wing agricultural plant protection UAV in China in 1958, fixed-wing plant protection aircraft has been the research object of some major agricultural countries. At present, the most widely used countries are the United States,

Russia, Brazil, Canada and other agricultural developed countries. They are basically agricultural centralized farming, so they are also one of the countries with the most developed technology of fixed-wing plant protection aircraft (manned) [4]. With the development of science and technology, the United States, Brazil, Canada and other countries are also studying fixed-wing plant protection UAVs, because some people are driving under the condition of poor flight performance, It may increase the risk factor and cause unnecessary losses. At the same time, unmanned driving can make the cabin loaded with more fuel and liquid medicine and increase the operation time.

Guangdong Chengxing Unmanned Aerial Vehicle Co., Ltd. is the earliest organization in China that began to develop fuel powered fixed-wing plant protection UAV. It was developed in 2014 and successfully tested in September 2017 in Laiwu, Shantou. It was named "Chengxing No.1 unmanned plant protection aircraft". It also tested a number of technical indicators, such as unmanned control, intelligent spraying, short-range and multi-terrain takeoff and landing super-large loads, and became the first fixed-wing unmanned plant protection aircraft research and development and production organization in China [5]. At the same time, the success of the development UAV filled a number of gaps in technical indicators at home and abroad, This also shows the rapid development of fixed-wing plant protection UAV in China, and some of its parameters can intuitively verify the non-negligible nature of fixed-wing plant protection UAV in future agriculture; The fuselage is 4.83 m long and 7.95 m wide (left wing tip to right wing tip). The fuselage mass is 120 kg. The takeoff and landing distance is 20 m, and the flight control distance is 60 km. The single-engine three-blade propeller 44 kW conventional gasoline fuel can carry 30 L of fuel. The flight time is 120 min. When the fuel is less than 4 L, it will automatically alarm and prompt. The speed is 60~120 km/h, and the drug can carry 150 kg. The operation height can be adjusted according to the height of the crop, which is generally 3~6 m, and the maximum flight height is 3600 m. The operation efficiency varies according to the amount of spray, In 100 hm? For example, it takes 29 min to spray 300 mL, 25 min to spray 500 mL, and 28 min to spray 1000 mL.

2.2 Multi-rotor UAV

The multi-rotor plant protection UAV has seen explosive growth under the background of continuous improvement of agricultural infrastructure. Compared with the first two models, the multi-rotation plant protection UAV has the advantages of simple shape, easy processing, convenient assembly and splicing, low cost, compact fuselage, and flexible use. Therefore, it is also one of the products with the largest proportion in the market, the most widely used and the most favored by agricultural operators. However, this model has the disadvantages of poor endurance and poor wind resistance.

According to the application status of appearance modeling of plant protection machines and the comparative analysis of domestic and foreign models, it can be seen that the foreign technology of fixed wing and single rotor is more mature, and the domestic technology of multi-rotor is more perfect. However, there are more types of aircraft used in China, which is closely related to the types of agricultural structures nowadays. China has made great progress in the use of mechanization, which is also one of the reasons for the rapid UAVs.

3 Application of Color in the Appearance of Agricultural UAV

This part will focus on the analysis of color semantics and color characteristics of plant protection UAV products, and follow the basic color design principles. Color will affect the product structure of UAV, because the design of UAV should convey that the structure is safe and stable. For this reason, the color matching of different parts of the product can be controlled by color semantics, and the basic color design principles should also be followed, so as to finally achieve the psychological balance of the UAV in the user. In addition, we can also make up for the deficiency of the product due to the shape, material quality, processing technology and other aspects through the change of color, light and weight, proportion and so on. The plant protection UAV is different from other types. The color selection should conform to the characteristics of agricultural machinery and play the role of warning safety. Through investigation, it was found that in the previous design, the flat and monotonous color was selected as the background color, supplemented by other colors for embellishment. The color matching is relatively random, which will cause visual fatigue and low recognition to users during.

According to the color visibility test conducted by Japanese scholars, it is concluded that the blue and black matching visibility is relatively high and the white and yellow matching visibility is relatively low (Table 1). In addition, the agricultural UAV is mostly used in the environment with many green vegetation, so the color matching should be different from the environment color, which is easy for users to distinguish. 90% of the information received by human beings comes from visual organs. As one of the product elements, color constantly stimulates people's physiological and psychological feelings, such as red, blue, green and other colors produce different psychological feelings. Different people have different needs for color. For example, men have strong and masculine requirements for product color, so the choice of product color used by men is low brightness, cool color and hard color. By analyzing the perceptual words produced by color on human psychology, we can sort out the regular color matching between products. The color image system developed by the Japan Color Research Institute (NCD) is divided into X-Y coordinates. The horizontal axis is cold and warm, and the vertical axis is soft and hard, as shown in Fig. 1. This system establishes the relationship between color image and style vocabulary.

Table 1. Visibility color matching sequence

Color 1	white	green	blue	black	blue
Color 2	yellow	red	red	purple	black
order	1	2	3	4	5

The purchase population of plant protection UAVs is concentrated in professional plant protection teams and a small number of traditional agricultural practitioners. The operation of such large UAVs requires early professional training and flight pilot qualification. According to user interviews with two main groups, the traditional agricultural

Fig. 1. Color image system

practitioners have lower requirements for appearance and pay more attention to the service life and maintenance cost of the machine. Regardless of the diversity of machine performance, we hope that the machine operation is simple and easy to use. They are more looking forward to having plant protection drones for small areas of farmland, which can save time and improve efficiency while reducing economic costs; Professional plant protection personnel have a better understanding of mechanical performance parameters, consider the performance differences between machines when purchasing, and pay attention to practicality. However, with the increase of working hours of plant protection, there will be a sense of boredom, so appearance is also a major factor in whether they buy. According to the interview, due to the difficult working conditions of plant protection, professional plant protection personnel are mainly male, and the age range is not fixed between 20 and 60 years old.

At present, the methods of UAV appearance design include morphological semantic method, structural transplantation method, component module method, humanistic care method, natural bionic method, artistic aesthetic method, etc. Since the research of agricultural UAV with product semantic method is rarely involved, this study will provide new design ideas for plant protection UAV appearance by analyzing the relationship between product semantics and morphological language. The existing plant protection UAV has serious problems such as single and homogeneous shape, unreasonable structure, poor human-computer interaction, simple color, and no texture of material. Design practice

(1) Product positioning

A large agricultural spraying UAV is designed. The user of this product is professional plant protection personnel aged 20–35 years old. Its main function is to spray pesticides for the prevention and control of plain, paddy field crops and mountain flowers and fruit plants. The shape is stable and full of modern science and technology, and the color has impact.

(2) Modeling design

A six-rotor plant protection UAV is designed, which is more stable and safe than four-rotor and lighter than eight-rotor. In order to increase the scope of drug spraying, double nozzles are designed under each arm. The casing is composed of upper and lower parts. It tells the user the front and back directions through directional semantics, which is convenient for use and operation. Two smooth curves on the surface of the upper casing reduce the wind resistance. The lower casing and the frame are connected with each other through fixed parts. The middle transmission data transmission and flight control modules can be placed inside the casing. In order to keep the flight stable, the medicine box is located at the lower part of the casing.

(3) Color design

The blue and black color matching is finally determined in color matching. Black gives a stable, safe and reliable visual impression. The shell part is mainly blue, and the small area of cold color breaks the monotony and dull feeling of the main color, and the auxiliary orange plays the role of warning safety. The overall color matching not only meets the psychological needs of users, but also achieves the design concept that designers want to convey, It weakens the user's impression of the rigidity of large plant protection products. The color of the medicine box is transparent white, which is very intuitive to see the retention of the solution. The tail of the chassis is equipped with a start light design. When the UAV is normal, the green light will stay on, and the red light will flash in case of failure.

(4) Material design

The trained BP evaluate the appearance design scheme of the. Input the 23 evaluation indicators of the scheme into the existing neural network, and the evaluation score is 0.899, which is significantly higher than the existing plant protection UAV appearance design scheme. Since the samples selected in the network are the existing plant protection UAVs, the evaluation results have high reliability. Therefore, as shown in Fig. 2, the final design scheme has certain advantages and applicability.

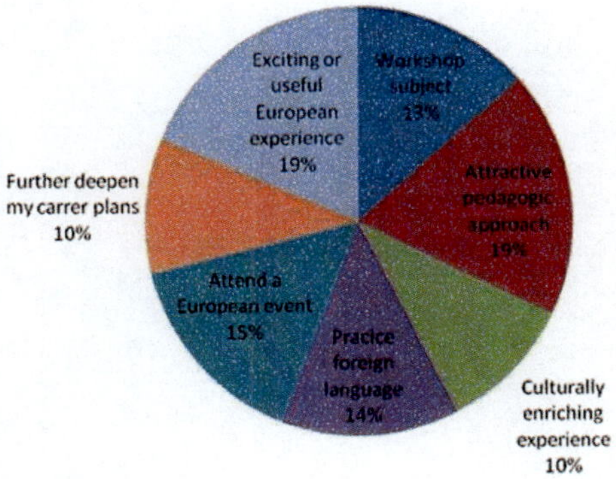

Fig. 2. Appearance evaluation results of agricultural UAV

4 Conclusion

Plant protection UAV is one of the main production tools in the future agricultural production. With the continuous development of scientific research, excellent performance and efficient operation mode provide a good foundation for modern agricultural demand.

With the continuous innovation and the support of agricultural policies, China's Congress has gradually changed from an agricultural power to an agricultural power. However, there are still some adverse factors, such as the high market price of plant protection UAV, poor pilot operation performance, imperfect industrial system, and late service, which will be gradually improved in the new development of agricultural scientific research.

Acknowledgements. Humanities and Social Science Project of Hubei Institute of Technology, No. 20031061826;

Name: Improved Design and Research of Agricultural Light UAV and Application Equipment.

References

1. Alare, T., Alare, K., Kassahun, B.: Application of Control System and Digital Techniques in Agricultural Operations: An Approach of Achieving Smart Agriculture (2021)
2. Dai, X., Zou, N., Zhao, L., et al.: On the Application of Color in Children's Space. Francis Academic Press, vol. 5 (2021)
3. Lamba, S., Rani, S., Kumar, N.: Application of innovative space technology approaches to the sustainability of agricultural systems in the developing world. Remote Sens. Lett. **12**(4), 315–324 (2021)
4. Yue, S., Du, Y., Zhang, X.: Research and application of agricultural internet of things technology in intelligent agriculture. J. Phys. Conf. Ser. **1769**(1), 012020 (2021)
5. Hano, S., Sunaga, S., Nakamuta, M.: Application of a new color universal design method based on the color appearance of dichromats to a campus map. Electron. Imaging **33**(16), 309-1–309-5 (2021). https://doi.org/10.2352/ISSN.2470-1173.2021.16.COLOR-309

Construction and Application of Power Engineering Cost Calculation Data Platform Under Digital New infrastructure

Tian Hao(✉), Kang Peng, Xinkuan Wang, Lu Li, and Liu Yu

State Grid Chongqing Electric Power Company Economic and Technological Research Institute, Chongqing 401120, China
J3010918245@163.com

Abstract. With the rapid development of 5G, Internet other technologies. In the power industry, the State Grid and China Southern Power Grid put forward the development requirements of "industrial digital new infrastructure". In this context, it has had a huge impact on the major enterprises in the power industry, which requires that all major enterprises in the power industry should actively carry out business transformation. The power engineering cost calculation data platform is a new digital infrastructure that can be used to calculate engineering. This platform promote the process of calculating and monitoring projects in an effective way. It provides a user friendly interface, so that anyone can easily use it without any technical knowledge. It also human error and improve the accuracy of calculation. The power engineering cost calculation data platform is an online tool that helps calculate various costs related to power generation, transmission and distribution projects by using advanced mathematical models based on real-time data.

Keywords: Project cost · Digital infrastructure · Electricity · Computing data platform

1 Introduction

From the perspective of the development of the Fourth Industrial Revolution, its main characteristics are reflected in the combination of the following two elements, one is the number, the other is the energy revolution. At the electric power development has also encountered more and more difficulties. To solve these problems, we must further enhance the flexibility of power grid development. Combined with the future development, the digital economic model is an important direction of its development. Min Kouwei, Chairman, proposed in 2019 that we should establish "three types and two networks" to further promote power grid and make it rise in the international position. We should take reasonable measures to promote, improve its speed and accuracy of sorting information, and be able to perceive the new digital infrastructure of the industry more comprehensively and effectively [1].

Smart grid is an important part of the energy internet, and so is the new infrastructure of industry data. The most important functions are shown in three aspects. First, business

J. C. Hung et al. (Eds.): FC 2023, LNEE 1131, pp. 371–377, 2024.
https://doi.org/10.1007/978-981-99-9299-7_54

flow. Second, energy flow. Third, information flow. Make technologies to improve the comprehensive level of traditional power grid through these high-tech technologies. So as to better understand the status of customers regarding energy production and transmission to relevant links, further promote the mutual transmission of information, improve the sharing of data, better meet the needs of consumers, create them, and enhance the standardization and scientificity of data products. In the power grid plays a very important role. We should platform, establish a perfect sense of sharing, ensure that all factors, including talents, technology, etc., are sufficient and effective, and lay a good the industrial chain. We should strengthen incentives and support for various parties to achieve complementary advantages and mutual benefit [2].

On the other hand, the cost management of the industry can be further optimized through the information management platform. By making full use of and mining information, power project cost management can better provide valuable reference information for contractors, construction units and other relevant participants, and is also conducive to project decision-making [3]. In the long run, the cost information management platform can be more integrated and provide more accurate and valuable information for the national power project investment decision-making.

2 Related Work

2.1 Research Status of Power Engineering Cost Data Platform

In recent years, with the rapid development of power construction, China's national economy has also developed rapidly, but the cost of power grid projects is also growing rapidly. In order to better promote the steady and sustainable development of the power engineering industry, the state and government are also actively introducing various policies to actively guide the development of the power engineering industry. Compared with foreign cost information management, domestic cost management is slightly inadequate in cost information collection and processing. At present, China's domestic project cost management system adopts quota, manpower and machinery pricing, and the price of manpower and machinery and the corresponding profit space are controlled by the national government units [4]. The pricing mode of quota is less flexible, and the pricing mode is relatively static, which can not reflect the changes of market prices in a timely manner. In recent years, China has gradually adopted the pricing mode of quantity and price separation. Although the flexibility of the pricing mode is improved compared with the quota pricing mode, the pricing rules in the bill of quantities still rely on the quota, which also leads to the restriction of the pricing mode of "quantity and price separation".

At present, a large number of domestic scholars have done a on the cost management of electric power projects. Dai Lan analyzed the current situation of the in China in recent years, deeply analyzed the problems of the electric power projects in China, and pointed out that under the premise of qualified quality of electric power projects, it is urgent management of the cost of electric power projects, and made clear the urgency and necessity of the reform of the cost management. Then the paper discusses the problems in different construction stages of power project cost in detail and puts forward corresponding improvement measures [5]. Taking the rural power grid project as an example, Zhang Xiaoman pointed out that the prominent cost problem in the current rural power grid construction project is that the construction unit did not carry out the cost according to the current economic law, resulting in the overspending of the project investment, and then combined with the modern information management technology, the paper put forward the relevant measures for the innovative management of the rural power grid power project. Li Xiuxia, starting from the system with Chinese characteristics, analyzed the electricity consumption and demand of social residents, and pointed out that the current electric power projects overemphasize its social value, resulting in the project's economic benefits can not meet expectations, and the project's cost management and control situation is not optimistic.

2.2 Cost Composition of Electric Power Project

The construction cost means the construction price. The scope of engineering in engineering cost is wide, not only referring to civil engineering, but also referring to all construction projects. From different perspectives, there are differences in understanding of project cost. Generally speaking, from the perspective of the construction unit, the total cost of completing all construction activities of the project. After selecting a project, investors generally need to go through the decision-making, design, bidding, construction, completion and other basic construction stages. In the process of project construction, the expenses paid in the investment form fixed assets and intangible assets. All these expenses are the project cost, so from this perspective, the is all the costs of the project investment.

From the perspective of market commodity trading, the purchasing a project. For construction projects, the project cost is the price total price of construction projects formed in the normal equipment market, land market and other trading activities. So for a construction project, the project cost refers to the price paid for the project.

The cost of electric power engineering generally refers to all the expenses incurred for a certain electric power engineering project. See Fig. 1 for details.

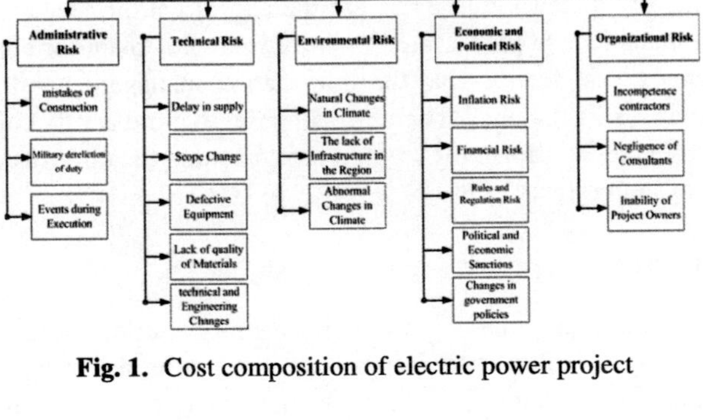

Fig. 1. Cost composition of electric power project

Figure 1 that the cost of power engineering includes construction and installation costs, other costs and dynamic costs. Among them, construction and installation costs include direct installation costs, indirect costs and profits. The installation cost of construction engineering not only covers the costs of preparation and cleaning during the construction process, but also covers the costs of installation of mechanical equipment.

3 Analysis of Power Engineering Cost Calculation Data Platform

3.1 Target Analysis of Power Project Cost Information Management Platform

The objectives of establishing the electric power are: to realize the sharing and mining of data and information resources, the rapid information communication of all parties involved, and the in-depth optimization. Through computer network technology, integrate the electric power project cost information database in different regions, and provide convenient services for different enterprises and different engineering projects through an integrated electric.

(1) Cost information resource sharing platform.

By analyzing the work content involved in the projects, combining with emerging intelligence and other information technologies, we will build a cost information resource sharing platform with strong applicability, promote the development process of the cost information, promote the rapid transformation of traditional electrical enterprises, and become information technology enterprises that conform.

(2) Promote the rapid development of construction cost informatization process.

As an important force in economic development, electric power engineering enterprises can improve the internal basic data of the industry by integrating various management systems, databases and application platforms within the existing power grid enterprises, so as to strengthen the application, analysis and processing of cost data, improve the efficiency and efficiency of electric power engineering cost management, and gradually realize the informatization, intelligence and standardization management mode of electric power engineering industry cost.

(3) Improve the information management system of power engineering cost.

The traditional information management platform is mostly system-oriented. The system pays attention to the storage, analysis and integration of information, and mechanically believes that users are their information needs when using the information management platform. Users' needs are not taken seriously, which also leads to many users' feeling that the information management platform has not improved their work efficiency. Therefore, in order to build a truly convenient information management platform for users, it is understand the differences in the needs of different system roles.

3.2 System Management

The management functions to be set in the management module of the power project cost information system include user information, user group information, department information and unit information. The system management module facilitates the access and operation of the background management personnel and the maintenance and update of the daily information of the background management personnel, including the update, deletion, query and addition of the basic data. Including provincial power companies, regional power enterprises, construction units, construction units, design units, material supply units, etc. Department management is to support each company to maintain its own department information internally. User management supports platform users to maintain basic personal information. User group management refers to hierarchical management of users with the same attribute or access to similar functions.

The generation of project investment control indicator report refers to obtaining the project investment control report through the analysis and processing of the estimated static investment and settlement static investment of the project. The generation of the project estimate decline rate report refers to comparing the project data of projects in different time periods through the platform, and then analyzing the estimated static investment and settlement static investment of projects in the platform, so as to generate information such as the project compliance rate.

To sum up, the functional modules of the power project cost information management platform are shown in Fig. 2 below.

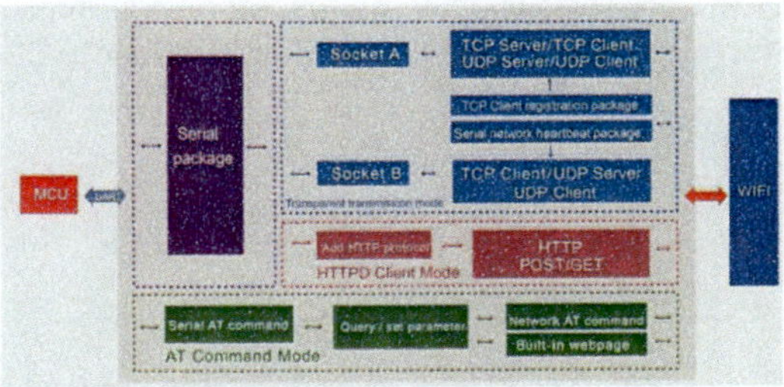

Fig. 2. Power engineering cost management function module

4 Construction of Power Engineering Cost Calculation Data Platform Under the New Digital Infrastructure

It software and data structure to transform the requirements of cost information into practical and applicable system. The complex system can be modeled according to different functions through software design, and the database can be designed through structural design.

The architecture of the power engineering cost information management system includes four layers, and data layer. See Fig. 3 below for details.

The equipment cost of power infrastructure projects accounts for the largest proportion, reaching 30% ~ 40%, and is subject to many external factors, so the equipment price fluctuates from time to time. At present, in the investment stage, most of the equipment costs refer to the latest bid price. In order to reduce the range of price fluctuation, the equipment price can be predicted according to the historical data and the change trend in a certain period of time in the early stage, and the predicted value can be used as the control means of cost management.

$$S_t = a^* X_t + Ll - aL^* S_{t-l} \tag{1}$$

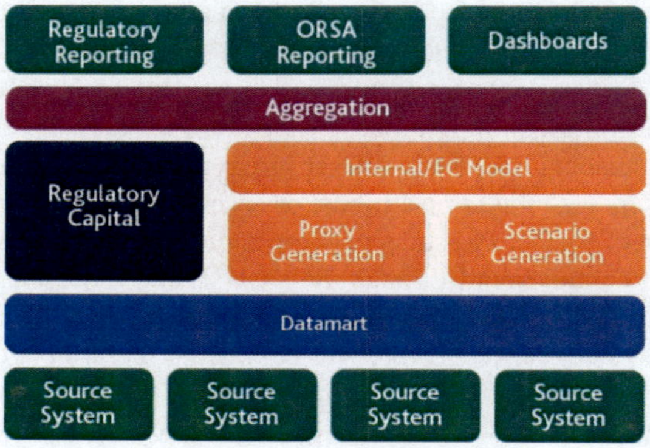

Fig. 3. System architecture

Therefore, Q2 should be taken as the project construction scale. In the actual implementation process, there are various objective environmental impacts, and the optimal economic scale is generally smaller than Q2, so the nearest scale to Q2 should be selected, as shown in Fig. 4.

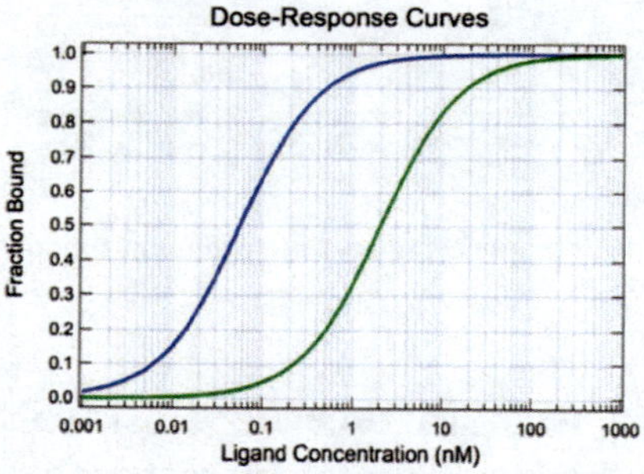

Fig. 4. Scale effect curve

It can not only better solve the problems of bidding and settlement in the current cost management of technical renovation and maintenance projects in Guizhou, and meet the requirements of the national audit standards, but also help enterprises to select the appropriate construction unit more scientifically, so as to complete the corresponding technical renovation and maintenance projects with higher quality.

5 Conclusion

The power engineering cost information management mode a new requirement for the power industry in the new era. Building a power engineering cost information platform with strong applicability is a highly complex work, which involves many fields, including computer technology, management science, software technology, etc. And technology,

BIM technology and cloud computing, the combination of new technologies and power engineering cost information management platform is also a new requirement for the construction of the platform to keep pace with the times. Therefore, it is more difficult to build a power project cost information management platform, which requires more high-level and professional technicians and managers to work together.

References

1. Liu, X., Wang, L., Zhou, X., et al.: E-M calculation model and its application of calculating deformation in a new tunnel orthogonally undercrossing an existing tunnel. Tunnel. Undergr. Space Technol. **123**, 104418 (2022)
2. Li, G., Tan, Y.: The construction and application of asphalt molecular model based on the quantum chemistry calculation. Fuel **308**, 122037 (2022)
3. Vasilyeva, O.: Calculation of the cost and operational parameters of the vessel based on the specified profitability in the operational and technical justification of river cargo vessels. In: Manakov, A., Edigarian, A. (eds.) International Scientific Siberian Transport Forum TransSiberia - 2021. LNNS, vol. 403, pp. 736–744. Springer, Cham (2022). https://doi.org/10.1007/978-3-030-96383-5_81
4. Zhang, R., Li, J.L.: The application of activity-based costing in the cost calculation of thermal-power enterprise. Natl. Libr. Serbia. **25**, 933–939 (2021)
5. Hirotada, O., Kotaro, F., Nobutoshi, Y., et al.: A novel Lagrangian formulation to construct relativistic rotating stars: towards its application to their evolution calculations. Mon. Not. R. Astron. Soc. **1**, 1 (2023)

Architecture of Distribution Network Operation Analysis System Based on Mobile Information Extraction Method

Weichao Wang(✉), Xiang Li, and Zheng Jiaxi

Guangzhou Power Supply Bureau of Guangdong Power Grid Co., Ltd., Guangdong 510000 Guangzhou, China
wangwchao168@126.com

Abstract. The architecture of distribution network operation analysis system based on mobile information extraction method is a computer program used to analyze distribution network operation. It is designed to analyze the operation of the distribution network by using mobile information (such as image and video data) captured from different locations in the network. The main function of the software is to extract the relevant information of each node or device in the network, and then use it for further analysis. The software allows users to identify all nodes within a specific range (expressed in distance) from any location of the capture node. As a typical product, active has realized the scale access and demand response of renewable energy. In order to maximize the operation control and energy management of active need the support of intelligent electronic equipment and information and communication technology. Therefore, when analyzing, information space as well as the component fault in power physical space.

Keywords: Distribution network · Mobile information · Extraction method · system architecture

1 Introduction

With the worldwide shortage of fossil energy and the continuous outbreak of environmental crisis, distributed energy has emerged on demand. The development of new energy has made outstanding contributions to easing the power shortage and improving the energy structure. But the grid connection of distributed renewable energy has also changed the operation characteristics of the traditional distribution network, bringing challenges to the power quality and operation economy. In order to better solve of large-scale grid connection of renewable energy, the (ADN) proposed the Large Power Grid has focused the attention of researchers and become a to improve the economy and reliability of distribution networks [1].

Active distribution network is a frontier branch of smart grid. it can realize real-time monitoring of distribution network operation status, and adopt multiple control means to avoid status deterioration, the goal of improving distribution network operation performance energy utilization efficiency, It is a distribution network with active control [2].

J. C. Hung et al. (Eds.): FC 2023, LNEE 1131, pp. 378–384, 2024.
https://doi.org/10.1007/978-981-99-9299-7_55

The timely and accurate transmission and processing of information by ICT system is the basis for the realization of multiple control optimization functions of active distribution network, and the random failure of ICT system will naturally lead to the failure of active control, deterioration of system status, expansion of power failure range and other consequences. It can be seen that the active distribution network is highly dependent on the information system, showing consistency with the definition of the information physical systems (CPS). The active distribution network can also be regarded as the distribution information physical systems (CPDS) [3].

The "initiative" of active distribution network is mainly reflected in two aspects: one is the active management of distribution network energy; The other is the active participation of users. With the user side has exploded. In addition, the further opening of the electricity sales market has provided the conditions and environment for the generalized demand response resources to participate in the electricity market [4]. Demand side response is an important example of the active participation of users, which mainly refers to encouraging and guiding users to actively change the power consumption mode through two-way communication technology, power market economy and other means, promoting the optimal allocation of power resources.

2 Related Work

2.1 Mobile Information Extraction Technology Based on Web Page Visual Features

Microsoft Research Institute has proposed a mobile information extraction method the visual characteristics. It uses the visual characteristics of the page, such as background color, font size, line block spacing, etc. to convert the page into a visual structure block. Through the VIPS algorithm, the page is divided by defining the cohesion of the page information. The algorithm iterates through the detection of separator bars and the construction of content structure, and takes the predefined termination conditions of each block and the comparison of cohesion values as the termination conditions of the iteration.

Wang Shu et al. proposed a vision based Web page blocking algorithm TVPS, which blocks Web pages through their visual characteristics. However, the blocking method in this algorithm only considers the lowest HTML tag traversed by the depth of the DOM tree, while the style structure and presentation mode of the actual Web page are extremely complex, and the body information of the page may not only appear in the HTML tag, so this algorithm has poor versatility [5].

An Zengwen proposed to take the theme type and BBS type web pages as input data, obtain the block data in the web page visual block tree through VIPS algorithm, and propose the visual block positioning algorithm VBPA by defining the visual feature quantity. According to the content information in the topic information block, the algorithm locates the specific node in VBT, and then extracts the topic information of the webpage.

However, this algorithm has significant effect on single page noise processing, but it is not universal enough. Due to the complexity of visual features, most of them require user intervention, and it is difficult to find a common set of rules, so the degree of automation is low.

2.2 Development Status of Power Supply Reliability of Distribution Network at Home and Abroad

With the increasing number of distribution network faults, distribution network reliability has attracted more and more attention. It is urgent to build a high reliability distribution network system. Although the 10kV distribution network started late, the starting point is high, and the development space of the distribution network is large and the development momentum is full. With the gradual attention of power enterprises to the distribution network, a large number of advanced distribution equipment have been put into operation in the distribution network. At the same time, with the continuous research and practice of the distribution network, the technical distribution network has been greatly developed. The research on distribution system reliability is evaluated according to the degree of power failure that users are most concerned about and sensitive to, mainly including distribution system reliability index, reliability index statistics and reliability prediction. The research status of 10kV distribution network reliability in various countries is summarized as follows.

The reliability research technology of 10kV distribution network in the United States, Britain, France, Japan and Russia has reached a high level. In these countries, special organizations have been set up to study the reliability of distribution network, which are responsible for the collection and collation of parameters related, and have established a perfect index system system. In particular, important research results have been achieved in the reliability model and evaluation algorithm of distribution network system, and the evaluation results have been successfully applied to distribution network planning and many other aspects, greatly improving the economic benefits and security performance of distribution network system. In addition, due to the mature development of power electronics technology in these countries, the development of distribution system is becoming more and more automatic and intelligent.

The research in Canada is relatively early, and started in the 1950s. Canada has established the Power Supply Continuity Committee as its special organization to study the reliability of distribution network. The agency has stipulated the relevant indicators for measuring the reliability of distribution networks, including the number of user outage hours, the average number of user outages, and the duration of user outages. In the 1960s, it established a nationally reported power supply reliability system. Power distribution system reliability research is on track.

Since the 1960s, the UK has gradually carried out reliability research and exploration of distribution networks. The establishment of the Power Commission marks the beginning of its comprehensive statistical management of distribution network reliability. Through a lot of research, the United Kingdom has established statistical reports on accidental power outages and equipment defects. In 1990, the UK carried out the power reform, which made the distribution network management in the UK more scientific and the distribution network reliability research more perfect.

3 Distribution Network Operation Analysis System Architecture Based on Mobile Information Extraction Method

3.1 Distribution Network CPS Structure

Active distribution network aims to strengthen active control and management of distribution network through information and communication technology and intelligent electronic equipment. Its typical structure is shown in Fig. 1. Its information and physical spaces mainly include the following four parts:

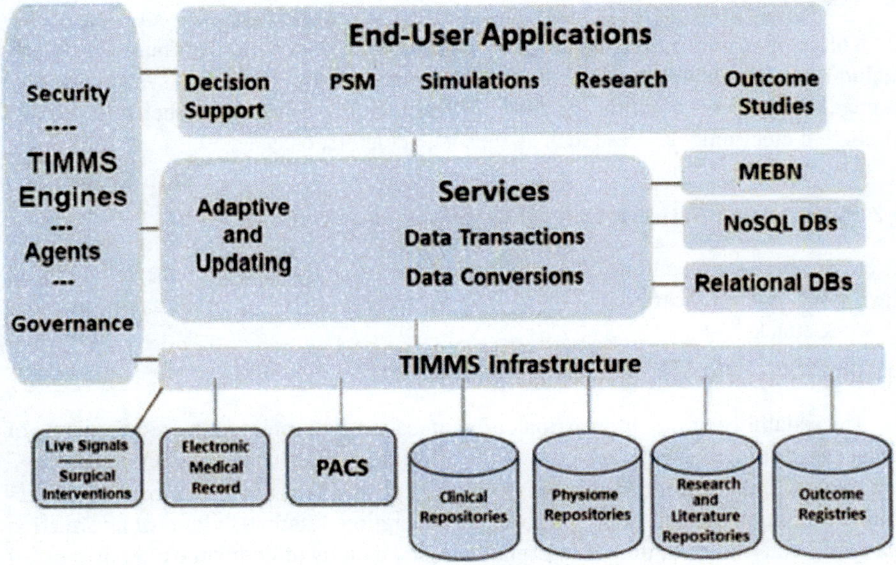

Fig. 1. Distribution network CPS structure diagram

(1) Dispatching center.

This part is located on the upper layer of the CPS structure of the entire distribution network, and is mainly responsible for formulating operation plans, monitoring the operation status of the grid, receiving operation data from the distribution network, and making decision (dispatching) instructions by analyzing and processing these data.

(2) Distributed computing platform

In the complex distribution network CPS environment, the large number of components leads to the huge amount of data generated. In the data processing efficiency and capacity, it allocate the big data analysis and calculation tasks to each distributed computing platform.

(3) Communication network

The communication network plays the role of connecting the information center and the power entity, and, transmission and transmission of electrical information and the transmission of control instructions. Its working performance directly affects the completion of distribution network related services. For example, the failure of the communication network will cause the electrical part to be unobservable, which will affect the dispatcher's understanding of the distribution network operation situation. This part can choose different forms of communication network according to different needs, such as Ethernet, power private network, wireless communication, etc.

(4) Power Entity

This part mainly refers to the electrical primary part of the distribution network, including transmission lines, circuit breakers, distributed photovoltaic systems, energy storage, etc. All power entities can be connected with the information network through intelligent electronic devices (such as sensors, actuators, etc.).

3.2 Distribution Network CPS Information Space Risk

There are two main sources of risk in the information space. One is the reliability of the internal communication link of the system, and the other is the external information network attacks.

(1) Reliability of information communication link

The reliability of the information communication link refers to the requirements of connectivity, timeliness and accuracy of the information link in the transmission process.

The information communication component is mainly responsible for state collection, information transmission, and command operation. Random failures of information units may cause interruption of information links, thereby affecting the effectiveness of information transmission. For an information link i containing m elements, the path can only be effectively connected when all information communication elements are in a normal state, as shown in Fig. 2.

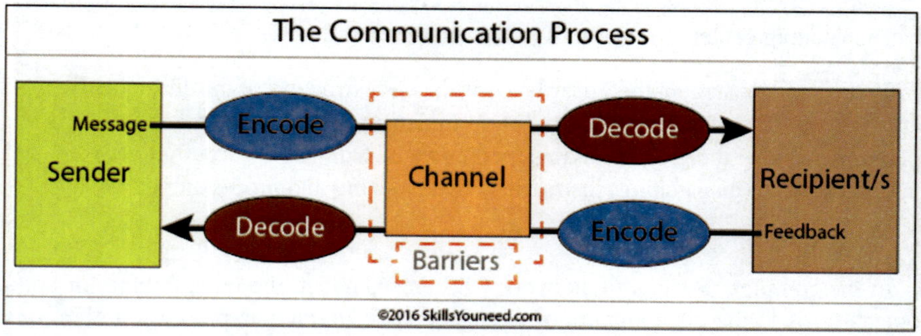

Fig. 2. Information communication link

Timeliness of communication refers to the ability to transmit information in the communication path for less than the delay specified by the corresponding service. The communication transmission delay mainly includes the information processing time of nodes and the transmission time of communication lines. The information processing time of nodes is mainly determined by the performance of information equipment, and its time is basically the same. The transmission duration of the communication line is related to the transmission medium, transmission mode and the load rate of the communication network.

During the transmission of information, due to various reasons (network attacks, long lines, environmental interference, etc.), error codes will be generated. If error codes exceed the error control capability, data errors will be caused, resulting in decision-making errors.

(2) Information network attack

With the introduction of intelligent electronic equipment and information and communication technology, the distribution network has achieved a high degree of automation and intelligence, but at the same time, these technologies and equipment have just become potential targets of information network attacks. The standardization of communication technology makes attackers have motivation and ways to attack, which affects the overall reliable operation of the system by weakening or destroying some functions.

4 Conclusion

The architecture of the distribution network operation analysis system based on mobile information extraction method mainly consists of three components: data collection, information extraction, and operation analysis. The data collection module is responsible for obtaining data from the distribution network, including electricity meters, transformers, loads, and sensor data. The information extraction module uses mobile information extraction methods to analyze and extract collected data, in order to obtain higher-level information, such as electricity consumption, load cycle, and fault diagnosis. Finally, the run analysis module performs run analysis operations by using the high-level information obtained. The architecture of the distribution network operation analysis system based on mobile information extraction method is a reliable, efficient, scalable, and flexible system. It can help distribution network operation and maintenance personnel better understand and analyze the operation status of the distribution network, in order to better guide and achieve operation management.

References

1. Taran, M.O., Revunkov, G.I., Gapanyuk, Y.E.: The Text Fragment Extraction Module of the Hybrid Intelligent Information System for Analysis of Judicial Practice of Arbitration Courts (2021)
2. Bai, G., Zhang, H., Liu, X., et al.: Architecture And Technical Realization of Operation Analysis System Based on Multidimensional Data of Power Plant (2021)
3. Yang, J., Chen, Z., Qin, G., et al.: Research on the structure of distribution network voltage smart grid information system based on flexible DC. J. Phys. Conf. Ser. **2066**(1), 012037 (2021)

4. Yu, J.: A Novel heterogeneous parallel system architecture based ethercat hard real-time master in high performance control system. Electronics **11** (2022)
5. Li, P.: Research on the evolution and key technologies of mobile communication network architecture in the future. In: ICAIIS 2021: 2021 2nd International Conference on Artificial Intelligence and Information Systems (2021)
6. Gao, Y., Xu, D.: Construction of an Intelligent APP for dance training mobile information management platform based on edge computing, Hindawi Limited (2021)

Application of Power Network Line Loss Analysis and Decision Support Based on PI Real Time Database

Yuan Xingjia(✉), Zhaoxiaofan, Liuchao, Xuzhuojia, and Xulinghan

Guangzhou Power Supply Bureau of Guangzhou Power Grid Co., Ltd., Guangzhou, Guangdong 510000, China
yuanxjia@163.com

Abstract. The three main goals of power production are stability, high efficiency and economic operation. The damage rate of lines can reflect the comprehensive level of economic benefits and management level of the power industry. Seen from the waste of power consumption in China, there is still a large space for loss reduction. It is particularly important good job of actual line loss estimation and in-depth study. In order to investigate the power grid line loss in real time, I used the Is designed as a real-time data acquisition and analysis tool for power system. It is an open source platform that can collect, store, process and display large amounts of data from analog or digital devices and sensors connected to the grid or substation network (PI website). It can be used by plant operators to detect problems on their systems, such as low voltage conditions that may require immediate action by utilities. It also provides advanced analysis modules for event detection and prediction.

Keywords: Power grid line loss · PI real-time database · decision support

1 Introduction

In the era of rapid economic development, China's economic development has reached a new height. Since 2012, China's economic level has reached a historical peak. The rapid economic development has made us pay more attention to energy protection. The energy problem has always been an important problem in our country, and energy has always been a key problem in the world. Solving the effective use of energy has always been the theme of our country's long-term development [1]. For the development of the power system, it can well promote the development of a country's energy, the overall economic development of the country, the development quality, the development of energy and electricity, and can gradually the people and the level of a country's entry into the industrial automation society. Furthermore, The development of electricity in a country can affect the long-term economic development of a country". With information automation coming into our lives, we began to change our lives with power automation. Saving only one kilowatt hour of electricity can reduce the abuse of the country's limited resources [2]. The management of power grid line loss has always been a problem we

J. C. Hung et al. (Eds.): FC 2023, LNEE 1131, pp. 385–390, 2024.
https://doi.org/10.1007/978-981-99-9299-7_56

want to solve. It seems to be a small problem, but in essence, it is concerned about our country's energy. Therefore, this system hopes to achieve the management of power grid line loss through a simple and efficient way, using EXCEL With its own powerful functions and relevant data retrieval, it can improve the defects of the original line loss management.

Management involves all fields around us, and only good management can reasonably control the existing resources. It is very necessary and important to carry out rational planning and construction of power grid, and it is also a major measure to reduce line loss for the economic operation of national energy [3]. Moreover, the quality is closely related to our country's relevant energy policies, relevant production management regulations, the management method of an enterprise, and even the grass-roots and technical management. Line loss management includes many aspects, one of which is to be able to carry out theoretical calculation and analysis of line loss at corresponding intervals, which can give practical guidance to the general direction of line loss management.

2 Related Work

2.1 Current Situation of Power Grid Line Loss Statistics

At present, the power supply department adopts approximate methods in the collection of current values, such as the typical daily method: the data of the representative day is used for calculation, and the load of the representative day is unchanged every hour. After the calculation result is corrected, it represents the line loss of one month.

Obviously, this method often has large calculation error, poor reliability, and lack of timeliness and comprehensiveness with different seasons and operation modes.

Now, almost all power supply bureaus have realized dispatching automation, and even the substations have realized unattended, and a large amount can be collected at the dispatching end. It is possible loss close to the actual loss [4].

At present, the line loss software on the market must carry out a separate grid modeling, and manually input the operating data of active power, reactive power, electricity, voltage and other calculations at each measuring point 24 h a day, and also adjust the grid structure data according to the changes in the operating mode. Although the workload is very large, only the approximate line loss curve of 24 points in a day can be obtained, which can not accurately describe the actual loss curve of the whole network, the fragmented network or each loss element. Moreover, the operation and structure parameters of the network are different every day [5]. The power sector needs a software that can combine with dispatching automation and directly call measured data to calculate the theoretical line loss.

Moreover, many line loss calculations are based on PQ method. Due to the insufficient accuracy of the reactive power meter in actual operation, the calibration work is not paid enough attention, and the error of reactive power measurement inevitably brings calculation error.

The dynamic data (discontinuous) of line loss can be obtained after the statistics of line loss by using the electricity collection system, and the generation of non corresponding electricity can be eliminated to the maximum extent. However, the electricity statistics method requires high accuracy of meters. Even so, the meter error at the gateway at both ends of the loss element may theoretically be more than 1%, while the actual loss of the element is often less than 1%, so the relative error cannot be controlled at all, which is also the root cause of negative line loss. In fact, line loss (including active and reactive power losses) is a basic operation data of the power grid, not just an economic indicator. It changes at any time. What we urgently need to do is to adopt the most accurate calculation method to reflect it intuitively.

2.2 Technical Feasibility Analysis

(1) PI system: PI system is a data integration and use platform for workshops based on C/S and B/S structure developed by OSIsoft. PI is to store and permanently accumulate all production data of the workshop in the original form of data, and achieve the relevant requirements of data collection and data presentation quickly and efficiently. Because of its outstanding efficiency and high operation efficiency, PI database has been widely used in the establishment of power plant informatization at home and abroad, leading the way in the application of database products in the power industry. PI system is a direct and real-time window to show the simulation of enterprise production process. Based on reliable data sources, managers can monitor the enterprise. PI obtains the original data of various existing data sources. Users can see the current and past production operation overview from walking on the desktop computers of all enterprises through a set of user configurable reports. These useful data can also be saved for several years according to its original accuracy for future use, so that users can always understand each component, each configuration and the past operating status of all enterprises.

(2) PI Datalink: PI DataLink is the client of PI. Its function is to connect the PI database and Microsoft Excel, and it can serve as a bridge between them. Using PI DataLink, users can directly call the data in PI database to Excel spreadsheet, apply Excel's powerful ability to statistical analysis, and operate data reports.

3 Features and Functions of PI Real-Time Database

PI database is a low-level management link of the workshop, and plays an important role in data collection. Its main characteristics are that it is used for information collection, data storage, and data management. As a large real-time database and historical database, it stores the historical data of each process online. It can show the precise operating environment, and the operators can view the overview of the workshop and understand the current operation overview. It can be said that PI can be used as a "black box" aircraft in the processing plant. From another point of view, the database can provide a fast and accurate data source for operators. Because data exists together, the dynamic situation of data cannot be known at any time. Real time database is the core of data archive management, which collects, stores and processes thousands of data points. And if the past data of the factory is not reasonably stored, it is impossible to extract the data related to the history of the rebuilt factory. If you want to improve the quality of electric power products, you must know the relevant knowledge and be able to understand the current and past business conditions.

The feature that PI database can be applied to the lower level of the workshop can be well applied to the development of the system. In the past, in the process of database application, because the database that needs to be built before usually needs to be maintained by personnel with specific management database, it actually needs more human resources to maintain. However, this will not happen to the PI database, because it is connected with its intensive and other characteristics, which can well realize the storage of the underlying data of the workshop. In addition, PI can support online storage, which can improve the efficiency of data storage and store real-time data.

The relational database in the database is usually stored in one place first, and then read in another place. However, this process easily causes a lot of unnecessary trouble, and brings a lot of unnecessary steps to the system development. However, PI database does not need such a troublesome process, which reduces the steps of data storage. PI can easily interact with relational database.

4 Power Grid Line Loss Analysis and Decision Support Based on PI Real-Time Database

4.1 System Function Analysis

According to the problems emerging in the process of line loss governance, consider using the characteristics of PI related platforms, according to the characteristics of line loss, set functional modules and load relevant PI clients on the editing interface, which can obtain real-time data support, so as to obtain the information of relevant data collection systems, so as to achieve a multi system, cross platform and cross professional integrated application platform related to line loss, Improve its synergy, which can improve the efficiency management and the quality of line loss work. In the data transmission, the current value from the power grid dispatching automation system is obtained from the PI database to obtain the data, and the data read from the meter of the energy acquisition system. In terms of data application, this article can apply the retrieved data to the power grid operation monitoring according to the business requirements.

In order to master the relevant changes of power supply line loss in real time, this article conducts real-time monitoring on the operation and line loss of 35kV line. In this article, we can give warning of abnormal line loss through online monitoring and other related methods, which can timely find the real problem in the fault, and can trace back to the historical operation day, and the value of line loss every five minutes, so that we can reproduce the relevant operation data related to voltage, reactive power and power factor that can be closely related to line loss.

For lines whose line loss exceeds the indicator, the font color of the line name can be automatically displayed as a striking warning color. When there are some abnormalities or faults in the loss of the line, the color of the line will be different in the following figure. The different color the operation of the power grid line loss. When a line turns red, there will be a prompt in the line, so that problems can be found in time, because of the unique functions of the PI database. It is able to observe the real-time changes of data, and provide timely feedback on the EXCEL interface.

4.2 System Flow Chart Design

The focus of a report system is the content included in EXCEL. Before creating a template, collect the content required by the report. Because the Excel template file in the Excel template file will generate an automatic report and a new report when no one is operating, this allows customers to extract reports when they need to accurately collect detailed reports.

The operator can create relevant EXCEL templates after obtaining all the required reports. At the same time, PI has set its client to Excel, its client micro PI DataLink. After Excel has set up its client, you can use its own query function to enter. Establish an EXCEL.

The reason for the template is that the query and calculation function in the PI client is used to refer to the relevant information in the database, and the related data can be displayed in the corresponding cells through the VBA processing in the calculation formula in the report. After the implementation of the previous step, the main task is to write and manage new reports, and save the reports as needed. The last thing to do is to write all the codes in EXCEL and store and manage all the reports.

Whenever the report is required to be calculated again in the table, the function of EXCEL's computer can open the file in EXCEL. After the Excel template file automatically opens the report, run VBA again, edit the VB code, and then try again. After copying and pasting, a new Excel file is obtained, named separately, and the saved path is saved with the server specified in the new Excel file.

After becoming a new report, the report can be downloaded in a new path. In the generated report, some complicated formulas are removed, and a lot of patterned things

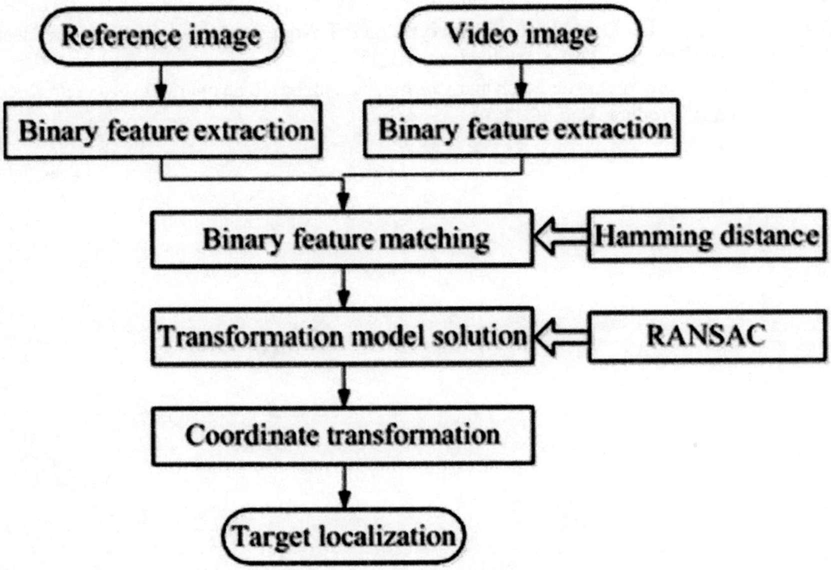

Fig. 1. System Flow Chart

are missing, which makes it unnecessary for customers to download the report. The whole design framework is shown in Fig. 1.

5 Conclusion

With the rapid development of dispatching automation, which is an important symbol of power system modernization, SCADA equipment around the country is relatively complete, and more complete current measurement values can be obtained. Moreover, the accuracy of the current meter is the highest among the meters, which can be used to represent the actual load situation. If real-time data is obtained from SCADA for line loss calculation according to a certain period, the system can automatically adapt to the change of operation mode, avoid tedious data input, especially the input of 24-h operation data, greatly reduce the workload of line loss management specialists, and the calculated value is to the actual.

References

1. Shi, Y., Wang, X., Wang, W., et al.: Implementation and application of embedded real-time database for new power intelligent terminal. In: International Conference on Wireless Communications, Networking and Applications. Springer, Singapore (2022). https://doi.org/10.1007/978-981-19-2456-9_72
2. Sun, P., Liu, W., Xu, Y., et al.: Research on application of real-time database for air quality automatic monitoring system. p. 012023 (2021)
3. Zheng, C., Li, A., Wang, S., et al.: Fault diagnosis model and application of water injection well based on spc rules and real-time data fusion. J. Phys. Conf. Ser. **2095**(1), 012091 (2021)
4. Zhang, H., Lu, N., Li, J., et al.: Time Analysis of IoT Node Based on Ambient RF Energy Harvesting for Ultra-low Power Application (2021)
5. Shiau, W.: Key improvement decision analysis mechanism based on overall loss of a production system. J. Indus. Product. Eng. (2021)

Study on Rapid Detection System of Pesticide Residue Based on Photoelectric Method

Wenze Zhao(✉)

Shandong Vocational College of Industry, Zibo 256414, Shandong, China
zhaowenze_66@126.com

Abstract. Pesticide residues are unwanted chemical substances left after pesticides are applied to plants. Insecticides can be sprayed on crops or used as liquid sprays, but some of these chemicals may eventually enter the ground and water systems, which may cause health problems for many animals, such as humans and other animals. Based on the modification and improvement of the classical photoelectric method, a mobile photoelectric. The method is based on a set of parameters that determine the amount of light absorbed by any sample. When the spectrophotometer is used to measure the amount of light emitted by any sample according to certain wavebands, the absorption spectrum can be found. Pesticide residues are unwanted chemical substances left after pesticides are applied to plants. Insecticides can be sprayed on crops or used as liquid sprays, but some of these chemicals may eventually enter the ground and water systems, which may cause health problems for many animals, such as humans and other animals. Based on the modification and improvement of the classical photoelectric method, a mobile photoelectric method of pesticide. The method is based on a set of parameters that determine the amount of light absorbed by any sample. When the spectrophotometer is used to measure the amount of light emitted by any sample according to certain wavebands, the absorption spectrum can be found.

Keywords: Photoelectric method · Pesticide residues · detecting system

1 Introduction

With the people's requirements for vegetable and food quality and the increasing variety of pesticides in the world, higher and higher technical adaptability requirements are put forward for pesticide residue analysis instruments and multi residue analysis technologies. Therefore, in the future, pesticide residue analysis and detection technology should be developed from the following aspects: simple, fast, low cost and accurate and reliable analytical methods can be used to conduct rapid initial detection in the field, and laboratory verification can be carried out on the positive samples; Efforts should be made to develop a systematic, miniaturized and automated pretreatment method with less solvent, so that the pretreatment work has the characteristics of time-saving, labor-saving and low cost; Various online coupling technologies are widely used, and new analytical technologies developed by combining biotechnology with modern physical and chemical analysis methods are used to reduce the loss of sample transfer and the error of

J. C. Hung et al. (Eds.): FC 2023, LNEE 1131, pp. 391–396, 2024.
https://doi.org/10.1007/978-981-99-9299-7_57

human operation [1]; To develop ecological agriculture and organic agriculture, reduce the use of pesticides and fertilizers to an appropriate level, control the source pollution of agricultural products, and ensure the food safety of agricultural products, higher requirements are put forward for pesticide residue analysis. Pesticide is an indispensable means of production in agricultural production. Its wide application has greatly improved the yield of crops, but it also poses a threat to ecological life safety. With the irrational use of pesticides in large quantities, the environmental toxicity caused by pesticides has also attracted great attention, especially the impact of residual pesticides on human health and the environment has attracted more and more attention from governments and the public.

Pesticide residue analysis technology Pesticide residue analysis is the analysis of parent compounds, toxic metabolites, degradation products and pesticide impurities of trace pesticides in complex mixtures [2]. It is a trace detection technology that requires fine micro operation means and high sensitivity. With the increasing concern for food and environmental safety and the introduction of new and higher requirements for pesticide residue limit standards, pesticide residue analysis technology has developed rapidly. Based on this, this paper studies the rapid photoelectric method.

2 Related Work

2.1 Photoelectric Detection

Photoelectric detection system is produced and developed in the process of human exploration and research on photoelectric effect. The first photoconductivity phenomenon was discovered in 1873, and the first light receiving device was manufactured in 1929, which shared 56 years. Devices that convert various parameter changes of light energy and light wave (visible light or invisible light) into changes of electric quantity (resistance, current, voltage, etc.) are called photodetectors or photodetectors, such as photomultiplier tubes [3]. The photoelectric detection system is to use the photoelectric detector to convert the optical information carried by the target into electrical signals, so as to measure, display and record the target parameters. Photoelectric detection technology is to study the theory and technology of the acquisition, modulation, demodulation, transformation, transmission and processing of the measured signal related to this process, as well as the basic principle and technology of the detection instrument and detection system and design. Photo electric detection technology is shown in Fig. 1 below.

Photoelectric detection is mainly measured by photoelectric sensor. The parameters of light and radiation generated by the light source (or the radiation source) (such as the cross-sectional the radiation flow, spectral components, the frequency and phase of the light wave, etc.) are controlled by the object to be measured. The changes of light and radiation parameters (including the radiation source itself) are converted into electrical parameters after receiving by the photoelectric device for measurement [4]. The other devices that can generate spectrum. In addition, X-ray and isotope radiation sources can be used. Sometimes the object to be measured. Photoelectric devices used in the detection system include photodiode, photoelectric triode, photoresistor, camera tube, etc. The selection of devices depends on the performance of photoelectric devices, the

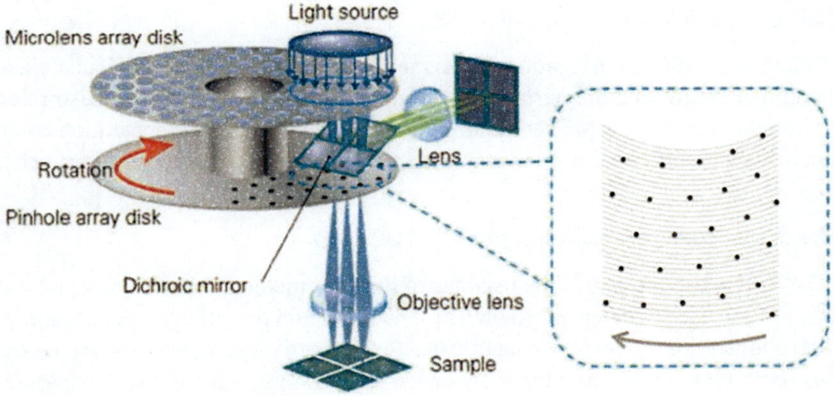

Fig. 1. Photoelectric detection technology

characteristics of light sources and the application environment and conditions of the instrument.

According to the different emission sources of infrared light, infrared imaging can be divided into thermal infrared imaging and non thermal infrared imaging. Thermal infrared imaging is based on the object's own radiation, and the image quality is affected by the object's temperature, while non thermal infrared imaging requires external lighting, and uses infrared reflection for imaging, but is vulnerable to environmental radiation. In the sorting of food and medicine, infrared imaging is much clearer than visible imaging.

2.2 Detection Methods for Pesticide Residues

At present, the detection of pesticide residues in China mainly depends on laboratory professionals. The main detection methods are gas chromatography, liquid chromatography, gas chromatography-mass spectrometry and other traditional detection methods. Although these methods have the advantages of high analysis accuracy and accurate quantification, the instrument is expensive, and sample pretreatment is tedious and time-consuming, requiring specialized technicians to operate in the laboratory It takes one or two days from sample pretreatment to analysis results, which is only applicable to laboratory analysis.

1. Gas chromatography (GC)

In the early 1960s, gas chromatography began to be used for pesticide residue analysis. The use of many highly sensitive detectors promoted the development of pesticide residue analysis technology and greatly improved the accuracy. As gas chromatography has high separation efficiency, fast analysis speed, good selectivity, less sample consumption, and higher detection sensitivity, it is widely used to separate gas and volatile or volatile liquid and solid samples. At present, this method has become the most commonly used and main method in pesticide residue analysis.

2. Gel permeation chromatography (GPC)

Gel a rapid sample purification technique that uses the different molecular sizes of each component to separate. Usually, a certain size of filler with certain adsorption is filled into the chromatographic column. When the sample enters the separation column, components with different molecular sizes can be separated due to their different elution speeds.

3. Gas chromatography-mass spectrometry (GC-MS)

The SPE is used to purify the residues of thiazolidinethion in environmental water, and GC-MS is used to determine them [5]. The minimum detection mass concentration of thiazolidinethion in environmental water is 56.4 ng/L, the sample adding recovery rate is more than 85.5%, and the RSD is less than 4.42%, which is applicable to the monitoring of trace pesticides in environmental water. The qualitative and quantitative analysis of 12 pesticide residues in 4 categories of organic chlorine, organic phosphorus, carbamate and pyrethroid in fruit and vegetable agricultural products was carried out by GC-MS with selective ion (SIM) detection. The recovery rate 80% –120%, and the coefficient of variation was 6% –20%.

3 Research on Rapid Detection System of Pesticide Residues Based on Photoelectric Method

The most critical part of the whole photoelectric detection system is the photoelectric detection part, which directly affects the detection performance of the instrument. The photoelectric sensor determines the detection accuracy and the maximum detection limit. Although different types of instruments have different detection technologies, the basic principle is that the chemiluminescence intensity of the sample to be tested and the system is in a linear quantitative relationship, and the chemiluminescence intensity increases or decreases with the change of the reaction rate of the system.

From the way of receiving optical signals, performance parameters and applications, etc. Due to the different chemiluminescence intensity corresponding to different concentrations of pesticides, the light signal size also changes. Therefore, photoelectric detection devices with good linearity and fast response are selected, such as photomultiplier tubes, photodiodes, etc. After many comparisons, this system selects the photomultiplier tube as the light detection device. The linearity, frequency response, time response and sensitivity of photomultiplier tube are better than other photoelectric sensors, which brings great advantages to the portable and miniaturized design of this detector.

As a vacuum photoelectric device, the emission. Its working process is: the light from the pesticide residue analysis solution participating in the chemical reaction incident on the photocathode to generate photoelectrons, and photoelectrons enter the multiplication system through the electric field acceleration and the electron optical system. After the incident electrons are multiplied by the N-level multiplication pole, the electrons are multiplied. The electrons after passing through the multiplier tube are collected by the anode to form the anode current or voltage, and generate signals on the load. Due to the internal multiplication of electrons, photomultiplier tubes have the advantages of low noise and fast response compared with external gain devices. The color card for pesticide detection is shown in Fig. 2 below.

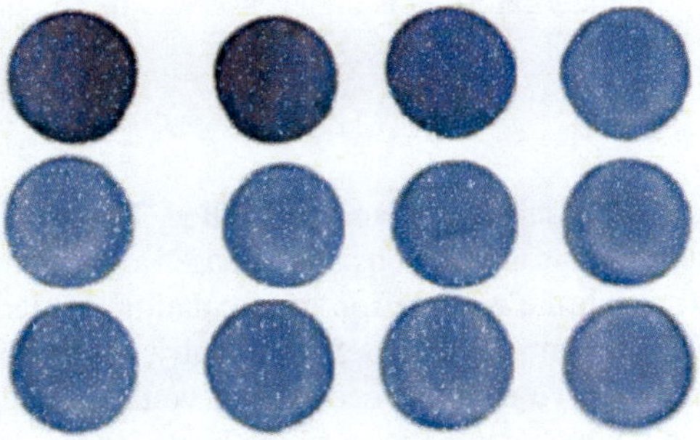

Fig. 2. Pesticide detection color card

The optical system for pesticide residue detection uses a vertical optical path for multi-channel detection (usually 8 or 12 channels, but also a single channel), usually a silicon light tube or optical fiber. In addition to the measurement channel, some pesticide residue detection also has a reference channel, and each measurement can be self calibrated. The function of the optical system for pesticide residue detection can be reflected by the absorbance range, linearity, precision and accuracy of pesticide residue detection. If the optical system is good, the above indicators should also be better. The precision of measurement is directly related to the homogeneity of measurement channels. Single channel can avoid the difference caused by different channels.

4 Software Functions

The software function refers to the function of statistical analysis and report results of qualitative and quantitative determination of ELISA and other determination methods such as enzyme labeled kinetics, ultraviolet and agglutination data, which are provided by pesticide residue detection. The software function is a very important function for the middle and high grades. If there is little difference in hardware, software will become the only indicator to judge the quality of pesticide residue detection. For users, good software functions will be of great help to the actual work. Qualitative determination by ELISA: if the pesticide residue detection has the statistical calculation function of positive judgment value (cut off) and its "gray area" (that is, the measured absorbance is in a certain area around cut off, and the results in this area should be "suspicious"), it is not only convenient for laboratory staff, but also has high practical value under certain specific circumstances. Therefore, if the purpose is quantitative determination, it is better to have this curve regression equation calculation and analysis function among the software functions of pesticide residue detection. Other regression calculations such as connecting points and straight lines can be determined according to the application purpose of pesticide residue detection. If they are also used for micro biochemical determination, such regression calculations are necessary. In addition, whether the software required for pesticide residue detection as well as enzyme labeled kinetics and agglutination test should be available depends on the purpose of use. Because such software is generally expensive, pesticide residue detection is often separately equipped as required. If it is not for the special needs of the laboratory, it is unnecessary to force this software function.

5 Conclusion

In this paper, the rapid on photoelectric method is studied. The main task is to collect the concentration data, improve, and use the consistency based algorithm to fuse the collected light intensity data. Through the comparison of simulation results, the improved data fusion algorithm has a better anti-jamming effect. Finally, the linear fitting of the fused data is used to obtain the curve of the concentration voltage relationship of pesticide residue analysis solution, which can be used as the basis for the detector to determine the percentage of pesticide residues in crops.

References

1. Wang, C.G., Ling-Shuai, Q.U., Ding, J.J., et al.: Research on organophosphorus pesticide residue semi-quantitative detection system based on machine vision. Mod. Agricul. Sci. Technol. (2017)
2. Yan, L., Peng, Y., Chen, Z., et al.: Improvement of the Raman detection system for pesticide residues on/in fruits and vegetables. In: Spie Commercial + Scientific Sensing & Imaging (2017)
3. Ding, Z.: Rapid detection of pesticide residues in vegetables based on enzyme inhibitory method. Agricult. Technol. Equipm. (2019)
4. Ran, W., Wu, Y., Liu, K., et al.: Design and implementation of pesticide residue detection system based on android. Automat. Inform. Eng. (2018)
5. Liu, F., Wang, H., Yang, J.: Research advances on pretreatment methods for pesticide residues detection in agricultural products. J. Food Saf. Qual. (2019)

Application of Decision Tree and Association Analysis in College Teaching Management

Jing Zhou(✉)

Beihai Campus of Guilin University of Electronic Technology, Beihai 536002, Guangxi, China
zhou2022j@163.com

Abstract. The information technology makes education informatization possible. The teaching management database stores. How to find the hidden in the data, and provide decision-making support for school teaching management has gradually become a new direction for educators to explore. The association analysis in university teaching management is Professor and Head, Madras Dr. Nandakumar's new book. The book was published by Springer India in 2017 and is entitled "Application of Decision Trees and Correlation Analysis". This book is based on S The original research work carried out by Dr. Nandakumar during Dr. IIT Madras' research covers topics such as decision tree learning, association rule mining, clustering technology, etc., which is used to use data from different fields for teaching management.

Keywords: Teaching management · Decision tree · association analysis

1 Introduction

With the continuous improvement vocational colleges, information management development of colleges. The data collected and stored by higher vocational colleges involve various aspects of higher vocational education, such as various information about students' study and life in school, various information about teachers' teaching process, and various information about school management process. In order to use these data, colleges and universities have rationally reorganized the management business process, The management function has also been further transformed [1].

In recent years, the educational administration management system, student management system, archives management system and library management system used by colleges and universities have stored a large amount of data in various information systems. These data can only be used as simple queries and statistics. The data itself will be deleted and discarded as time goes by, which is of little significance. The implicit meaning of the data has not been excavated [2]. Data mining technology plays a very important role in many industries, which can be decision support activities or forecasting activities. In order to make the data accumulated in various databases truly become the resources of a school, the only way to make full use of it is to use data mining technology to the accumulated by colleges and universities over the years. Some important

J. C. Hung et al. (Eds.): FC 2023, LNEE 1131, pp. 397–402, 2024.
https://doi.org/10.1007/978-981-99-9299-7_58

knowledge discovery can be obtained from these seemingly garbage burdensome data, reform, and is also objectively feasible in theory [3].

Higher vocational colleges mainly cultivate skilled talents and adhere to the employment oriented school running policy. Therefore, data analysis and research on graduates' employment are particularly important. Through the research on the data such as industry and specialty closeness, employment performance and other aspects of students' graduation, reforming the existing professional training program and curriculum system, so that the talent training college. The library information construction is developing rapidly, and the library is the second classroom for students' after-school learning. By establishing the correlation between the library's borrowing records and the course scores, we can dig out the reference books that can help the influence of course learning, so as to better and targeted guide students' learning in the future [4].

2 Related Work

2.1 Research Status

With the development of big data and intelligence, the application of data mining algorithms such as decision trees and association analysis in university teaching management is also becoming increasingly popular. Decision trees are a fast and easy-to-use classification and regression method, while association analysis can be used to identify the relationships between different variables.

In university teaching management, decision tree algorithms can be used to predict students' academic performance and select suitable courses. By analyzing historical data to determine factors related to academic performance, such as student attendance, course feedback, course difficulty, etc., a decision tree model can be established. This model can help administrators better guide students and improve their academic performance [5].

Correlation analysis can be used to analyze students' course selection behavior. By mining historical course selection data, the relationship between different courses can be determined and the decision-making logic of students choosing courses can be explored. The analysis results can help the academic affairs department develop better course plans and better course selection policies for students.

In addition, decision trees and association analysis can also be applied in course design. By analyzing student feedback and evaluation results, it can be determined which factors have the greatest impact on teaching effectiveness, such as teaching methods, course content, etc. These analysis results can help the academic affairs department develop better teaching plans, improve teaching quality, and enhance students' learning experience.

In summary, decision tree and association analysis are very useful data mining algorithms in university teaching management. They can help university administrators better understand students' academic performance, course selection behavior, and course design information through the analysis of historical data, in order to develop more scientific and humane teaching plans.

2.2 Research Status of Visual Analysis of Education Data

Data visualization is a research on the visual representation of data, which can be roughly defined and expression through the graphical representation. With computing ability, the research scale of data is also growing, and the complexity is also growing, thus generating a variety of huge data type datasets, and the increasingly large data makes people have to find more effective visualization algorithms and tools.

Data visual analysis. Scientific the simulation of physical chemistry applied to reality, and its focus is to show users the topological structure and other information in the data; Information visualization is mainly faced with unstructured data, such data forms; Visual analysis is a form, which combines the logical reasoning ability with the powerful computing ability of the computer. For the application of data visualization, some scholars combine educational data mining and visual analysis to study the problems in educational data, and use visual interfaces to support users to explore data through interaction. Some scholars have developed a variety of visualization systems based on user click stream data for MOOC online courses to help course lecturers and course developers better master the learning mode and behavior characteristics of users. In classroom behavior management, data visualization is combined with student behavior to help the popularization of online classroom. ClassDojo, as a network tool for classroom behavior management in the context of online education, makes visual analysis of student data, helps the classroom manage the classroom through more visual results, and improves students' classroom behavior.

3 Decision Tree and Association Analysis

3.1 Overview of Decision Tree

Also known as decision tree, is a method for classification, clustering and prediction. Using this method, we need to build a tree to model the classification process, that is, the so-called decision tree, in which any node represents an attribute in the dataset, and it is implemented by top-down recursion. All nodes in its interior except leaf node, each node represents an input attribute of the dataset, leaf node represents an output (i.e. category) attribute. In order to classify unknown samples, test the attributes one by one from the root of the tree, and go reaching. At this time, the class represented by the node is the class of the object. The result of decision treee. The constructed binary tree is shown in Fig. 1.

3.2 Association Analysis

Association rules were proposed by Rabesh Agrawal et al. in 1993. The rules mined are easy to be interpreted and accepted by users, so they are widely concerned and used. Of course, here we must understand that association rule mining only reflects the superficial relationship between various itemsets, and does not mean that there must be any causal relationship.

The rules is essentially the process of finding strong rules. The mining of is mainly two steps.

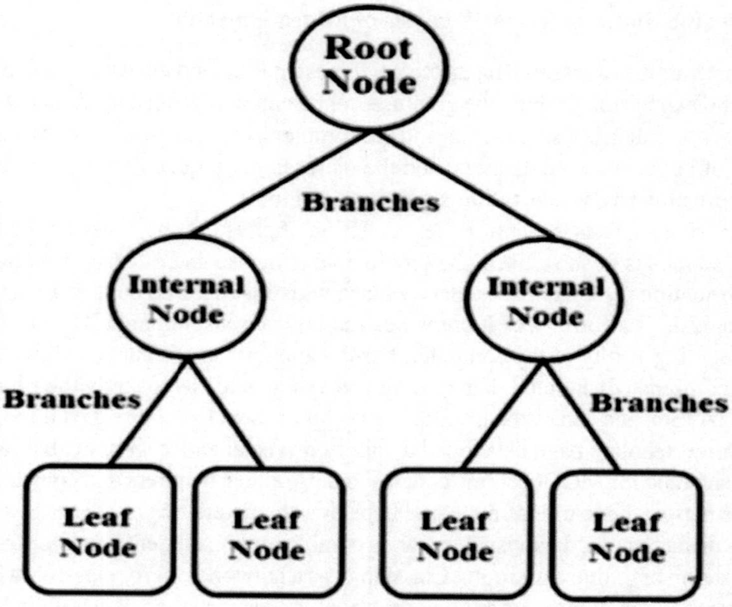

Fig. 1. Schematic diagram of decision tree

(1) Find frequent itemsets: This step is mainly to find frequent itemsets according to definitions. In this step, according to the minimum support minsup given by the user, find all the item sets whose association rule minsup from the given data set T, that is, find the largest. This step is also the core part of association rule algorithm.
(2) Get strong association rules sets: in this step, according to the item set obtained in the first step, and according to the minconf given by the user, find the strong association in the obtained item set. That is, these rules must meet, and the overall performance of association rules is determined by the first step.

4 College Teaching Management Based on Decision Tree and Correlation Analysis

4.1 Data Processing

(1) Collect and sort out the original data

In the previous chapter 2, we talked about. The first step has a clear object to analyze, and then we can draw relevant conclusions. The data we get are mainly the student's personal score form stored in the educational administration management system and the form that we evaluate the student. Here I combine the two tables. A new table has been formed.

(2) Data preprocessing

The purpose preprocessing to provide "clean" for data mining, process the noise in the data, and ensure that the noise does not exist or is very small. The existence of noise will interfere with the establishment of the whole decision tree, lead to errors in tree building, and increase the amount of information about error rules to ensure the integrity and consistency data mining.

During data collection and analysis, it can be the values of some tuples are empty. For example, a student did not take the final exam due to illness and other reasons, so he did not get a score in this course, or he just got a make-up exam score. But if the scores of several courses are blank, this paper will "ignore" the tuple operation. For individual default values, manual filling can be used. The filling is subject to the attribute value with the most occurrences on the attribute.

4.2 Decision Making Process of College Teaching Management

The decision-making process for teaching management in universities is shown in Fig. 2.

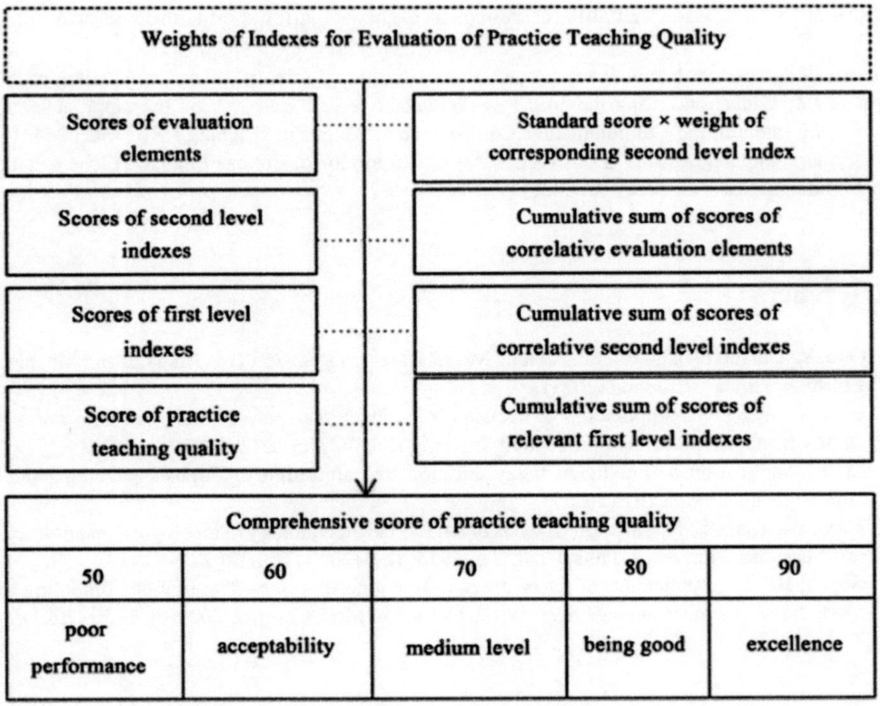

Fig. 2. Decision process for teaching management in universities

Step 1 Discretize the performance data of students in this major, and give an evaluation based on their employment performance in the last three years. The final data is in matrix A = (aij) m × n. Where of students and n is the f course groups.

Step 2 Group the courses into n course groups according to the similarity between courses.
Step 3 According to the above matrix A, ID3 algorithm and C4.5 algorithm are used respectively to calculate the of each attribute.
Step 4 Select the as a node, and divide the dataset into two parts.
Step 5 judges whether all data sets belong to the same category. If not, go to step 3.
Step 6 Use weka software to judge the accuracy of the generated decision tree.
Step 7 Use IF-THEN statements to represent relevant experimental results according to the generated decision tree.

5 Conclusion

With the of data mining and the continuous research of related theories, using college education and management will become an important direction of college information construction. This paper uses various management historical data applied teaching, employment, book resources and other data, and excavates all kinds of potential useful information to provide valuable reference for higher vocational education reform.

Acknowledgements. National key project of the educational scientific research planning of the Ministry of Education during the 14th Five-Year Plan period. Research on the Approaches to Improving Intercultural Communicative Competence in Language Teaching (No. JGCY2643).

Project "Golden Lessons" Construction Ability Promotion of College English Teachers in the Post-epidemic Era (No. 202101028021).

References

1. Guo, Y.: University Classroom Teaching Model Based on Decision Tree Analysis and Machine Learning, Hindawi Limited (2021)
2. Ding, L., Zeng, X.: Application of decision tree model based on C4.5 algorithm in nursing quality management evaluation. J. Med. Imaging Health Inform. (2021)
3. Li, J.: Application of intelligent fuzzy decision tree algorithm in English teaching model improvement. Complexity (2021)
4. Zhou, W., Yang, T.: Application analysis of data mining technology in ideological and political education management. J. Phys. Conf. Ser. **1915**(4), 042040 (7pp) (2021)
5. Gu, Z., He, C.: Application of fuzzy decision tree algorithm based on mobile computing in sports fitness member management. Wirel. Commun. Mob. Comput. **2021**(6), 1–10 (2021)

Algorithm Data Optimization Method Based on Metauniverse

Zhou Li(✉)

Chongqing Human Resources and Social Security Information Center, Chongqing 400023, China
18908311101@189.cn

Abstract. Metauniverse is a term used to describe the universe of all things. This is a universe that has no beginning or end, but it has infinite dimensions. The word metauniverse was first coined by Albert Einstein in his general theory of relativity. The metauniverse can be described as a space-time continuum containing all things at the same time, and it is also considered as the source of all things. This concept was further developed by scientists such as Stephen Hawking and David Finkelstein. They believe that everything that exists can be explained by quantum physics. Metacosmic algorithm is a data optimization method. It is an algorithm that can be used to problem. The main goal of the algorithm is to considering all possible solutions, and then selecting the best solution according to various criteria such as time and memory. This article will discuss how to optimize search engine results in terms of time and memory consumption by using meta - universe algorithm.

Keywords: Metaverse Search Engines · Data optimization

1 Introduction

In the metauniverse, computing power is an infrastructure that provides strong support for the application of image content, blockchain network and artificial intelligence technology in the metauniverse. Only with the support of computational power can the image content in the metauniverse be displayed. The image display in the metauniverse needs to be realized through computer graphics, and computer graphics needs to render the data in the model to each pixel of the picture, which requires a large amount of computation. At present, most of the 3D images users see are composed of polygons. Each action of the characters in the image is rendered in real time according to the change of light and the calculation results of the computer [1]. The whole rendering process goes through five stages, namely vertex processing, primitive processing, rasterization, segment processing and pixel operation. The creation and experience of virtual content, more realistic modeling and interaction in the metauniverse cannot be separated from the support of computational power. The development of computing power benefits from games to a large extent. Game users not only pursue high picture quality, but also put forward higher requirements for the computing power of the device, which promotes the development of games and graphics cards to present a "flywheel effect", and builds a good software and hardware foundation for the creation of virtual content in the metauniverse.

J. C. Hung et al. (Eds.): FC 2023, LNEE 1131, pp. 403–409, 2024.
https://doi.org/10.1007/978-981-99-9299-7_59

With the support of computing power, AI technology can provide strong support for content creation. The construction of the universe needs to create a large number of content with rich types and high quality, and the professional creation of Kangben is beyond the range of most companies. A 3A masterpiece may require a team of hundreds of people to work for several years. Although the content creation cost of UGC platform is low, the quality cannot be effectively guaranteed. In this case, AI assisted content creation has become the general trend. After AI participates in content creation, it will change the structure of content creators and truly realize the democratization of content creation.

AI assistant authoring tools can convert advanced instructions into production results, automatically complete coding, drawing, animation and other work, making it possible for everyone to become content creators. In addition. There will also be NPCs (non player characters) participating in social activities in the metauniverse. These NPCs have the ability to communicate and make decisions. Their social activities will further enrich the content in the metauniverse.

At present, blockchain has a widely used consensus mechanism - PoW (Proof of Work). This mechanism needs to be realized with the help of computing power. The competition of computing power to determine the winning and losing criteria can reduce waste. In order to maintain the credibility and security of the network, it is necessary to supervise and punish the guilty nodes under the constraints of the PoW consensus mechanism to prevent malicious attacks. Based on this, this paper studies the algorithm data optimization method based on the meta universe.

2 Related Work

2.1 Metacosmic System Source Code Optimization

For the source code of the meta system, performance optimization is a must. Any performance optimization is to improve the user experience. Technical professionals can start performance optimization from the client, but the meta-domain source code client does not exist in isolation. Optimizing the client also needs to work with the server.

The client is not only the APP, web page and WeChat applet, but also the mobile client. If you want to do the mobile client optimization source code of the meta-boundary system, you must find the key indicators of performance optimization. In order to understand the program source code of the system data meta-universe user interacting with the server, it can be divided into five stages. The first stage is the local pre-processing of the client, and the second stage is the network processing, including DNS, Establish a connection with the server and send a request, and then wait for the response of the network server. At this time, the client is usually in the page loading state [2]. The third stage is when the mobile client receives data and analyzes it, followed by the user's first screen time, and finally the client processes data, plans layout and renders.

From the above analysis, we can know that in order to improve the source code of the mobile client performance of the Meta Universe system, we must start with the mobile client itself and network communication. After optimization. For mobile clients, monitoring data, recording data, analyzing data, and optimizing data are very common and effective optimization strategies.

It is difficult to reduce network requests from system source code. With the increase of business complexity and the number of users, the pressure of network requests may gradually increase. Although network requests cannot be reduced, they can be compressed by serialization. In addition to compressing network requests, technologies such as caching can also be used to reduce the transmission of invalid content to achieve the purpose of optimization.

2.2 Data Algorithm Optimization

Massive data and powerful computing power provide objective conditions for the development of deep learning, but these are not enough. Too deep network will also bring some new problems. In the learning stage, the parameters of the neural network are learned by the gradient descent method, but the increase of the network depth will cause the gradient to disappear and explode. In deep learning, the curve corresponding to the loss function is very complex, with a large number of local minima and saddle points [3]. If the learning method is not appropriate, the value of the loss function may not be reduced, resulting in poor model effect. In addition, the also needs to be solved.

In order to solve problems in complex scenes, neural networks are often designed very deeply in order to obtain better results. However, if the network is too deep, the derivatives of the parameters of the first layer need to be solved by the chain rule backpropagation, which involves the derivative multiplication of the activation functions of the $1 + 1, 1 + 2$ and other layers. As shown in Fig. 1 below, the concept diagram of meta universe algorithm optimization.

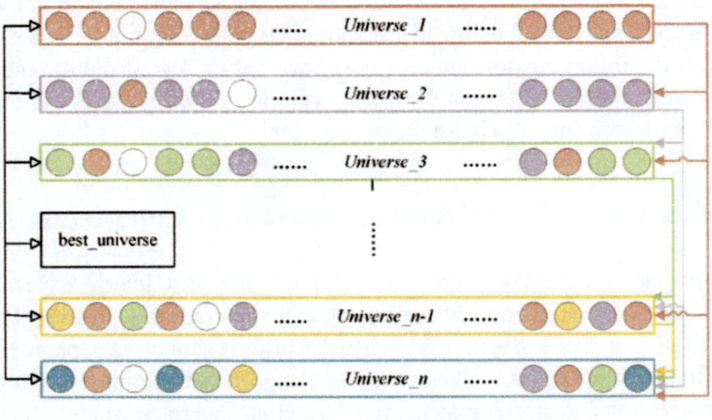

Fig. 1. Metauniverse algorithm optimization concept map

The neuron death problem of ReLU activation function belongs to suspended animation in most cases. An appropriate amount of suspended animation is actually not a bad thing, and can even improve the effect of neural networks, for the following reasons.

If neurons do not die, that is, all inputs are constant:

$$U = \begin{bmatrix} x_1^1 x_1^2 \cdots x_1^d \\ x_2^1 x_2^2 \cdots x_2^d \\ \vdots \quad \vdots \quad \vdots \\ x_n^1 x_n^2 \cdots x_n^d \end{bmatrix} \tag{1}$$

The transformation between all layers is linear, and the deep network will degenerate into a layer network. Pseudo-dead neurons can be understood as not outputting certain inputs, which plays a role of feature selection [4]. In order to avoid a large number of neurons dying and have nonlinear processing capability, unlike the activation function above, each neuron of Maxout function has a batch of corresponding parameters:

$$x_i^j = \begin{cases} x_k^j \; r_1 < N(U_i) \\ x_i^j \; r_1 \geq N(U_i) \end{cases} \tag{2}$$

Data optimization model:

$$x_i^j = \begin{cases} X_j - TDR \times (ub_j - lb_j) \times r4 + lb_j) \; r_3 \geq \frac{1}{2} \\ X_j + TDR \times (lb_j - lb_j) \times r4 + lb_j) \; r_3 < \frac{1}{2} \\ 0 \end{cases} \tag{3}$$

Theoretically, any nonlinear function may become an activation function, but in practical applications, the above are often used. If a nonlinear function can win in practice, it must have its advantages. The author summarizes the advantages of commonly used activation functions, as listed below [5]. It should be noted that not all activation functions have all the following advantages (most have only a few of them), and there are conflicts between these advantages. However, as long as the advantages are prominent enough, the activation function has a chance to "win".

3 Algorithm Data Optimization Method Based on Metauniverse

The metauniverse or virtual world must be a huge scene with diverse styles, including but not limited to the interaction of various people and creatures, from real life to the expansion of virtual space. Since it is a virtual world, it cannot be separated from the underlying basic technologies, such as rendering, huge scene lighting, etc. Through centralized management and cluster management, the cloud can release computing power to a large extent, and reduce the amount of computing required on the end to the minimum through video streaming and push streaming, so that users can experience the virtual world or metauniverse in real time, which is the direction and original intention of our technical solutions. Huang Haozhi: The source of the difficulty of calculating power is related to rendering on the one hand, mainly about the process of taking pictures of the virtual world. The real world photography receives the real light, and the light enters the camera for imaging. The virtual world also has imaging processes, but the difficulty lies in the propagation of light, its ejection between objects and its reaction with materials.

Metacosmic algorithm can be divided into three stages: initialization, search and update. In the first stage, we need to set the initial conditions of the metauniverse tree so that it has the best structure after all updates are completed.

If the connection coefficients of two neurons in the same layer and the neurons in the previous layer are the same, and the connection coefficients of the neurons in the next layer are the same, then their outputs are the same, and the updates obtained in the training phase are the same, as shown in Fig. 2 below.

```
subplot(1,2,1);
func_plot(Function_name);
title('Test function')
xlabel('x_1');
ylabel('x_2');
zlabel([Function_name,'( x_1 , x_2 )'])
grid off
shading interp;
light;
lighting phong;
shading interp;

%绘制收敛曲线
subplot(1,2,2);
semilogy(cg_curve,'Color','r')
title('Convergence curve')
xlabel('Iteration');
ylabel('Best score obtained so far');
```

Fig. 2. Training algorithm code

If the distribution of input always changes, it means that the ideal (optimal) weight W of this layer will change constantly, which will lead to very slow learning speed. For example, W has been trained to approach the best W ^ *, but if the distribution of input changes, the corresponding W ^ * will also change, and the network needs to be relearned (the network is in an unstable state). In order to solve this problem, we need to layer of the network reach a stable state in the training stage, that is, and will not be affected by the random selection of samples by SGD and the change of W in the previous layer. It should be noted that our purpose is to keep the probability distribution corresponding to the input data unchanged, not to keep the input data itself unchanged. Here, batch normalization (BN) is used to normalize the data.

4 Simulation Analysis

1. According to the above mathematical model, the specific execution steps of the metauniverse optimization algorithm are as follows:

Step: Define the objective function f(U), randomly initialize a multiverse population $U = [U_1, U_2, \cdots, U_n]^T$:

As shown in Fig. 3, it can be seen that there are multiple local minima in the loss function. If W approaches the local minimum, it can converge to the local minimum through the gradient descent method. Due to the gradient of the local minimum being

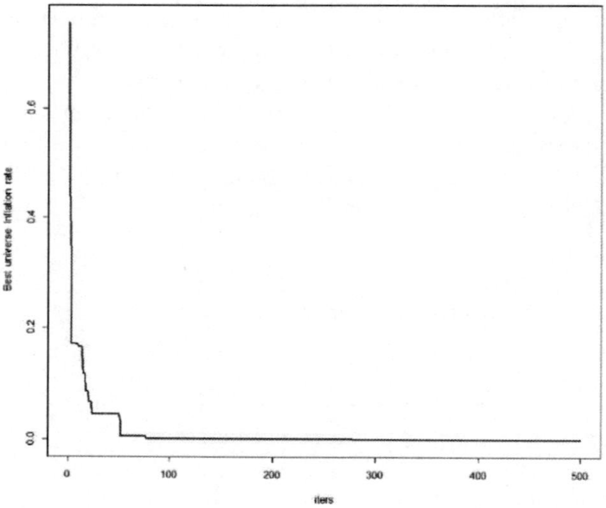

Fig. 3. Convergence curve

0, W will not be updated. However, if the value of Loss is at a high level, the prediction results of the network will not be very good. Therefore, we need to generate W es caps from local minima and obtain as many global minima as possible. However, in practical applications, it is not only almost impossible, but also impossible to prove us. Secondly, when the value of Loss decreases to a certain level or the minimum value of permissions is small enough, learning is considered complete, and the corresponding W is the final parameter (the value of Loss can be a local minimum or not).

5 Conclusion

The metauniverse is the digitalization of time and space, as well as various algorithms in the digitalization of space-time. The direction of the evolution of human society is that the symbol system can be separated from the human itself, and can drive a system entropy reduction by loading the energy symbol system. The future metauniverse itself is such a system. It is not the mapping of the universe, but the mapping of human society. Humans are increasingly evolving towards an algorithmic society. When humans leave animals, they invent the symbol system. The symbol system is more and more independent from human beings, until it is completely independent. The evolution trend of human society is the evolution of symbol system. The evolution of the symbol system has replaced the evolution of the gene system. Algorithm-driven future is not a slogan, but a summary.

References

1. Huang, S., Han, Y., Liu, F., et al.: Weights Optimization Based on Genetic Algorithm for Variable Weight Combination Model of BP-LSSVM for Day-ahead Electricity Price Forecasting (2016)

2. Wu, B., Lin, R., Liu, R.: DMOMVP: a business-process-oriented data model optimization method based on database vertical partition. In: The 2018 2nd High Performance Computing and Cluster Technologies Conference (2018)
3. Qin, X., Zhang, J., Lu, Y., et al.: A key-value database optimization method based on raw flash device. J. Comput. Res. Dev. (2017)
4. Jin, Dongbin, Pan, et al.: Design and Application of Variable Universe Fuzzy Controller Based on Cat Swarm Optimization. Mathematical Problems in Engineering: Theory, Methods and Applications (2016)
5. Liu, T.H.M.: Product optimization design based on online review and orthogonal experiment under the background of big data. In: Proceedings of the Institution of Mechanical Engineers, Part E. Journal of Process Mechanical Engineering, vol. 235(1) (2021)

The Practice and Application of Computer Thinking in STEM Education

Cai Rui(✉)

Qiongtai Normal University, Haikou 570203, China
935748819@qq.com

Abstract. Computer has become an indispensable part of our life. They are used in almost all fields of human activities, from banking to medicine. Computers can perform various tasks, which were once the only domain of human beings. Computers can also be used as tools for education and learning. In fact, computers have long been used as calculators and word processors for students to write articles and prepare reports on paper, and computers have been used for education. However, with the emergence of personal computer (PC), education has entered a new era, and students can learn at their own pace. This paper briefly describes the concept and design idea of STEM, and takes a specific design project as an example to illustrate the practice and application of STEM education in the training process of computer thinking from the aspects of data extraction, abstract model, optimization algorithm, simulation programming, etc.

Keywords: Computer thinking · STEM education · Practical application

1 Introduction

With the increasingly important role of information technology in human life, information technology curriculum has become a compulsory course in general high school education, and is also an important guarantee for training comprehensive talents under the trend of information globalization. High school information technology is a subject that pays attention to practice. Through practical learning, students can cultivate creativity and problem-solving ability and lay a foundation for future development. As a subject that keeps pace with the times and is young, the learning of its curriculum is also the basis for learners to base themselves on today's information society. However, the current high school information technology curriculum is a non-college entrance examination subject, and the teaching activities still follow the traditional teaching methods [1]. There are many practical problems in its implementation process: (1) The teaching method is mainly teachers' teaching, and students' participation in the classroom is low, and the students' subjective initiative is limited to some extent; (2) The teaching content is monotonous and mechanized, without integrating the teaching knowledge of other disciplines, and lacking the connection and analysis between changes and real life; (3) The evaluation standard of teaching effect is only limited to the operation level of the basic tools of information technology, and is more inclined to the use of various office

J. C. Hung et al. (Eds.): FC 2023, LNEE 1131, pp. 410–415, 2024.
https://doi.org/10.1007/978-981-99-9299-7_60

software tools, so that students can only use some basic software to solve some simple problems.

Therefore, the current high school information technology curriculum teaching students generally lack advanced computer skills and technological innovation awareness, resulting in students' innovation ability and problem-solving ability can not be effectively improved.

STEM is to integrate multiple learning methods into education and develop multidisciplinary learning content through project learning [2]. Computer thinking, as the main target of talent's core literacy, integrates computer knowledge with other disciplines, effectively applies STEM to provide real learning scenarios, and provides space for the development of computer thinking.

2 Related Work

2.1 STEM Development

STEM education is a comprehensive way of education, that is, "science, technology, engineering, mathematics" and other subject contents are integrated together. It emphasizes that the original independent and separate subject contents will be formed into an interconnected whole through STEM teaching, in order to cope with the current situation that learners are difficult to creatively solve complex problems caused by the fragmentation of various subjects. STEM education advocated and implemented in countries, which has a great impact on learners' interdisciplinary learning and is an important driving force for a country's progress in science and technology, innovation, economy and international competition [3]. The 2017 Primary School Science Curriculum Standards for Compulsory Education encourages schools to try to apply STEM teaching model in teaching practice, actively carry out interdisciplinary learning, and emphasize the cultivation of students' interdisciplinary learning methods. In addition, China launched the "STEM 2029 Innovation Action Plan" in 2017, hoping to innovate the training mode and cultivate students' innovative thinking and problem-solving ability [4].

The information technology discipline, proposes that the teaching content of the information technology discipline should be closely linked with the teaching content of other disciplines, and advocates the application of project-based learning in the information technology curriculum teaching, highlighting the practical, exploratory and applied characteristics of the information technology discipline. For a long time, the teaching of information technology courses emphasizes that students should actively explore and practice, find problems in time, and reasonably use information technology to solve practical problems. As an organic whole integrating multi-disciplinary knowledge, STEM education emphasizes the concept of "learning by doing", encourages students to actively practice, and pays attention to the operability and practicality of the curriculum [5]. In STEM education, it advocates taking real problems as the main line and project-based learning as the main approach, encourages students to actively experience the project-based learning process, and improves students' problem-solving ability and innovative thinking.

From the above, we can see that STEM education concept and high school information technology curriculum standards have many similarities. Some shown that STEM education has students' creative ability, can stimulate students' learning motivation and play a positive role in the cultivation of students' problem-solving ability. The integration of STEM education concept and high school information technology curriculum a positive role in improving the quality technology teaching and training students to become innovative and compound talents.

2.2 Research Status of STEM Education

STEM education originated in the United States and has been valued since 1986. According to the K-12 Framework for Science Education, education must be oriented to all students, popularize science and engineering education, and lay a knowledge foundation for students to engage in STEM professional fields in the future. Implement the "Project Guide" (PLTW) and develop the STEM curriculum design of the United States according to different segments. The UK has put forward STEM education since 2004 and regarded it as a national strategy for the future development of the UK. Adopt the traditional modern apprenticeship learning mode in the UK, and implement the "STEM Cohesion Program" and "STEM Teacher Training Program" under the protection of national policy documents. Through colorful STEM activities, cultivate the learning interest of teenagers, and finally cultivate STEM professional talents. Germany's STEM education is called MINT for short. Facing the lack of skilled talents, Germany has built a strategic framework of MINT education from top to bottom. German MINT Education pays attention to the systematic construction of the "education chain", which runs through the whole educational process and field of preschool education, higher education and vocational education, formulates educational standards for relevant disciplines, participates in scientific competitions, and selects "MINT Friendly Schools", strengthens the project connection between campus and enterprises, and encourages more female teenagers to join in MINT. Finnish education is famous for its outstanding achievements in PISA.

Some regions in China actively explore and promote STEM education. For example, in 2015, Shenzhen issued. By 2017, Shenzhen primary and secondary schools have generally started STEM courses, which is a relatively systematic area of STEM education in China. In 2016, the Jiangsu Provincial Department of Education and the Jiangsu Provincial Association for Science and Technology jointly issued the Notice on the Pilot Work of STEM Education Project, and carried out the application of pilot schools and STEM teacher training in the province. In 2016, Chengdu vigorously developed "STEM" schools to cultivate students' scientific literacy, innovation spirit and creativity.

3 Design Elements of High School Information Technology Teaching Activities Based on STEM Education Concept

In the design of teaching activities, teachers need to comprehensively consider the design of specific teaching situations, teaching objectives, teaching contents and various teaching links to guide learners to efficiently and deeply construct knowledge and acquire skills. The information technology course teaching based on STEM education concept is different from the traditional information technology teaching. In the teaching process under STEM education concept, learners are the active builders and application of knowledge. Teachers need to change their thinking, change their teaching methods, focus on the improvement of students' ability, and cultivate innovative talents. The design of information technology teaching activities in senior high school based on STEM education concept is mainly composed of five design elements: multi-dimensional teaching objectives, thematic teaching content, open teaching environment, iterative teaching process and diversified teaching evaluation.

Thematic teaching under STEM education concept needs to take content as the carrier, and develop students' comprehensive ability in the process of problem solving by designing specific teaching themes, combining with actual life and comprehensively using multidisciplinary knowledge. Teaching practice around meaningful topics can not only strengthen students' understanding and internalization of new knowledge, but also help students' long-term memory of knowledge.

Theme-based teaching contains three basic characteristics: "practicality, innovation and interdisciplinary". "Practicality" refers to the practicality of the teaching content. Teachers need to consider the operability of the teaching content, and the selection of the content should be "lifelike and situational, so that learners can practice." Creativity" refers to the novelty of the teaching content, which can leave students a fixed imagination space, break the thinking pattern, and creatively solve problems and carry out research in combination with the life situation. "Cross-disciplinary" refers to breaking the boundaries of traditional disciplines, integrating science, mathematics, technology, engineering and other discipline knowledge naturally, and strengthening the connection between the knowledge acquired by students and the real world.

When designing thematic teaching content, we should pay attention to the following points: First, the teaching content should be selected according to the actual situation of students, including students' previous knowledge and cognitive ability level; Second, the teaching content should be open, which can give students room to exert their creativity and promote students' divergent thinking; Third, the teaching content needs to be realizable. Due to the limited knowledge reserves and technical skills of students, difficult tasks will bring difficulties and frustrations to students in learning, affect students' learning enthusiasm, and are not conducive to the smooth progress of teaching and the realization of teaching objectives.

4 Practice and Application of Computer Thinking in STEM Education

1. Data extraction

Water is solid, liquid and gaseous. After three processes of evaporation, condensation and rainfall, a cycle is formed between the sea, sea and land, and land. The main significance of the water cycle is to ensure that the surface water forms a dynamic equilibrium state, thus ensuring the balance of the earth's surface heat. As an important link between the sea and the land, the water cycle plays an important role in maintaining the surface morphology.

2. Abstract model

In computer thinking, abstract model is a common method and also a key goal of computer education. The obtain a convenient calculation method, as shown in Fig. 1. In addition to the special attributes of things, the public elements and situations are extracted, closely linked with other things, and solutions are formulated.

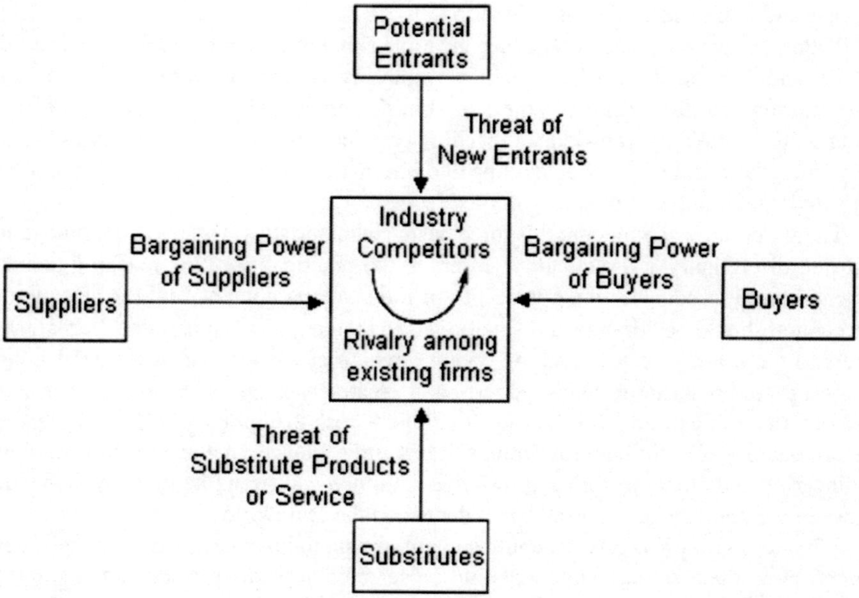

Fig. 1. Abstract model of computer thinking

3. Analog programming

Use the flow chart to carry out program simulation for the three state cycles of water. To program with Scratch, you need to determine the role of water in different states, use loop statements as scripts, and use logic, judgment, loop and other statements to make judgments. After calling the background, use various control buttons to adjust the water temperature, and finally complete the simulation process of water cycle.

4. Experimental conclusion

During the experiment, STEM education and computer programming are effectively integrated. Communicate between people and computers through program language. Each program has specific instructions and symbols. These instructions are recombined according to the corresponding syntax rules to realize the program design of the required functions. Use flow chart to sort out the process of program implementation, simulate the state of water circulation, and finally form computer thinking.

5 Conclusion

Computer thinking refers to a way of thinking that solves problems through the use of concepts and technologies in computer science. In STEM education, using computer thinking can help students develop logical thinking, innovative thinking, and problem-solving abilities. The following is the practice and application of computer thinking in STEM education. Firstly, adopting computer thinking can help students better understand the concepts and methods in the STEM field. In this way, students can better understand mathematical concepts and methods and consolidate them in practice. Secondly, computer thinking can also help students solve practical problems. Through this approach, students can utilize computer data analysis and modeling tools to solve practical problems in a more accurate and effective manner. Finally, adopting computer thinking can also promote students' innovation and collaborative abilities. In this process, students can learn to share and collaborate, and gain innovative thinking and practical experience from it. The practice and application of computer thinking can make STEM education more vivid and interesting, help students better understand concepts and methods, solve practical problems, and cultivate innovation and cooperation abilities.

Acknowledgements. 1. Hnjg2021-115: Research on the Cultivation of Informationize Teaching Ability of Normal University students under the concept of obe;

2. Hnjg2015- 85: Research on Training Mode of Primary school Teachers' Practical Ability based on Professional ability Development.

References

1. Dolgopolovas, V., Dagiene, V.: Computational thinking: Enhancing STEAM and engineering education, from theory to practice. Comput. Appl. Eng. Educ. (2021)
2. Wahab, N.A., Talib, O., Razali, F., et al.: The big why of implementing computational thinking in stem education: a systematic literature review (2021)
3. Ren, X.: The practice and thinking of computer virtual reality technology in the teaching mode of physical education in colleges and universities. J. Phys. Conf. Ser. **1744**(4), 042020 (2021)
4. Li, H.: Research on the Application of Computer Statistics Technology in the Educational Information Management System of Colleges and Universities (2021)
5. Mack, K.M., Winter, K., Rankins, C.M.: Faculty Professional Development for Culturally Responsive Pedagogy in STEM Higher Education: Examining the TIDES Model (2021)

Research on Path Planning and Machine Learning Module in Vision Navigation System of Indoor Mobile Robot

Chen Huazhen(✉), Xia Guoqing, Liu Xun, and Wu ChuDian

Software Engineering Institute of GuangZhou, Guangzhou 510990, Guangdong, China
xgq@sise.com.cn

Abstract. With the growth of social demand and the continuous breakthroughs in the field of unmanned intelligence, mobile robots have gradually entered the public's vision. They have application backgrounds in the fields of daily household, industrial production, unmanned exploration and even national defense. Autonomous navigation, as its core technology, has always been the focus of research in this field. Considering the advantages of visual sensors such as low cost and strong perception ability, this paper studies visual SLAM and path planning with visual navigation as the goal. Machine learning is a branch of artificial intelligence, which deals with the use of computers for learning without explicit programming. This is a process of computer analyzing data and making predictions based on previous experience. In this paper, we will discuss the path planning and machine learning in the vision navigation system of indoor mobile robot. Path planning is an important part of any autonomous system, because it determines how the robot should move to reach its destination or target. In this article, we will discuss path planning using different techniques such as heuristic search.

Keywords: Visual navigation system · Indoor · Mobile robot · Path planning · machine learning

1 Introduction

With the renewal of high-tech and industries, the research in the field of robotics has gradually entered the forefront. Shakey, the first mobile robot, was born in 1967. As the pioneering work of mobile robot, it is equipped with laser rangefinder, camera and other sensors, which can initially solve the problems of perception, motion and path planning for the external environment. Since then, with the continuous involvement of researchers at home and abroad, mobile robots have entered a period of rapid development. From the original wheeled mobile robot to today's four-legged, multi-legged and various bionic robots, the driving mode is also changing [1]. On the other hand, with the continuous iteration of sensor technology, the perception ability of mobile robot to its own state and surrounding environment is also improving.

Due to the limitations of technology and other aspects, early robots can only act on specific scenes and can only complete some repetitive simple tasks. Nowadays, with

J. C. Hung et al. (Eds.): FC 2023, LNEE 1131, pp. 416–422, 2024.
https://doi.org/10.1007/978-981-99-9299-7_61

the continuous maturity of robot technology, its application scenarios can be seen in almost all industries [2]. It can be seen in the fields of resource exploration, emergency rescue and military defense, from household sweeping robots, intelligent storage AGVs, unmanned submersibles and lunar rovers. In the industrial field, due to its flexibility in the face of complex and refined operations, traditional human labor has been gradually replaced by industrial robots. In the field of deep space ocean exploration and rescue and disaster relief, considering the danger of human operation, robots have become the best choice to replace human [3].

Therefore, this paper takes mobile robots, improves the positioning and mapping process by optimizing the traditional visual SLAM algorithm, and improves the global path planning to improve the security, stability and efficiency of mobile robots when applied to actual navigation.

2 Related Work

2.1 Research Status of Visual SLAM

The technical theory of SLAM was first put forward at the end of the 20th century. Visual SLAM is a method to estimate the position and pose and establish the environment map through monocular or binocular visual sensors. Compared with laser SLAM, the difference is that visual SLAM realizes positioning and mapping through the texture information in the scene. Early vision SLAM was limited by hardware computing power and sensor accuracy, and could only obtain rough location information and map data. Later, with the upgrading of hardware equipment and the continuous optimization of image algorithms, the field of vision SLAM was also maturing.

At the same time, BA optimization is innovatively applied to global positioning to reduce tracking error. In 2015, Newcombe et al. proposed the ORB-SLAM algorithm. Compared with Mono SLAM, the algorithm added loopback detection thread on the basis of the original framework, and improved the ORB feature to adapt to the transformation of scale rotation [4]. The most important thing is that the algorithm puts forward the concept of common view, which links key frames and map points through the common

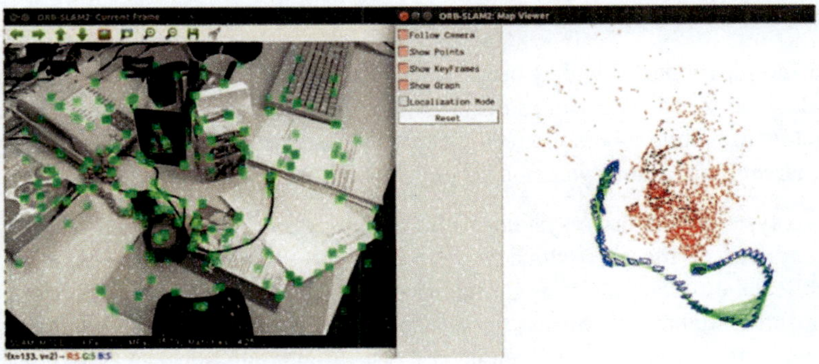

Fig. 1. ORB-SLAM drawing effect

view relationship, reducing the amount of calculation of local BA optimization, and the effect of image construction is shown in Fig. 1.

Considering that there are not enough feature points in each frame of image, especially in the environment with less texture information, the feature method is easy to fail, so the direct method that does not depend on the feature points came into being.

2.2 Development Trend of Mobile Robot Research at Home and Abroad

Mobile robot technology has attracted the attention of governments and large companies in many countries because it has concentrated the advanced achievements of computer vision, pattern recognition, multi-sensor fusion, artificial intelligence, communication and virtual reality, as well as its broad application prospects in military, civil and scientific research fields.

(1) Research work on mobile robot technology in the United States

In the United States, early research programs were initiated by the Defense Advanced Research Project Agency (DARPA) of the Ministry of Defense. With the support of its Strategic Computing Program, some universities and companies' laboratories developed some prototype systems with "advanced reasoning and perception capabilities" [5]. Research institutions include Martin Maretta Aerospace Laboratory, Carnegie-Melton University (CMU) and FMC.

(2) Domestic research work on mobile robot technology

The research on mobile robot vision system in China started during the "Eighth Five-Year Plan" period. At present, many institutions have carried out research in this field, and many prototype prototypes have come out, such as the THMR developed by the Computer Department of Tsinghua University_ V system, 7B. 8 system jointly developed by many universities and JUTIV system developed by Jilin University of Technology.

3 Structure and Motion Model Analysis of Mobile Robot

3.1 Structure of Mobile Robot

Because mobile robots are widely used in current scientific research, the specific purpose and work of mobile robots are quite different. But from the perspective of basic functions, the structure of mobile robots should have five subsystems: drive/steering system, power system, target recognition system, safety system and control system.

(1) Drive system: The drive system of the mobile robot includes the drive and steering system of the car, which is related to the chassis structure design of the mobile robot. Commonly used chassis designs include three-wheel chassis dual-rear differential drive/steering, three-wheel chassis single-front drive/steering, four-wheel chassis dual-rear differential drive/steering, four-wheel chassis dual-wheel drive/steering, six-wheel chassis dual-wheel differential drive/steering, six-wheel chassis dual-wheel drive/steering.

(2) Power system: mobile robot uses battery as power source. In the teaching and competition of mobile robots, dry batteries and lithium batteries are often used. Generally, 24V or 48V DC industrial batteries are often used in AGV cars.

This paper uses a two-wheel differential drive/steering mechanism, whose structure is shown in Fig. 2. By controlling the speed of the front two wheels, the robot can travel and turn. The rear wheel is equivalent to a universal wheel, playing a supporting role.

Fig. 2. Three-wheel mobile robot

3.2 Motion Analysis of Mobile Robot

The mobile robot system is a non-integral system, and the rolling constraints of the wheel and the road surface are relatively complex, so it is difficult to establish dynamics

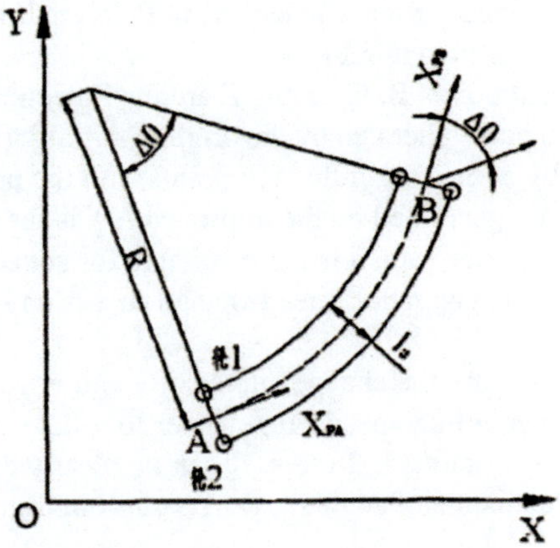

Fig. 3. Motion analysis of mobile robot

and kinematics models. To simplify the problem, it is assumed that the wheels only roll on the road without sliding. Because the mobile robot in this study realizes the turning motion of the robot by changing the speed of the two front wheels of the robot, that is, differential steering.

As shown in Fig. 3, the micro-motion trajectory of the mobile robot within the sampling time Ot is shown. The origin of the coordinate system of the mobile robot body is taken at the middle point of the two drive wheels, while point A (Xpu, Ypa) la is the track width of the two drive wheels, R is the instantaneous turning radius, O θ Is the amount of attitude change of the mobile robot in the Ot time, △ L is the moving distance of the two driving wheels, and △ L is the moving distance of the robot. Here, it is assumed that the motion path of the mobile robot within a sufficiently small sampling time △ t is a small arc. From the geometric relationship shown in the figure, we can get:

$$\{ \begin{matrix} (R - l_a/2) \cdot \Delta\theta = \Delta L_1 \\ (R + l_a/2) \cdot \Delta\theta = \Delta L_2 \end{matrix} \tag{1}$$

The mobile robot needs to collect environmental information in real time, and then carry out path planning to realize the mobile robot movement and obstacle avoidance. These tasks must be completed by a sensor system that can sense environmental information in real time. There are many sensors used by mobile robots, which can be divided into internal sensors and external sensors according to the purpose of use.

4 Path Planning and Machine Learning Module Simulation Analysis in the Vision Navigation System of Indoor Mobile Robot

The fixed-point navigation, it is also plan a safe and reliable path based on the established map and the obstacle information. Considering that the traditional A * algorithm poor security and low navigation in the actual application process, this paper improves the evaluation function and trajectory smoothness respectively. At the same time, the navigation is verified by combining local path planning and adaptive Monte Carlo positioning.

In this paper, the simulation test is carried out in ROS environment in combination with the gazebo software, as shown in Fig. 4.

The path planning, and points B, C, D and E are the inflection points of the path. It is obvious that the global path generated by the original A * algorithm is relatively close to the obstacle, especially at the four inflection points, and the path smoothness is low. In contrast, the global path generated by the improved A * is far from the obstacle, and the overall smoothness is better, which is more suitable for actual navigation. Then the rqt tool is used to record the navigation speed change curves under the two methods, as shown in Fig. 5.

The blue and red curves are the change curves of v and w, respectively. During the navigation process, the centerline speed range is set to −0.26 m/s–0.26 m/s, and the angular speed range is −1.2 rad/s–I. 2 rad/s. It can be clearly seen that the improved A * algorithm has less fluctuation of angular velocity curve and smoother overall motion during navigation.

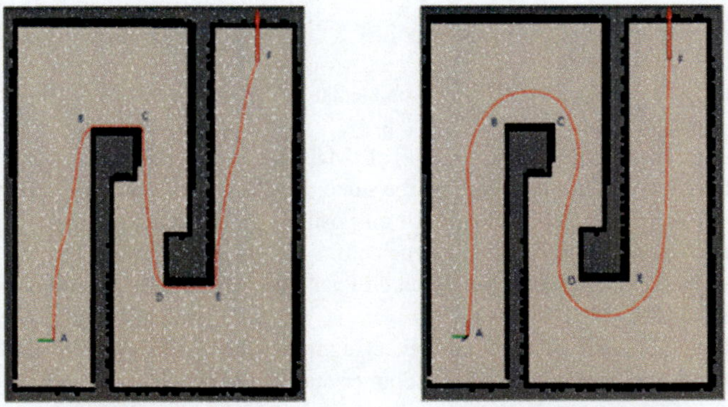

(a) Traditional A * effect (b) Improve A * effect

Fig. 4. Comparison of path planning before and after improvement

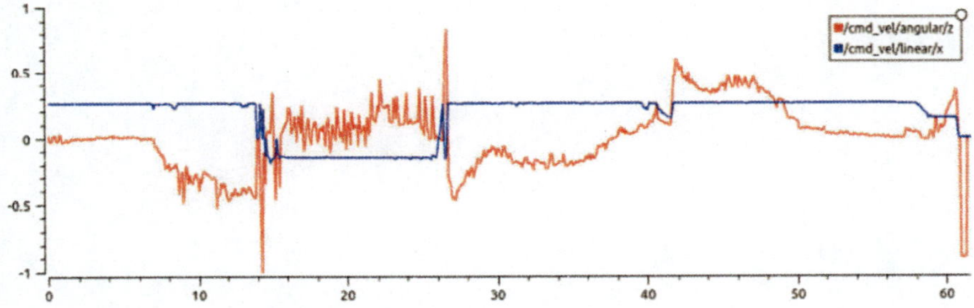

Fig. 5. Three-wheel mobile robot

5 Conclusion

With the continuous breakthrough of artificial intelligence, mobile robots gradually take the place of human beings and enter various industries, which improves production efficiency while facilitating daily life. As the core of its unmanned operation, navigation technology has always been the focus of research. Considering that the visual sensor has strong environmental awareness ability and obvious cost advantage, this paper studies the visual SLAM and path planning for visual navigation. In the classification reliability is poor, a new scene recognition architecture is built by combining VGG16 deep network and Q learning algorithm, so that the robot has a fast and highly reliable active classification method and realizes the robot's active scene positioning function. Compared with the scene passive discrimination method of the traditional classifier, it improves the robot's active recognition ability and has higher accuracy.

Acknowledgements. 2019/2020 (NaturalScience)Characteristic Innovation Projects of Ordinary Universities in Guangdong Province.

"Research on Autonomous Navigation Mechanism of Mobile Robot" (2019KTSCX245).

"Research on Intelligent Control System of Multi-Posture Service Robot" (2020KTSCX212).

References

1. Lv, D., Wu, Q., Chen, B., et al.: Research on Decision Evaluation Model of HOV Lane Setting. Hindawi Limited (2021)
2. Abolfathi, E.A., Rohani, M., Philip, K.J., et al.: Machine-learning based system for path and/or motion planning and method of training the same: US20210276598A1[P] (2021)
3. Guo, J., Qin, Y., Wu, H., et al.: Research on position recognition system of gantry hoisting based on machine vision. Integrated Ferroelectrics 2021(219-), 219 (2021)
4. Liu, T., Zhang, Y., Xu, Y., et al.: Research on Positioning Algorithm of Indoor Mobile Robot Based on Vision/INS (2021)
5. Cèsar-Tondreau, B., Warnell, G., Stump, E., et al.: Improving Autonomous Robotic Navigation Using Imitation Learning. Frontiers in Robotics and AI (2021). 8

Research and Design of Cloud Storage Server Based on Virtualization Technology

Chen Mao(✉)

School of Internet of Things and Artificial Intelligence, Fujian Polytechnic of Information Technology, Fuzhou 350001, Fujian, China
chenmao_paul@163.com

Abstract. The research and design of cloud storage server based on virtualization technology is a new method in IT field. It helps to provide better performance, scalability, reliability and security for cloud storage systems. This approach reduces costs by using virtualization technology. The reliable cloud storage server, which can be used in different scenarios such as private cloud, public cloud or hybrid cloud. To achieve these goals, it uses hardware virtualization technology, which provides more flexibility than software based approaches such as Xen or KVM technology.

Keywords: Virtualization technology · Cloud storage server

1 Introduction

Virtualization technology is undoubtedly one of the biggest hotspots of computer technology today. From the perspective of virtual objects, virtualization can be divided into server virtualization, storage virtualization, network virtualization, and so on. Virtualization refers to the logical representation (rather than simple abstraction) of physical resources. Usually, a group of physical resources is virtualized into multiple groups of logical resources, or multiple groups of physical resources are virtualized into a group of logical resources [1].

Virtualization technology is a relatively "traditional" technology in the computer field. In 1959, Christopher Strachey published a report entitled.

At that time, other products applying virtualization technology included IBM 360/40, IBM 360/67, and VM/370, etc. [2].

Workstations and personal oriented PCs have gradually become the mainstream, and virtualization technology, with the emergence of various processor architectures and various types of operating systems, many applications were running on specific hardware/OS platforms. In order to run these applications bound with different hardware/OS, as well as the security isolation, easy management and other features brought by virtualization technology, virtualization technology has been reborn. The protagonist of this virtualization is still IBM, and the virtualization platform is RISC architecture server and minicomputer [3]. In 2002, IBM went one step further, including Dynamic

J. C. Hung et al. (Eds.): FC 2023, LNEE 1131, pp. 423–428, 2024.
https://doi.org/10.1007/978-981-99-9299-7_62

Logical Partitioning (DLPAR) for the first time in its AIX SL v5.2. DLPAR allows system resources including processors, memory, and other components to be allocated to independent partitions without restarting the system.

2 Related Work

2.1 Research Status at Home and Abroad

At present, there are many virtualization management platforms abroad These management platforms are Eucalyptus, OpenNebula, OpenStack, OpenQRM, XenServer, Oracle VM, CloudStack, ConVirt, etc. I will take several of these products as examples.

OpenStack is an IaaS system compatible with Amazon EC2. OpenStack includes OpenStackCompute and OpenStack Object Storage. The advantage of OpenStack Object Storage is that it can use general the new server. In this way, there is a problem. The system will generate a lot of data during operation. If data redundancy is adopted to ensure data security, it will take up a lot of system space. Secondly, OpenStack Object Storage has a disadvantage.

OpenQRM virtualization environment, including the basic layer (framework layer) and plug-ins. The basic layer (framework) is used to manage different plug-ins, and the management of virtual resources (computing resources, storage resources, image resources) is implemented through plug-ins. The framework of OpenQRM is similar to the interface in the Java language. It defines a series of methods for virtual machine resource lifecycle management, such as creating, starting, and shutting down virtual machines. On the basis of frameworks, OpenQRM implements different plug-ins for different virtualization platforms (Xen, KVM) to manage different physical and virtual resources. When new resources need to be supported, just write new plug-ins for OpenQRM to seamlessly integrate into the original environment. When the target node to be managed provides SSH login mode, the OpenQRM plug-in logs in to the computing node through SSH and directly runs the corresponding virtualization management command on the computing node [4]. When the target node to be managed provides an HTTP/HTTPS/XML-RPC remote call interface, the OpenQRM plug-in manages the target platform through the remote call interface provided by the target node. But the disadvantage is that OpenQRM uses plug-ins to meet different virtualization platforms, which is not reasonable enough. Because new plug-ins are created every time a new virtualization platform is encountered, which is not easy to manage. The scalability is also poor. Second, log in to the computing node using SSH. When the SSH protocol or port is disabled, the system will not be able to do any operations on the computing node, and the entire system will not work.

2.2 Storage Virtualization Technology and Its Development

There are many objects, and file/record virtualization. According to the classification method of SNIA (International Storage Industry Association), the classification of storage virtualization technology is shown in Fig. 1.

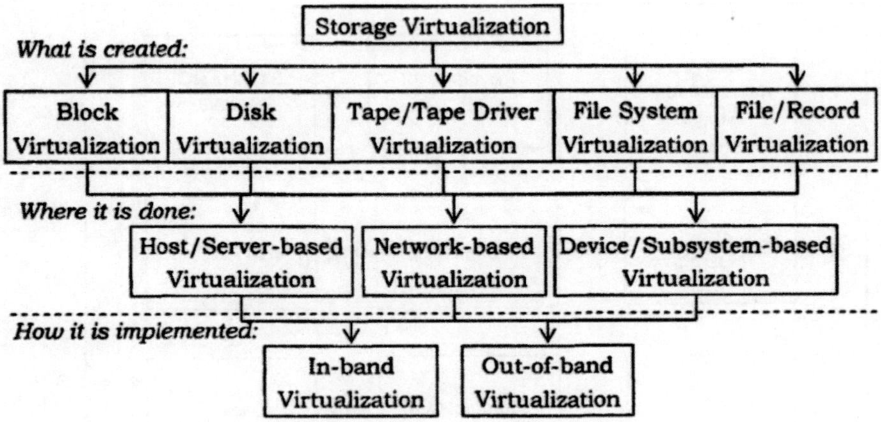

Fig. 1. Classification of storage virtualization technology

(1) Disk virtualization. Disk virtualization has appeared for more than 20 years and is widely used in almost all disk drives. Disk virtualization makes the physical characteristics of the disk such as cylinder, head and sector transparent, and provides users with a numbered, linear disk storage space, so that users do not have to pay attention to which sector the actual data is stored in when reading and writing, and can automatically detect the defects of the storage medium.
(2) Tape virtualization. As a backup medium, tape has the characteristics of low price and high reliability. Tape virtualization can be divided into tape media virtualization, tape drive virtualization, or tape library virtualization.

3 Storage Management in Virtualization Technology

There are certain conditions for virtualization technology to build a utility computing platform. Especially for the most critical problem of utility computing - how to automatically provide resources according to the dynamic change of demand. Mainstream virtualization technologies support the dynamic provision of CPU and memory resources for virtual machines. For example, in Xen, VMware, and MS Virtual PC, the CPU and memory owned by the virtual machine can be specified when the virtual machine is created. When the virtual machine is running, the size of these resources can also be dynamically changed through instructions or parameter settings. In this way, the virtual machine reduces the granularity of resource sharing and dynamic adjustment to below the device level, which is more consistent with the requirements of utility computing, so as to achieve better results. The storage management diagram in virtualization technology is shown in Fig. 2.

One of the shortcomings of current virtualization technology is that it does not support the sharing and dynamic adjustment of storage resources well. In Xen and VMware, only an entire storage device can be assigned to a virtual machine; In MS Virtual PC, the virtual machine uses an image file in VHD format as its virtual storage device. This image file supports continuous expansion with the increase of written data (similar to thin provisioning), but it cannot shrink.

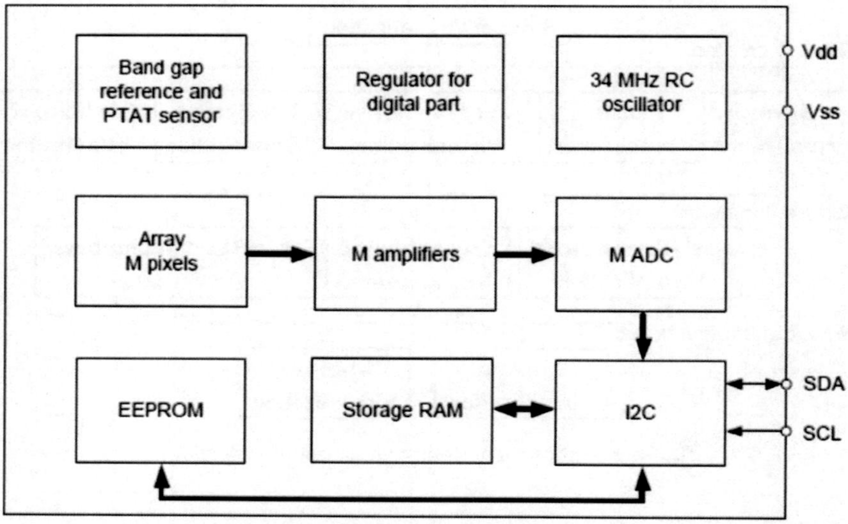

Fig. 2. Storage Management Block Diagram in Virtualization Technology

The virtual storage devices provided for Xen virtual machines can be in the following ways:

1) Physical hard disks or their partitions, including iSCSI disks or GNBD volumes. The hard disk of the virtual machine can be directly set as a hard disk or partition, and there is no need to organize the hard disk partition into another form. However, its disadvantage is that it is not flexible enough to dynamically change the storage space capacity with the needs of virtual machines. It is also only suitable for stand-alone environments and cannot meet the requirements of device sharing in large-scale systems.
2) Network storage protocol, including NFS and other network or cluster file systems. This method is suitable for some network storage devices. Its disadvantages are low performance (all data needs to pass through the network) and stability (related to network storage devices), and it is not universal. It is not suitable for the storage architecture of a distributed cloud storage center based on cheap heterogeneous servers.

4 Cloud Storage Server Based on Virtualization Technology

The goal of the whole system is to unify the management of SUN, HP and IBM servers and eliminate the heterogeneity of the three virtualization technologies. When there are new virtualization technologies, the entire system can be expanded very quickly. The system can realize smooth and convenient management and use of existing partition resources, simplify the creation and configuration process of new partitions, manage the entire life cycle of partitions, and implement partition operations such as querying, modifying, starting, stopping, deleting, restarting, and migrating partitions without changing the existing environment. At the same time, in order to monitor the operation of the entire ECS and maximize the use of ECS system resources. The system must realize real-time monitoring and data collection of ECS and partition resources.

The design of this system can not only meet the functional requirements and non functions of the system proposed in Sect. 3, but also focus on the key problems to be solved. The system adopts C/S architecture, and both the server side and the agent side need to achieve the characteristics of easy expansion. The system has designed a set of communication message protocols suitable for its own to ensure the robustness of the data link and poor data security. The overall framework of the system is shown in Fig. 3:

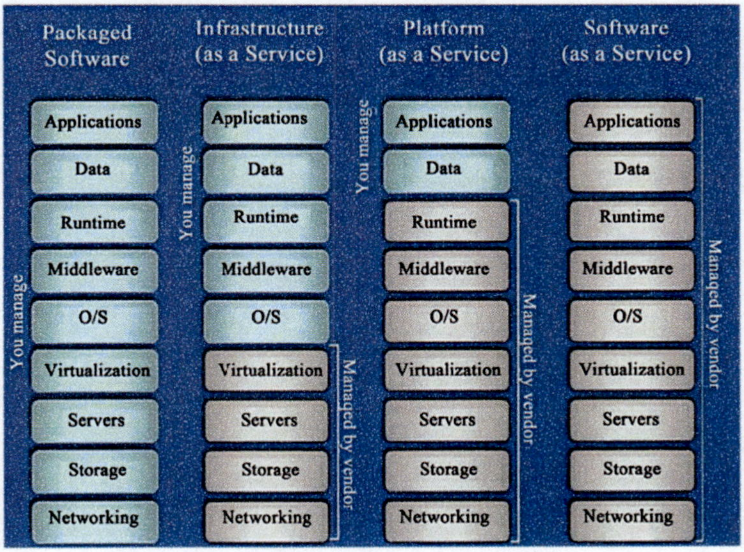

Fig. 3. Cloud storage server structure

5 Conclusion

The current system has been delivered to many domestic mobile companies and China Unicom for use. The operation is good, and the product greatly meets the needs of customers. However, the system still has a deficiency. The system needs to add KVM virtualization function. At present, all major telecom operators have put forward demands for convenient management of KVM virtualization technology. To make the system more perfect. The improvement of this system will increase the expansion of KVM virtualization technology.

References

1. Sawant, A.G., Rasal, P.R.: Research on the Virtualization Technology in Cloud Computing Environment. JETIR, 2021(6). www.jetir.org

2. Zhang, H., Zhenhua, H.E., Yalin, L.I., et al.: Research and application of high resolution fracture prediction technology based on ADM spectral inversion. Geophys. Prospecting Petroleum **55**(5), 737–745 (2022)
3. Baconguis, R.T.: Agricultural Technology: Why Does the Level of Agricultural Production Remain Low Despite Increased Investments in Research and Extension? Discussion Papers (2022)
4. Mccune, D.V., Dixon, S., Mohamed, M., et al.: Lets make space for young people to lead: integrating research and action programming in an arts and technology center: opportunities, challenges and lessons learned. J. Community Pract.Pract. **30**(1), 45–70 (2022)

Application of the Modified Apriori Algorithm in Addictive Behavior and Mental Health

Haiyang Ding[1](✉) and Qixuan Sun[2]

[1] Faculty of Psychology, Southwest University, Beibei District, Chongqing 404100, Chongqing, China
sean25@foxmail.com
[2] College of Computer and Information Science, Southwest University, Beibei District, Chongqing 404100, Chongqing, China

Abstract. In view of the practical application of Apriori algorithm in addictive behavior and mental health, the students of grade 2020 in a certain university are taken as the research object. We first outline Apriori algorithm, points out the defects of traditional Apriori algorithm, and second analyze four improved methods of Apriori algorithm. Finally, the practical application of the improved Apriori algorithm in the study of addictive behavior and mental health is explored.

Keywords: Apriori algorithm improvement · addictive behavior · mental health · practical application

1 Introduction

The emergence of addictive behavior represents the emergence of psychological problems. An individual's growth not only needs to take into account physical health, but also mental health occupies the same important position as the former. For colleges and universities, they need to be related and combined with various data and various elements in the process of carrying out mental health education. The application of Apriori algorithm can realize the purpose of analyzing and processing massive data, and find the correlation rules, which can be applied to mental health education and addiction behavior prevention and can play a good guiding role.

2 Overview of the Apriori Algorithm

In essence, the Apriori algorithm belongs to the category of the association rule mining algorithm, which is also the first algorithm and the most classical algorithm in such algorithms. The Apriori algorithm turns the item set relationship in the corresponding database into the corresponding rule. In general, the process of Apriori algorithm is completed through two process contents: first, to remove unnecessary intermediate results, which is also called "pruning"; second, the matrix operation, which is also called connection. The Apriori algorithm is shown in Fig. 1.

J. C. Hung et al. (Eds.): FC 2023, LNEE 1131, pp. 429–435, 2024.
https://doi.org/10.1007/978-981-99-9299-7_63

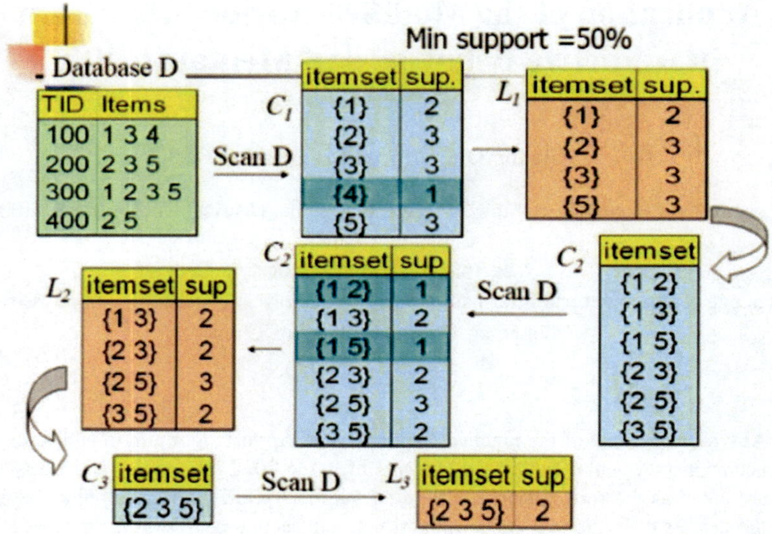

Fig. 1. Apriori algorithm

In the Apriori algorithm, the set of items is the so-called set of items, assuming that a certain item set is the set of K items, then the term set can be named the K item set. In general, the number of transactions containing a set of items is the frequency of the item set appears, the item set frequency. In terms of the progress process of the Apriori algorithm, when the item set can be satisfied with the minimum support degree, in this case, the item set can be divided into the category of the frequent item set. The frequent term set of Boolean association rules, which is also the most influential algorithm in this field. The two links of the plot downward closed detection and generating the candidate set. During the application of this algorithm, the key is the recursive algorithm based on the frequency set idea based on the above two links. Based on the perspective of the association rule type, according to different classification methods, the Apriori algorithm belongs to different types, which can generally be regarded as single-dimensional, single-layer, and Boolean association rules. In this case, each item set supported beyond the minimum support can be judged as a frequent item set.

At present, the scope of Apriori algorithms is expanding, and its applications are growing, such as cyber security, mental health and business [1]. Because this algorithm is mainly conducted through the layer by layer search iterative method, it shows relatively clear and simple characteristics, and there is no need to carry out tedious theoretical derivation in the practical application, and the implementation difficulty is relatively low. However, the traditional Apriori algorithm still has some more obvious defects and deficiencies. There are four aspects: first, when we apply the Apriori algorithm, which means that the application process is complicated, second, more unnecessary intermediate sets are generated, and the "pruning" method is needed to remove the unnecessary intermediate result; again, the Apriori algorithm depends on unique support; finally, the application surface of the Apriori algorithm itself is relatively narrow. The existence of these defects means that when the number of transactions in the set of the original items increases, the Apriori algorithm also works less efficiently in practice.

3 The Apriori Algorithm Improvement

It can be seen from the above content that the traditional Apriori algorithm still faces some problems and must be effectively improved. According to the current research methods in related fields, there are many forms of Apriori algorithm improvement.

First, the Apriori algorithm based on item set importance and interest. In this improved method, the application of Apriori algorithm will first take the user's point of interest as the basis, take a subset of the transaction set as the goal of data mining, and after realizing the generation of the project set, the corresponding weights of the elements are set. After this, when the support degree is calculated, the support degree calculation function with weights needs to be used as the basis for the generation of frequent sets of items, thereby, after which the association rules can be formed. This paper studies the modified Apriori algorithm in the application of the addictive behavior and mental health of college students.

Second, convert the transaction set to a Boolean matrix, transforming the whole process of mining the frequent set to operations for the matrix. In this case, the application of Apriori algorithm does not need to scan the database multiple times, but only once scan, that is, it can be abstracted as an operation against the matrix. When searching for the frequent item set, the improved Apriori algorithm can also be completed by the corresponding characteristics, matrix compression, thus relying on the matrix ranks of logic operation to complete the matrix class in the traditional.

Third, the Apriori algorithm based on division. In the process of implementation, this algorithm improvement method basically takes the concept of division and governance as the core, and divides the data transaction set originally in the overall state according to the corresponding conditions, making it into several transaction sets of disconnected and overlapping transactions. After this, the corresponding set of local frequentist items is generated by mining the form of divided transaction sets. Thereafter, these sets of local frequentist items can be integrated into a complete set of global sets of candidates. After that, a series of data mining work needs to be carried out to obtain the global frequent item set.

Fourth, the sampling-based Apriori algorithm improvement method. The theme of this improved method is to select some representative samples from the original transaction set, and to make these samples become a new transaction set, and then to apply the method of association rule mining to these newly generated transaction sets.

4 Practical Application of the Improved Apriori Algorithm in Addictive Behavior and Mental Health

4.1 Analysis of Addictive Behavior and Mental Health

From the perspective of meaning, addictive behavior refers to an extraordinary kind of habit and addiction. Some addictive behaviors are harmless to the human body, for example, some individuals are addicted to reading books. Most addictive behaviors will have a negative impact on an individual's development, such as drug addiction, smoking addiction, gambling addiction, alcohol addiction, abuse of prescription drug addiction, Internet addiction, and so on. Addicts may develop mental health problems when addictive behaviors are present. He Rihui, a domestic expert in addiction medicine and psychology, pointed out that addiction is not only a physical disease, but also a mental disease.

College students, for example, in recent years, college students' mental health problem has been widely concerned by the social from all walks of life, this group of addictive behavior and mental health problems comes from multiple levels, such as campus bullying, the pursuit of stimulation, etc., makes some college students appear Internet addiction, smoking addiction, alcohol addiction and other addiction behavior [2]. In addition to addictive behavior, some college students may be depressed due to school bullying, and their mental health problems are prominent.

Although the current Chinese colleges and universities are included in the compulsory course system of the college students' mental health education, but still a superficial phenomenon, some college students in the psychological crisis or addiction behavior, usually difficult to get the attention from the school, or some college students to their mental health problems hold a kind of escape, unwilling to admit attitude, make psychological problems are difficult to solve. Moreover, due to the complexity and concealment of an individual's psychological crisis problems and addictive behavior, the investigation and analysis of mental health problems in universities and colleges face great difficulty [3]. Generally speaking, Chinese colleges and universities have accumulated a number of data of students, including mental health test results data, student performance data and student status data and so on. In the current big data technology and information technology application depth increasingly extended background, colleges and universities in screening and analyzing students 'mental health problems and addiction, that can rely on such technology, set up mental health collection evaluation system, obtain students' mental health data, and from the database of student data administration system, student management system, while using improved Apriori algorithm, mining related rules, so as to screen and analyze students' psychological crisis and psychological counseling work.

4.2 Data Mining Objects and Results Analysis

In the improved Apriori research activities, selected a university of 2020 students, a total of 2345 students, the psychological assessment survey data were extracted, at the same time, the psychological assessment questionnaire distribution, a total of 2268 valid questionnaires. During the data mining period, the Apriori algorithm was used, including nine elements: depression, interpersonal sensitivity, compulsive symptoms, paranoia, fear, mental symptoms, anxiety, hostility, and somatization. After completing the mining, the data processing is also required. Although the Apriori algorithm used has been improved, there are also some useless data, missing values and duplicate values in the mined data, such as student student number, name and gender, so the original data needs to be cleaned. For example, the associations between the elements of a student's name and student number and psychological problems and addictive behavior are small, so they need to be removed during the analysis. The validation iteration analysis is shown in Fig. 2.

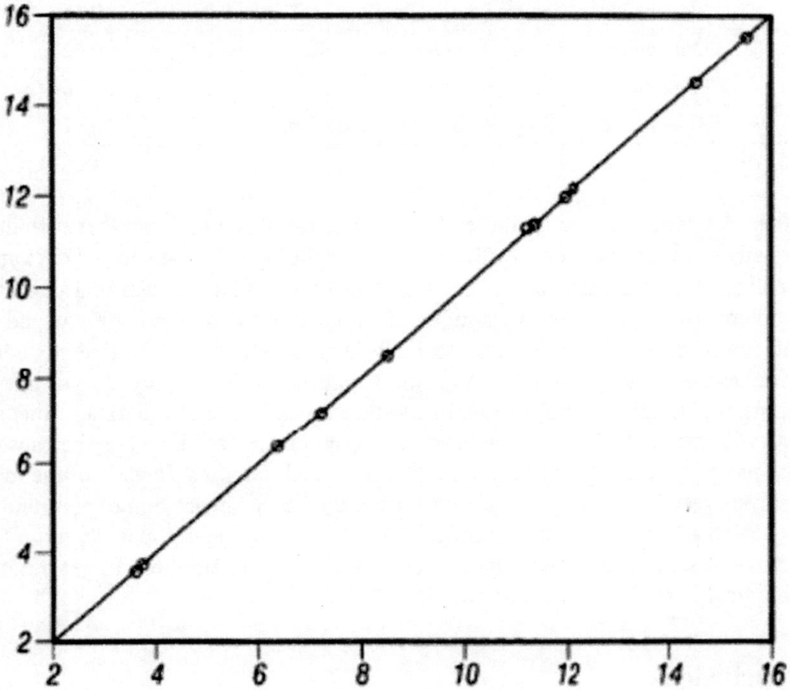

Fig. 2. Validation iteration analysis

In addition, since the selected objects are all grade 2020 students, the age difference between the students is small, so the student age data extracted from the system can also be removed. After the data cleaning, two results were generated, namely, the nine-dimensional association rule mining results shown in Table 1 and the association rule results between the mining properties and the compulsive psychological symptoms shown in Table 2.

Table 1. Mining results of nine-dimensional association rules

Number	Association rules	Support	Confidence
1	There is anxiety, hostility and coercion	0.21	0.94
2	There is anxiety, psychosis and depression	0.24	0.84
3	There is terror, there are relationships and depression	0.31	0.95
4	No compulsion, no paranoia, no human relationships	0.33	0.88

Table 2. Results of association rules between compulsive psychological symptoms

Number	Association rules	Support	Confidence
1	Student cadres, female without coercion	0.06	0.45
2	Only child, no depression in the countryside	0.09	0.42
3	High income, women have forced	0.14	0.66
4	Student cadres, low income without coercion	0.03	0.57

After analyzing the data in Tables 1 and Table 2, we can find that there are indeed some relatively close links between the different psychological dimensions. For example, some college students with compulsive symptoms tend to have concurrent psychological problems of interpersonal sensitivity, paranoia, depression, and anxiety, and such students are relatively more likely to have addictive behaviors. Some students who do not have sensitive interpersonal relationships, and they also usually do not have the symptoms of paranoia or compulsion. In addition, it can be seen from Table 2 that some students with higher family income have more obvious OCD, which may be caused by parents' more strict discipline for their children in such families. In view of the mining and analysis results of Apriori algorithm, have a certain impact on colleges and universities need to focus on implementing mental health counseling, pay attention to the psychological state of different types of students, and promote them to get out of the psychological obstacles as soon as possible.

5 Conclusion

To sum up, in the process of using Apriori algorithm to understand students' psychological state and investigate psychological crisis, colleges and universities can take the results and correlation rules generated by this algorithm as the guidance to carrying out mental health education. On this basis, schools can carry out crisis prevention education and response mechanism, and intervene in students' mental health problems in advance according to the discovered correlation rules, so as to ensure that the probability of addictive behaviors and mental health problems is effectively reduced.

Acknowledgements. In the process of completing this paper, I received the help and support of many people, which cannot be listed one by one, but have left important marks in my heart.

Firstly, I would like to thank my supervisor for his patient and meticulous guidance and support throughout the entire process. Through in-depth communication with him, my paper has been greatly improved and improved in various aspects.

Thank you to all readers. If you are reading this paper, then we have become friends for knowledge exchange. I hope you can be inspired by my work and share your practical experience and insights with me.

Once again, thank you all for helping me. Thank you for your patience and support. May we pursue the exploration and discovery of knowledge together.

References

1. Qi, W.: The application of incremental association rule mining algorithm in psychological crisis prevention of college students. J. Chifeng Univ. (Natural Sci. Edition) **35**(01), 66–68 (2019)
2. Hao, F.: Exercise prescription recommendation system for male compulsory isolation of drug addicts based on association rules. Shanxi University (2021)
3. sway. Psychological big data evaluation method based on association rule feature extraction. J. Zhoukou Normal Univ. **37**(02), 144–147 (2020)

Design and Development of Corporate Financial Risk Control System Based on Big Data

Qiang Du(✉)

Business School, SANDA University, Shanghai 201209, China
qdu@sandau.edu.cn

Abstract. This method has three important characteristics: (1) using large-scale and complex phenomena to obtain the main objective function; (2) Its design is based on the combination of fuzzy reasoning and evolutionary algorithm to optimize the value of the interdependent objective function; (3) The optimization mechanism focuses on "fuzzy trust" rather than simple "trust". In addition, a more detailed study was carried out by comparing with other results in related fields. Financial risk control is one of the most important problems that modern organizations must solve. Therefore, it is necessary to find effective risk control approaches and tools. Data can play an important role in this process because they are many different sources of information about business activities, and they can also help us make informed decisions when we use the appropriate methods.

Keywords: big data · finance · Risk control · system development

1 Introduction

Traditional financial risk control mainly uses financial data with strong credit attributes. Generally, it uses data of about 20 dimensions to identify customers' borrowing risks, including repayment ability and repayment willingness. Whether it is traditional finance or Internet finance, in common risk control processes, the credit information provided by customers when they enter the market is limited, and the volume of risk control data of business institutions is insufficient. Risk assessment based on internal risk data will be very one-sided, and it is impossible to comprehensively control the risk situation of a customer [1]. Therefore, the risk control data source based on the trinity of user input data, internal system data and external data is particularly important, an ideal risk control data system. The use of third-party data must be close to the business. First of all, we need to understand the business type, then sort out the risk category, develop risk control process, and build a risk profile to select appropriate risk control data.

In fact, risk control should run through the entire financial supply chain. Scenarios such as user account login, overdue loans, money laundering and anti cheating should be a relatively wide range of risk control applications. If transactions are involved, there should also be swipes. On the whole, we are also trying to reduce risks through data, not relying on the weather. Risk control is required on common platforms such as e-commerce, third-party payment, investment and financing. However, the whole

J. C. Hung et al. (Eds.): FC 2023, LNEE 1131, pp. 436–441, 2024.
https://doi.org/10.1007/978-981-99-9299-7_64

risk control is not just a matter of model algorithm, but also involves many operational strategies. For example, in the case of overdue loans, when monitoring the lender's business situation, specific operations should be involved to understand the situation [2]. If it does happen, how to solve the problem. In the case of money laundering, how to distinguish which are normal transactions and which are obvious cash laundering situations, these rules also need to be defined and learned.

2 Related Work

2.1 Risk Definition

The financial risk early warning mechanism has the dual role of financial management and financial diagnosis. Its significance lies in selecting a set of variables according to the relevant operation and management principles, establishing a set of early warning functions (Function i on), pointers (i ndi cat or) or benchmark values (crit i cal value). After using the early warning mode, if it is found that there is a situation that does not conform to the regulations or exceeds the warning range, it will give a warning, so that regulators can take regulatory measures as soon as possible.

If a person forgets the date of the IOU and repays it one day later, this situation cannot be simply defined as a bad user. When we cleaned up the data, we saw that there was such a category of "a small amount of overdue" in the customer's credit evaluation, which accounted for a considerable proportion, and played a significant role in the model, comparable to other categories of "good credit" and "bad credit". At the beginning, our technicians were confused about the classification of "a small amount of overdue", and could not understand whether the classification was good or bad. We didn't understand the meaning of this until we actually communicated with the risk control manager of a P2P company [3].

The emergence of financial risk is not accidental, it should have its internal and external causes of occurrence and change. The external causes of financial risks are mainly the changes. To explore these factors, we can examine them from the theoretical basis of financial risk early warning mechanism. Therefore, it is necessary to find a theoretical basis for the financial risk early warning mechanism.

Therefore, the perfection of risk control is closely related to user risk assessment. The more accurate the portrait is, the more accurate the risk assessment is. Accurate user portrait construction cannot be separated from comprehensive, efficient and stable data. The data application principles of financial institutions are internal before external, simple before complex, strong before weak, and rules before models. Internal data mainly includes user input data and user behavior data in the system, including order information, approval information, login information, repayment information, etc. External data classification only lists common data that can play a certain role in risk identification in the financial field. Including but not limited to credit data, report data, private credit, mobile user behavior data (including online consumption data, social data, etc.)

2.2 Financial Database Analysis

In a big data platform, the database occupies a very important position, and the design of the database will have an important impact on the running effect of the entire platform. In order to ensure efficient data storage and data integrity and consistency, database institutions are very important.

In this system, the user data information mainly presents the characteristics of huge data volume, multiple data types and scattered data. For a credit user, it generally includes at least basic information, Alipay information, billing information, JD information, communication information, etc. In general, there are hundreds of bills for a user. This leads to an explosive increase in the amount of data in the database. When storing bill data, the length of bill content varies, which is also one aspect of data fragmentation that is difficult to integrate. In terms of the above situation, in addition to the difficulty of data calculation, it also challenges the data storage.

At the same time, there are many users' personal information in the loan system, and the consequences would be unimaginable if they were stolen and used by criminals. Therefore, data security should be fully considered in database design. A secure database should have a user ID, and a user must enter the correct password in addition to having a legal identity. At the same time, different table permissions should be authorized for different product department personnel. If required, authorization can only be granted after the approval of the superior leader. When storing user sensitive information in the database, encryption processing should be carried out, such as adopting MD5 algorithm to encrypt user identity information, contact information, etc.

The database should be designed as predictably as possible to deal with possible table splitting, migration and other situations.

3 Design and Development of Corporate Financial Risk Control System Based on Big Data

OFBiz's Web application framework strictly follows the MVC pattern. The model in OFBiz's MVC implementation is assumed by its events and services that encapsulate business logic. Control is assumed by the controller, and View is assumed by traditional JSP, FreeMarker, JPublish, or BeanShell. Web server architecture is shown in Fig. 1.

The model of the entire OFBZ framework three types of elements: Custom Component, OFBiz Component, and 3rd PartyComponent. When the browser request of the client is submitted to the server, it is first parsed by the OFBiz component Control Servlet, and then transferred to the View JSP (page request) below or the Event (event request) above depending on whether it is a page request or an event request. The Control Servlet here is a group of objects that manage the web presentation layer [4]. It mainly assumes the role of Control in MVC mode, and its purpose is to completely separate the business logic from the presentation layer. Since the View JSP and Event components need to be developed according to different applications, they are not components provided by OFBiz, so they are called custom components. Some tags used for data display in JSP pages are components provided by OFBiz, which are called OFBiz components. No matter what kind of request is business related, it will be sent to the Service Engine. If it

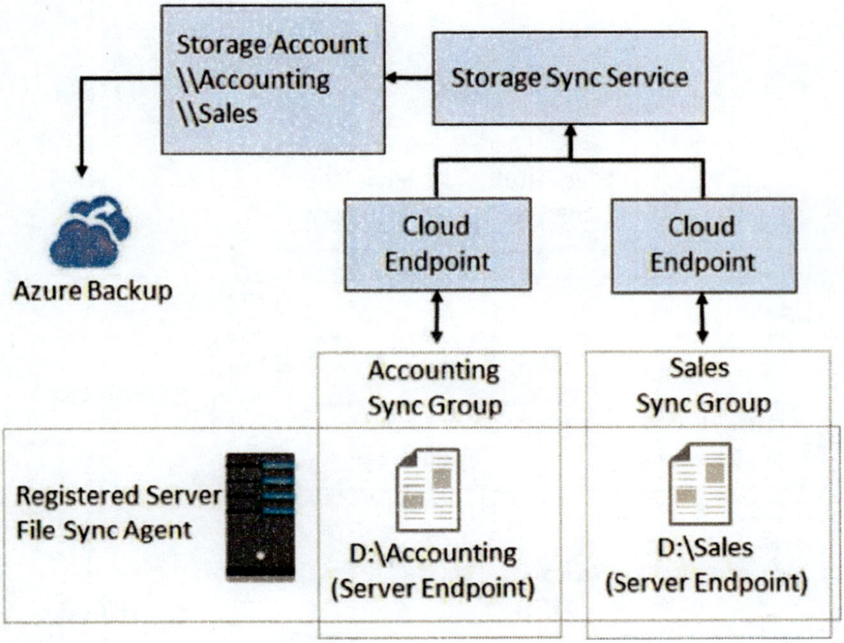

Fig. 1. Web server architecture

is a request that directly processes data, it will be transferred to the Entity Engine. These two components are the most important core modules in the OFBiz framework.

The risk control layer is located in the middle layer of the entire loan product platform, and risk decisions are made through interaction with the business layer and the data layer. Next, we will introduce the architecture of the entire loan product platform from top to bottom. The risk control layer is the main architecture layer of the risk control system.

4 System Development

The risk control system mainly provides risk decision-making, anti fraud decision-making, anti fraud services, monitoring services, offline computing, feature engineering feature extraction, etc. The anti fraud decision is called by the internal system, which mainly provides services for risk decision. Anti fraud service is to return the anti fraud results to external users or systems. Offline service mainly refers to offline computing when online data acquisition fails. as shown in Fig. 2.

Feature engineering feature extraction mainly requests WD data warehouse to obtain online calculated data and complete risk decision-making. The system package structure can be divided into the following eight layers:

(1) Process control layer: credit_ The model package is mainly responsible for controlling the invocation of specific models and anti fraud processes in risk decision-making. In this system, anti fraud is also regarded as a model, because anti fraud and

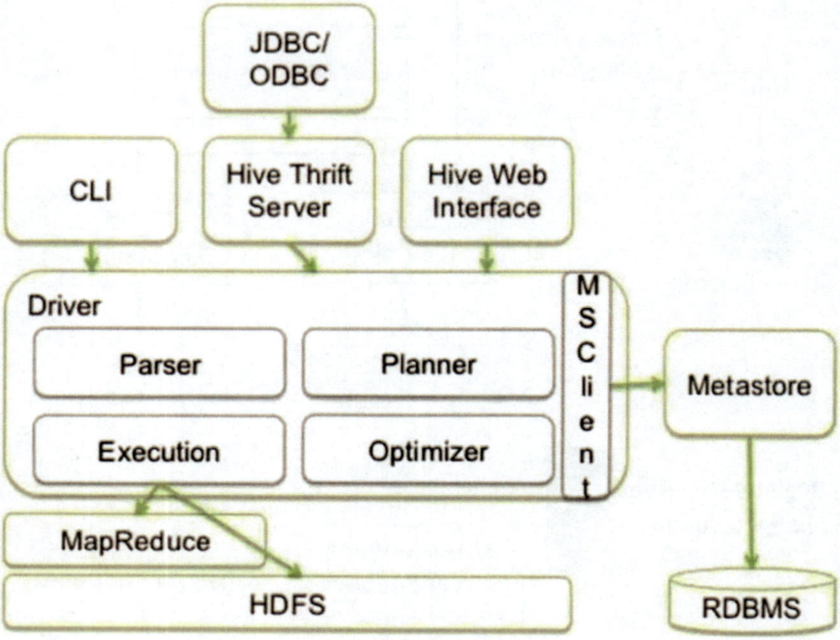

Fig. 2. Development Framework of Risk Control Decision System

credit model will also obtain data for simple calculation. Credit_ Each product in the model package corresponds to a class, and this product corresponds to the ksd_ multi_ credit。

(2) Model layer: model package and anti_ The fraud package mainly refers to the import and call of specific models and the implementation of anti fraud rules.
(3) Feature layer: feature package, mainly realizing feature acquisition and feature calculation. Each feature in this package corresponds to a feature calculation function.
(4) Feature preprocessing layer: data_ The source package mainly includes some basic functions for data processing, such as normalization calculation, slope of simple linear regression, intercept calculation, WOE calculation, etc.
(5) Online data acquisition layer: service package, mainly used for online risk decision-making to call the RESTful API of the WD data warehouse to obtain the required raw data.
(6) Online test layer: abtest package, which mainly realizes A/B test function and is used for calling in risk decision-making.
(7) Data monitoring layer: monitor package, which is mainly responsible for the monitoring of model results and anti fraud results after the completion of model credit. The data monitoring here is mainly for offline data monitoring.
(8) Offline data interaction layer: dao package, mainly responsible for the interaction between offline hive data warehouse and the system.

In risk decision-making, in addition to credit model, it also includes anti fraud module. Anti fraud in the risk control system is mainly achieved through anti fraud rules, because the anti fraud model takes a long time to build, and it is difficult to achieve practical results with constant changes in fraud means. It seems that anti fraud is just a dichotomy problem. In fact, if we regard each kind of fraud as a problem, we will face a multi classification problem. Therefore, this system has designed a more complex anti fraud process to reduce the loans of fraudulent users and improve the loan quality from the source.

5 Conclusion

The corporate risk control system based on big data and determine whether to use it. Unstructured information set containing quantitative and qualitative values and their relationships. The data set generated by the sensor forms another type of big data. Collecting large amounts of data requires expensive work, but it will bring huge benefits in terms of efficiency and accuracy. The modern business environment is characterized by an increasing concern for financial security, large amounts of data. Big data analysis can be used as a tool to prepare predictions, detect trends and predict future events with high reliability.

References

1. Gui, Y.: Analysis of Financial Risk Management of E-commerce Enterprises Based on Big Data (2020)
2. Lan, K.: Financial risk control of sports industry based on big data. J. Phys.: Conf. Ser. **1648**, 042068 (2020)
3. Lyu, X.: Research on internet financial risk evaluation under big data background:based on generalized DEA model and P2P network loan. J. Account. Econ. (2017)
4. Zhang, H., Khurshid, A., Wang, X., et al.: Corporate financial risk assessment and role of big data; new perspective using fuzzy analytic hierarchy process. J. Econ. Forecast. (2021)

Design and Application of Talent Training Program for Cross-Border E-Commerce of Agricultural Products Based on BP Neural Network

Yupeng Fang(✉)

Shandong Institute of Commerce and Technology, Shandong 250013, China
fangpeng315@126.com

Abstract. With the development of cross-border e-commerce, cross-border e-commerce for agricultural products has become an important market and application field. To ensure the sustainable development of cross-border e-commerce for agricultural products, it is necessary to cultivate qualified talents. This article is based on BP neural network and designs a talent cultivation plan for cross-border e-commerce of agricultural products, which has been verified in practical applications. The results indicate that the talents trained in this program can adapt to the current demand of the cross-border e-commerce market for agricultural products and have achieved good results.

Keywords: BP neural network · Electronic Commerce · agriculture products · personnel training

1 Introduction

Cross border e-commerce for agricultural products is an important field in international trade and a hot topic in the development of e-commerce. With the continuous development of the global economy, more and more consumers are paying attention to health, ecology, healthcare, and fresh food, leading to the rapid growth of cross-border e-commerce markets for agricultural products. However, due to the unique nature of agricultural products, the trade environment and legal regulations for cross-border e-commerce of agricultural products are relatively complex [1]. Therefore, the lack of professional talents is an important factor restricting its sustainable development.

In order to cultivate qualified cross-border e-commerce talents for agricultural products, this article designs a talent cultivation plan for cross-border e-commerce of agricultural products based on BP neural network and applies it in practice.

2 Related Work

The collation and refinement of the above documents provide the basis and reference for the follow-up research, but it still needs further consideration. Based on the above research, agricultural products, the existing research mainly takes a single influencing

J. C. Hung et al. (Eds.): FC 2023, LNEE 1131, pp. 442–448, 2024.
https://doi.org/10.1007/978-981-99-9299-7_65

factor or a specific angle as the entry point. As a complex system project, export has not only one influencing factor, the importance of each influencing factor needs to be clarified, and how to coordinate the development of each link still needs to be considered [2], This limits the persuasiveness of the research results to some extent. At the same time, in the research of existing literature, the phenomenon of theoretical accumulation and listing is serious, and there is little actual investigation and analysis of enterprise performance.

The e-commerce talent cultivation model refers to the cultivation of professional e-commerce talents through education, training, and practice within institutions such as schools or enterprises. The main modes of e-commerce training include "combining talent cultivation with practice", "mentor system", and "school enterprise joint training".

BP neural network is a feedback feedforward type of artificial neural network, which is currently one of the commonly used neural networks. It models the complex nonlinear relationship between input and output through information transmission and adjustment between multi-level computing units.

The talent cultivation plan for cross-border e-commerce of agricultural products based on BP neural network is divided into three levels: basic stage, professional stage, and practical stage. By offering relevant courses, students can understand the basic theories and current market demands related to cross-border e-commerce of agricultural products. Mainly including knowledge and skills in website construction, data collection, product display, online transactions, international logistics, etc. In the professional stage, the training program focuses on cultivating students' skills and practical experience in the field of agricultural e-commerce. Specifically, it includes knowledge and skills in marketing strategies and decision-making, international trade laws and regulations, market analysis and strategy, and agricultural product brand building.

The practical stage usually adopts the method of student internship and practice to help students apply their skills and knowledge in practical work. Students can choose to participate in work internships on actual cross-border agricultural e-commerce platforms, or participate in related international trade, market research, and other activities [3]. During the internship and practical process, students need to develop corresponding solutions and solutions based on the needs of the company or practical projects, in order to improve their practical abilities and operational skills. Students also need to improve their English proficiency and cross-cultural communication skills during the practical stage of knowledge application and ability enhancement.

This article selects undergraduate students majoring in e-commerce from an agricultural university as an example and uses BP neural networks for training and testing. The training dataset contains a certain amount of data on international trade, market analysis, marketing strategies, and other aspects involved in cross-border e-commerce of agricultural products. The test dataset includes data on students with initial abilities trained in the basic stage, as well as data on students trained in the practical stage.

By comparing the accuracy of test datasets and students' problem-solving abilities, it was found that the BP neural network-based cross-border e-commerce talent training program for agricultural products can successfully improve students' practical and problem-solving abilities. After training, students can better respond to the demand of the cross-border e-commerce market for agricultural products and improve their employment competitiveness.

3 Theoretical Basis of BP Neural Network

3.1 Basic Principle of BP Neural Network

This paper selects Back Propagation Neural Network to analyze the seller's credit risk model. Back-propagation neural network is also known as multi-layer perceptron, or BP neural network for short. It is a feedforward network composed of complex neurons (Hu Yuping, Wang Yudong, 2007). BP neural network is a biological mechanism function that imitates human thinking mode, processes data and information in the neural system, simulates classification, and finally gives feedback.

At the same time, the method of BP neural network to deal with nonlinear factors is to introduce activation function, as shown in Fig. 1, the corresponding data of the node of the previous layer is first linearly combined according to different weights, and its linear combination value is used as the dependent variable of the activation function.

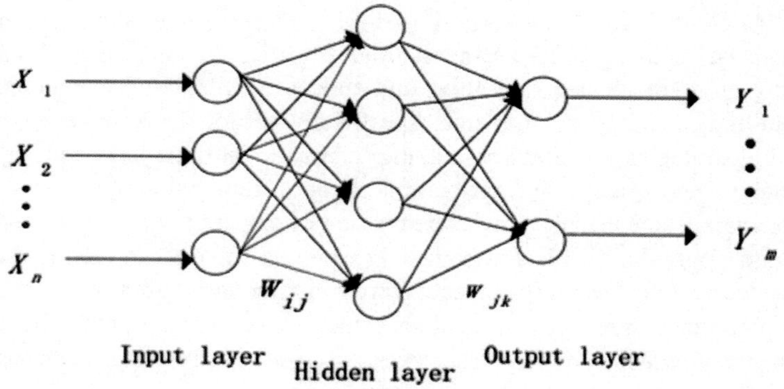

Fig. 1. BP neural network structure

There are four activation functions commonly used in BP neural network:

See Formula (1) for the expression. For any input, the sigmoid function is a nonlinear function. It can output any number between 0 and 1, so it can be used to represent the probability. However, when the input value approaches infinity or infinity, the curve of the sigmoid function tends to flatten, and its derivative value will become very small, which is easy to cause the phenomenon of gradient disappearance.

$$f(x) = 1/(1 + e^{-x}) \tag{1}$$

B. Tanh function: see Formula (2) for the expression of Tanh function. Similar to the sigmoid function, the Tanh function is also a nonlinear function, which can output any value between −1 and 1. It is applicable to more complex problems with different classification and low threshold [4]. However, if the data is relatively simple, it will lead to over-learning. Similarly, the Tanh function has the disadvantage of gradient disappearance similar to the sigmoid function.

$$f(x) = \tanh x = 2/(1 + e^{-2x}) - 1 = 2signoid(2x) - 1 \tag{2}$$

C. Relu function: see formula (3) for linear function. For the input value, if it is less than 0, the output is 0, otherwise the output itself has the efficiency of operation. It avoids the problem of gradient disappearance, and its operation speed is better than that of sigmoid function and Tanh function. However, when the activation function output of neurons in forward transmission is zero, the weight value will reach zero gradient, resulting in Relu necrosis phenomenon, resulting in permanent death of neurons in the network because they cannot play a role in forward transmission.

$$f(x) = \max(0, x) \tag{3}$$

3.2 Learning Algorithm of BP Neural Network

The relationship between simulated by machine learning. The model: it will conduct its own training and output the value of the optimized fitting effect, without manual weight assignment. Then transmits the data to the input layer. In the forward and backward propagation process, BP neural network will continuously adjust the connection weight to train the optimal model.

BP mapping an M-dimensional Euclidean space to an L-dimensional Euclidean space.

(1) Assuming that there are n training samples and each sample has m indicators, its initial information matrix can be expressed as:

$$X = (x_{ij})_{n \times m}, i = 1, 2, \cdots, n; j = 1, 2, \cdots m \tag{4}$$

(2) Weight adjustment

After can be obtained. The error is back-propagated and the connection weight w is adjusted:

$$\alpha_j^z = \alpha u_i z_i (1 - z_i) \tag{5}$$

Among them, η For learning rate, μ Is the momentum term, both of which are in the range of 0 to 1 open range. After that, input the next sample and return to (1) to continue training until all samples are trained. If E < Emin is met, end training, otherwise set E to 0 and retrain.

(3) Network inspection using inspection samples

In order to make the application process is shown in Fig. 2.

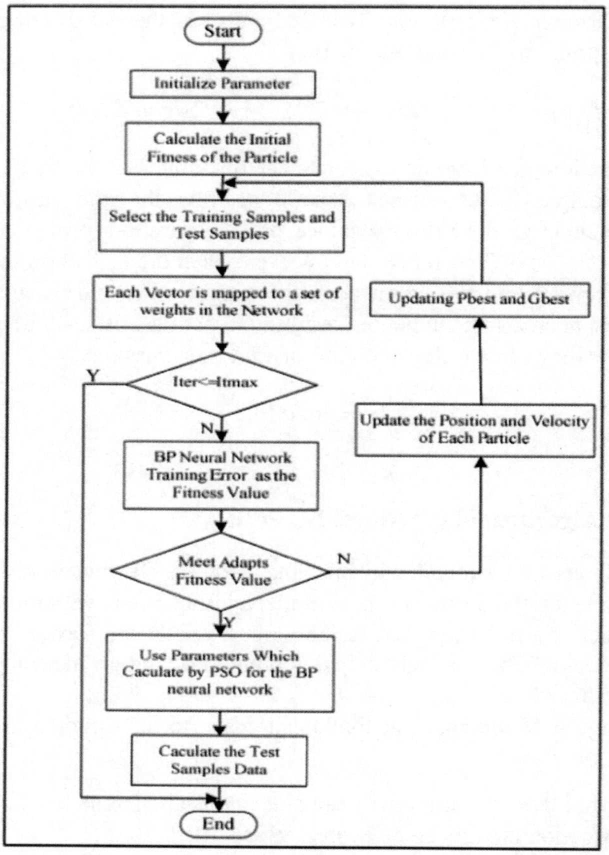

Fig. 2. BP neural network algorithm flowchart

4 Design of Talent Training Scheme for Cross-Border E-Commerce of Agricultural Products Based on BP Neural Network

With the penetration and application of the Internet, the population of netizens has expanded year by year in rural areas. Farmers have been "electrocuted" and began to try to sell their own agricultural products through We-media and e-commerce platforms at home and abroad. However, limited by their own cultural level, it is very are specialized in learning agricultural product-related knowledge, familiar with cross-border e-commerce operation procedures and have passed the English standard. At enterprises are faced with personnel loss caused by jet lag, unreasonable wage performance and other problems. Therefore, the lack of talents has an important impact on the profitability.

In the 1990s, people began to realize that human resources are the highest level of human capital and the most valuable capital to create profits. Adam Smith was the first economist who regarded talent as capital, and was regarded as the embryonic idea of talent resources.

Ye Zhonghai (1994) believed that talents are those who have specialized knowledge, high technology and ability in various social practice activities, can carry out creative work on their own, and have made some great contributions to understanding, transforming nature and society, and to human progress. Richard Forida (2000) believes that high-tech practitioners are the main representatives of "talents" and can be used as an alternative indicator of "talents". In terms of concept definition, he believes that "high-level human capital" is the main indicator of talent judgment, the most important factor to measure talent scale.

In the Global Information Technology Report (2007) released by the World Economic Forum, from the perspective of talent flow, the concept of talent should be "scientists, researchers, It engineers and scholars" [5]. According to the Global Competition Report for Talents released "high-quality labor" around the world, "the ability to acquire, understand and apply knowledge to promote technological and economic development" labor force, that is, the talent should have the above qualities.

Compared with the labor force, talent is the higher part of the labor force, which is the high-quality and high-level labor force required by the society. This is also the difference between talent and labor force. This labor force. Based on the above definitions of scholars and experts as well as different research institutions, this paper believes that talent refers to workers with certain professional knowledge or skills.

Through school-enterprise cooperation, introduce the projects and tasks of relevant enterprises in Fuzhou into the classroom, carry out the "order-based" talent training program, and establish the "Garden Expo Business Class" and other title classes through the "2 + 1 + N" school running mode. The school first relies on the off-campus training base, adopts the teaching mode of work-study alternation, employs the management personnel of enterprises with strong practical manpower as part-time teachers to teach and guide the training, and carries out the talent training through the whole training process in a hierarchical project way, completing the "2 + 1" learning stage. "N" learning stage refers to that after graduation, students' personal accounts will not be cancelled by the school, and they can permanently use the school's digital resources for independent learning, and communicate with teachers in real time to achieve lifelong learning.

5 Conclusion

This article designs a talent cultivation plan for cross-border e-commerce of agricultural products based on BP neural network and has been validated in practical applications. This program organically combines training content and practical experience from different stages, enabling students to adapt to current market demands and improve their practical and problem-solving abilities. The experimental results indicate that the talent cultivation plan for cross-border e-commerce of agricultural products based on BP neural network is feasible and effective. Further improvement and optimization are needed in the future to enhance the practicality and adaptability of the training plan, in order to meet more complex market demands.

References

1. Yuan, Q., Liang, Y., Zhu, X.: Analysis on the training mode of the rural e-commerce application talents——a case study of Suqian City in Jiangsu Province. J. Anhui Agricult. Sci. (2017)
2. Yuanren, W.U., School, B., University, Q.N.: A research on cross-border e-commerce promoting the transformation and upgrading of Quanzhou's Foreign Trade Enterprises. J. Quanzhou Norm. Univ. (2019)
3. Ting, X.U.: Research of Tianjin cross-border E-commerce development. J. Tianjin Coll. Commer. (2017)
4. Zhang, R., Zhou, R.: The research and application of the collaborative innovation team for the integration of electronic commerce and education in the Western Region. Logist. Sci-Tech (2018)
5. Zhou, Z., Zhong, X., Hongbo, L.V., et al.: Study on the training mode of E-commerce talents in applied colleges based on extension theory. Vocational Technology (2018)

Character Recognition System Based on Depth Neural Network

Feihang Ge(✉)

School of Information Science and Technology, Aichi Prefectural University, 1522-3 Ibaragabasama, Nagakute 480-1198, Aichi, Japan
fhge2018@163.com

Abstract. Deep neural networks are increasingly important tools for machine learning and artificial intelligence research, because they can learn very complex tasks without explicit programming. They are also flexible because they can be easily adapted to solve new problems. Character recognition involves the post perception process and is affected by many factors, such as the amount of visual information available and cognitive impairment. This method assumes that there are two types of representations in character memory: abstract forms, which are determined by font attributes, may vary with each letter, and specific forms or "embodiment" representations. The detection part adopts the horizontal text line detection method to detect the text, and in the text part, the system is analyzed and introduced in detail from the production of the model to the design and implementation of the neural network.

Keywords: Deep neural network · Character recognition · Character model

1 Introduction

Text is an important tool for human to exchange information. The way or carrier of text has changed a lot. Text no longer only stays in writing, but also appears in our lives in the form of signboards, banners, billboards, etc., or they are text information in pictures. Using computer to detect and identify these information will bring great convenience to our life [1]. For example, the automatic driving technology can identify various roadside signs, identify the license plates of parking lots, scan and enter ID card information, etc. In this information age with the high development of the Internet, manual recording of text information cannot be said to be outdated, but it is no problem to say that it is not efficient enough to keep up with the times.

For example, if the parking lot records the license plate number manually one by one, it will not only waste manpower, but also cause the parking people behind to wait for a long time, wasting the time of the parking people. Another example is that when teachers write on the blackboard in class, they usually take photos with their mobile phones when they have no time to record. For another example, the handwriting input method we used, the photo translation of Youdao Dictionary Baidu Dictionary, etc. are all based on text recognition in computer vision. However, with the development of various

J. C. Hung et al. (Eds.): FC 2023, LNEE 1131, pp. 449–455, 2024.
https://doi.org/10.1007/978-981-99-9299-7_66

technologies, we are not limited to simple and easy to achieve text recognition. We hope to further extract various text information in natural scenes. For example, traffic signs, signs of various shops, publicity banners, bulletin boards, warning slogans, etc. can be seen everywhere in our lives. These text information are different from our written texts. Because they are in the natural environment, various lights, different angles, rain and fog weather and other natural environmental phenomena have caused great interference to the detection [2]. At present, there are still many problems in the existing methods to recognize such text information, and text detection and recognition is still the character recognition system based on depth neural network.

2 Related Work

2.1 Deep Neural Network

As shown in Fig. 1, we know that the picture in the computer is actually one pixel after another, and the saved form can be seen as a two-dimensional matrix. The 55 green matrix in the figure can be seen as a 55 image, while the yellow 33 matrix is called the convolution kernel. Generally speaking, it can be seen as a filter, or a feature extractor. In actual use, the value in the convolution kernel is adjusted by the entire neural system through autonomous learning, without manual adjustment. We use this convolution kernel to scan an image according to certain rules. Each time the convolution kernel stays, each element in the convolution kernel is multiplied by the elements in the corresponding position of the image where the convolution kernel stays, and then all the elements are added. The values obtained are stored in a new matrix [3]. This process of multiplication and addition is called convolution. As shown in the figure, element position to get 4, which is stored in the pink matrix. When the convolution kernel sweeps the whole image in the way of step unit length, we will get a 33 matrix, which is the feature map corresponding to the convolution kernel.

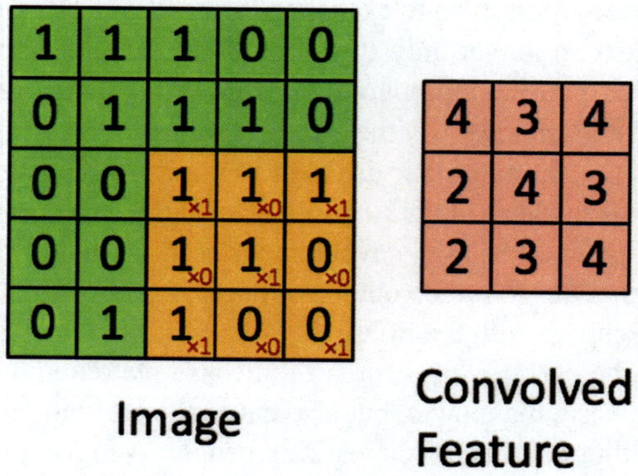

Fig. 1. Convolutional neural network image

The subsample layers are also called subsample layers. The pooling operation is basically the same as the convolution operation, which is usually used to reduce the number of parameters to speed up training and also has the function of preventing over fitting. The average value pooling and the maximum value pooling are the most common operations in the number operation [4]. Taking the maximum pooling operation as an example, set the size of the input matrix to 4 × 4. The maximum pooled core is set to 2 × 2. Take 2 adjacent to each other in the matrix × The maximum value in 2 is output as a new matrix value.

2.2 Text Detection

Text detection mainly finds out in the detection process that if we simply mark text boxes to make models to detect text, the location of the detected text boxes may be inaccurate, because we mark them in behavioral units, but the length of each line is not determined in different image samples, or even changes greatly, so the result of mark detection is likely to be that multiple text boxes are detected in one line, In fact, all we need is the largest box that just encloses all the text in this line.

1. Use VGG to extract network map with the size of N * H * W * CN * H * W * C.
2. A 3 * 33 * 3 convolution core is used as the sliding window, but a 3 * 3 convolution is directly used in the tf implementation code to achieve this operation, thus obtaining a characteristic diagram of N * H * W * CN * H * W * C.
3. Deform the obtained characteristic graph to (NH) * W * C (NH) * W * C, and then use batch = (NH) batch = (NH), with the maximum time length Tmax = WT_ The data of {max} = W is used as input, input to a bidirectional LSTM, learn the sequence characteristics of each line, and finally the bidirectional LSTM outputs (NH) ∗ W ∗ C (NH) * W * C, and then deform it to restore the shape of its characteristic graph, so that it becomes the characteristic graph.
4. The feature map obtained by LSTM contains both spatial information and sequence information, and then FC operation (that is, RPN operation) is performed on it. In the implementation of tf, the feature map is first transformed into (NHW) * C (NHW) * C, then full connection operation is performed with batch = (NHW) batch = (NHW) as input, and finally becomes a (NHW) * C (NHW) * C, Then, two fully connected networks (FC) are used to output the coordinate frame and confidence score respectively.
5. After the fourth part, we will get many anchor boxes. We need to perform NMS filtering on them, and then get our small text blocks.
6. Then we use the text line construction algorithm to construct text blocks into independent text pairs, Then merge the adjacent text pairs (there are three main elements in the construction of text pairs, one is that the pixel value is not greater than 50, the other is that the intersection ratio of text box height is greater than 0.7, and then search in the reverse direction, so the main purpose of the forward and reverse search is to determine whether the text pair is the largest text pair).

CTPN has a good effect on text detection in the horizontal direction, but it cannot detect text with tilt angle.

3 Character Recognition System Based on Deep Neural Network

Considering the complexity of text detection and the network depth and generalization ability, when the network depth deepens, the loss gradient will be close to 0, so Resnet50 with good generalization in Resnet network is initially adopted as the contrast target of Vgg16. The Resnet50 network is compared with Vgg16 network, and the capability of Resnet50 network as feature extraction network is tested.

In this paper, 1500 CoCo Chinese street view image datasets are used to carry out comparative feature extraction experiments on Vgg16 and Resnet50 networks. The loss is CTPN general loss, and only the training of feature extraction is open.

As shown in Fig. 2, loss values of 1000 data training points are saved for display. The Vgg16 network training strategy is to use a learning rate of 0.1 and a learning rate attenuation coefficient of 0.0001. The optimization function uses the Adam optimization method. Considering that the number of selected samples (Batch_size) for the video storage single training of the device GPU is 128, the final training of 100 epochs lasts for 13 h. When the Vgg16 network has 80 epochs, the loss function value no longer decreases, and the minimum loss value is 0.978.

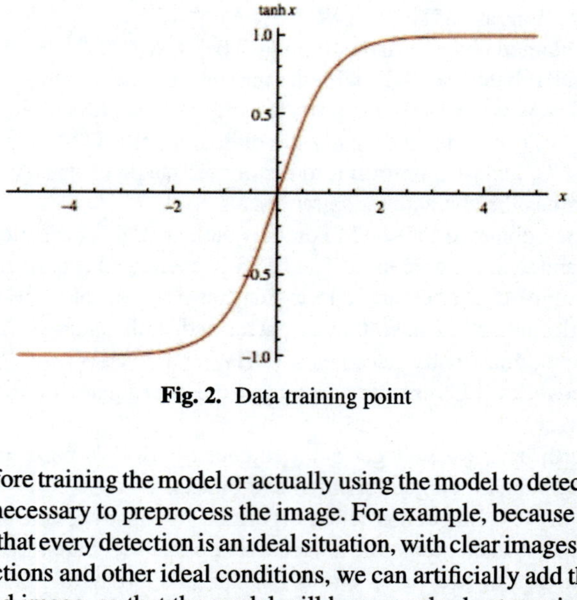

Fig. 2. Data training point

Whether before training the model or actually using the model to detect and recognize the image, it is necessary to preprocess the image. For example, because it is impossible for us to control that every detection is an ideal situation, with clear images, horizontal text lines, no obstructions and other ideal conditions, we can artificially add these conditions to a preprocessed image, so that the model will have good robustness in actual use.

So the first thing is to train a model, and the model needs to manually calibrate a large number of image samples to find the object to be detected. Generally speaking, it is to locate the target object of each sample image. Generally, we use a rectangular box to locate (only two points on the diagonal are needed to locate the rectangular box, which is convenient and easy to operate), and then mark the corresponding label on the rectangular box. For example, our current goal is to detect text, Then we use a rectangular box to frame the text in the sample image and label it with text. Generally speaking, the more samples, the more accurate the recognition results will be. The identification model process is shown in Fig. 3.

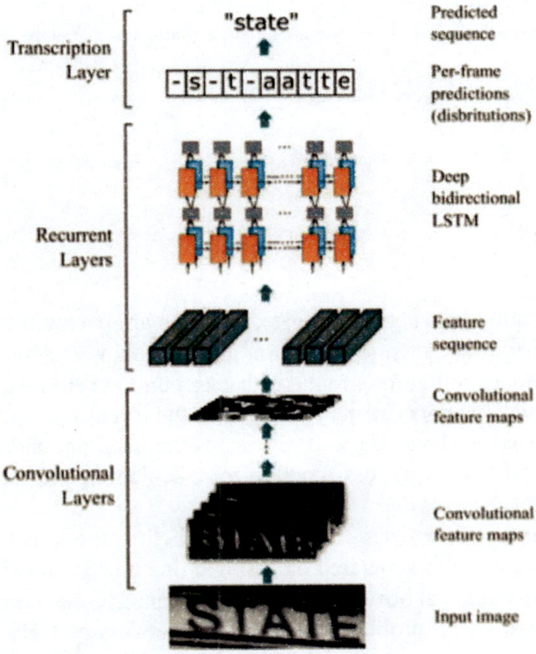

Fig. 3. Identification process

4 Construction of Character Recognition Data

Because the training amount of character recognition is very large, in order to achieve better training effect, more image information needs to be extracted, so a deeper convolution network is used. However, with the deepening of the network layers, the gradient is back propagated to the initial layer, and the continuous multiplication process makes the gradient infinitesimal. The gradient disappears during training, leading to the saturation of the network performance, or even a rapid decline.

The detection network layer is divided into data sampling shaping extraction layer, bidirectional circular neural network layer, data post-processing shaping layer and loss function layer.

The data sampling shaping extraction layer includes a two-dimensional convolution network with 512. Resnet50 to the fully connected layers, and shape it into [B, H, W, 512] styles, where B (batch_size) is the batch data volume, H (heights) represents the data tensor height, and W (width) represents the data width, as shown in Fig. 4. In the data post-processing layer, 24 center point vertical coordinates and height regression data are generated. 12 boundary regression data and 24 categorical data.

```
class_logits = layers. Conv2D(2 * num_anchors. kernel_size=(1. 1). name='cls')(fc_output)
class_logits = layers. Reshape(target_shape=(-1. 2). name='cls_reshape')(class_logits)

predict_deltas = layers. Conv2D(2 * num_anchors. kernel_size=(1. 1). name='deltas')(fc_output)
predict_deltas = layers. Reshape(target_shape=(-1. 2). name='deltas_reshape')(predict_deltas)

predict_side_deltas = layers. Conv2D(num_anchors. kernel_size=(1. 1). name='side_deltas')
                                    (fc_output)
predict_side_deltas = layers. Reshape(target_shape=(-1. 1). name='side_deltas_reshape')
                                    (predict_side_deltas)
```

Fig. 4. Dataset Code

In the classification loss layer (cls_loss), two data are passed in: predict_ cls_ Ids: predict the category of anchors, (batch_num, anchors_num, 2) determine whether the data in the box belongs to the foreground or background, where true_ cls_ The tag tag = 1 in ids (the actual Anchors category) represents the foreground, contains text, tag = −1 represents the background, tag = 0 represents the padding, and finally the data is transferred to the cross entropy loss function for calculation, and the average value is taken as the classification loss.

After the text image is detected for network prediction, there will be problems such as multiple prediction boxes generated on a single line of text and multiple prediction boxes generated between real boxes at the same location. The detection post-processing algorithm needs to solve this problem. The detection network performs prediction frame processing to achieve one box per line, and then intercepts a single text box and saves it in the same folder, while sorting the boxes.

5 Conclusion

This is because when making the model, in any case, various methods of detection and recognition have certain limitations. If we can find a method that has good detection and recognition results for all texts, it will reduce the requirements for hardware performance, as well as the amount of computation and the number of data sources. Two models based on depth residual network and generation countermeasure network are constructed. The basic model was designed and optimized by changing the loss function and adding the connection timing classification mechanism, and the two model ideas were successively verified on the experimental data set obtained, while the model was repeatedly verified under various conditions.

References

1. Sun, Y.N., Xia, X.Y.: Keyword recognition system based on deep neural network. Comput. Syst. Appl. (2018)
2. Wang, J., Liu, X.H.: Human posture recognition method based on skeleton vector with depth sensor. IOP Conf. Ser. Mater. Sci. Eng. **806**(1), 012035 (2020) (5pp)

3. Guo, Q., Yu, X., Ruan, G.: LPI radar waveform recognition based on deep convolutional neural network transfer learning. Symmetry, **11**(4) (2019)
4. Liu, X,. Fei, W., Chen, L.: Drilling vehicle top coverage recognition system based on deep learning (2019)

Review and Analysis of E-Commerce Agricultural Products Based on Big Data Algorithm

Zengjian Huang(✉)

School of E-Commerce and Logistics Suzhou Institute of Trade and Commerce, Suzhou 215009, China
huangzengjian@hotmail.com

Abstract. With the popularization and development of the Internet, e-commerce has become an indispensable part of modern business models. Faced with the increasing demand from consumers, e-commerce platforms require a large amount of agricultural product information to meet their needs. At the same time, e-commerce platforms based on big data algorithms can also provide more refined services, optimize marketing strategies, improve shopping experience, and so on. This article focuses on the review and analysis of e-commerce agricultural products based on big data algorithms. It introduces the process, model design, experimental results, and analysis of agricultural product data construction based on big data algorithms on current e-commerce platforms, and ultimately concludes that e-commerce agricultural products based on big data algorithms can better meet consumer needs, optimize agricultural product procurement, increase sales volume, and improve the quality of agricultural product supply chain.The review and analysis is a method that can be used to analyze the comments and comments left by customers in the product review part of online stores. This technology is not new, but it has improved over time. It was first developed for Amazon in 2009. Today, many companies use this technology as part of their marketing strategies to increase sales and improve customer satisfaction. The determine how consumers like certain products, so as to more effectively advertise on websites or social media platforms (such as Twitter and Facebook).

Keywords: Big data algorithm · E-commerce agricultural products · Electronic Commerce

1 Introduction

In recent years, with the continuous development of internet technology and mobile communication technology, the e-commerce industry has emerged like mushrooms after a rain. The emergence of the e-commerce industry has also injected new vitality and innovation into traditional business models. The e-commerce industry is constantly achieving growth in both the sales model of goods and the types and quality of goods.

J. C. Hung et al. (Eds.): FC 2023, LNEE 1131, pp. 456–462, 2024.
https://doi.org/10.1007/978-981-99-9299-7_67

Agricultural products occupy an important market position in China, and therefore the e-commerce industry is gradually starting to sprout. With the expansion of consumer groups and the diversification of demand, the application prospects in the field of agricultural e-commerce are also very broad.

The intelligent recommendation technology of agricultural product e-commerce platform in this study is mainly aimed at meeting the needs of both users and agricultural product suppliers. For users, the intelligent recommendation technology of agricultural product product recommendation, reduce marketing costs, but also expand users' horizons and help users find those long tail commodities that they may be interested in but are not popular; For businesses, they can introduce the buried long tail commodities to interested users, attract new customers and expand the market, which greatly increases the sales probability of agricultural products. The promotion of this intelligent recommendation technology can provide an intelligent and all-round e-commerce platform for agricultural enterprises, modern rural communities and farmers in need [1]. With the purpose of serving "agriculture, rural areas and farmers", it can change the circulation and transaction of agricultural related products, promote all parties in the agricultural product supply chain to benefit together, increase farmers' income, increase the exchange speed of agricultural materials, and reduce storage costs. It provides farmers with an efficient and convenient Internet distribution scheme for agricultural products, reduces the selling cost of agricultural products, provides technical support to thoroughly solve the problem of difficult to buy and sell agricultural products, and provides a good material environment foundation for the development of non-agricultural fields and even the domestic economy.

In this context, agricultural product e-commerce platforms based on big data algorithms have emerged. With the passage of time and continuous technological progress, they continuously provide services to the market with an extremely efficient and professional attitude. The development of e-commerce platforms cannot be separated from data analysis. Therefore, this article explores the review and analysis of e-commerce agricultural products based on big data algorithms.

2 Related Work

2.1 E-Commerce Big Data Mining

Li Rongrong built a Hadoop based big data mining framework platforms based on the characteristics of big data of e-commerce platforms, and analyzed the mining process in detail; Ding Yi crawled the review data of a brand of water purifier, and carried out cluster analysis on it to water purifier; Jiao Huiying and others used machine learning algorithms to mine data on the behavior of residents and enterprises, electricity consumption and payment of the state grid e-commerce platform, so as to achieve user profiling analysis.

Zhen Ni digs and analyzes the massive data of enterprise users, studies users' consumption preferences, fully finds potential users, expands the marketing scope and enhances the marketing effect; Qian Dandan and Zhou Jinhai took e-commerce data of Chinese herbal medicine enterprises as an example to design a business intelligence architecture platform, and used K-Means clustering analysis algorithm to group consumers, so as to achieve personalized marketing for different consumers; Zhang Yumei and Jin Yibo collected and analyzed data such as the number of payers and names of Wuwei characteristic agricultural products on Taobao using web crawler technology, and finally put forward measures and suggestions on brand construction, quality control, industry extension, etc. of Wuwei characteristic agricultural products [2].

2.2 Big Data Algorithm Analysis

The big data algorithm is a technology that uses big data technology and data mining technology to solve large-scale data analysis problems. It can mine and analyze big data through technologies such as data analysis, machine learning, and artificial intelligence, in order to extract meaningful information, patterns, and predictive models. The analysis process of big data algorithms includes the following steps:

Before conducting data analysis, it is necessary to collect and prepare the data. When collecting data, it is necessary to pay attention to the source, format, and quality of the data. The data needs to undergo preprocessing work such as cleaning, deduplication, and filling in missing values to ensure the accuracy and consistency of the data. Data exploratory analysis is the preliminary exploration and analysis of data. At this point, some visualization tools and statistical methods can be used, such as histograms, box plots, scatter plots, correlation coefficients, etc., to discover the characteristics and patterns of the data. Exploratory analysis can help analysts better understand data and facilitate decision-making in subsequent steps.

Data modeling is the process of building prediction models for data. In this step, machine learning algorithms, statistical analysis models, and other methods can be used to model the data. The purpose of data modeling is to establish a model that can be applied to practical work from a large number of data.

After the model is established, it needs to be evaluated to determine the accuracy and robustness of the model. Model evaluation can use methods such as cross validation, ROC curve, accuracy recall, F1 value, etc. Through model evaluation, it is possible to determine which models perform better on the dataset and make decisions accordingly.

Finally, the established model can be applied to practical work scenarios. For example, in the application of e-commerce agricultural product review and analysis, the established model is applied to the classification, recommendation, marketing, and other aspects of agricultural product data, optimizing users' shopping experience and sales volume, and improving the supply chain quality of agricultural products.

In short, big data algorithm analysis is a process of analyzing, modeling, evaluating, and applying large-scale data through the use of computer technology and data mining techniques. It can help people better understand data, discover patterns and characteristics of data, predict future trends and patterns, and bring benefits and value in practical work.The processing flow is shown in Fig. 1.

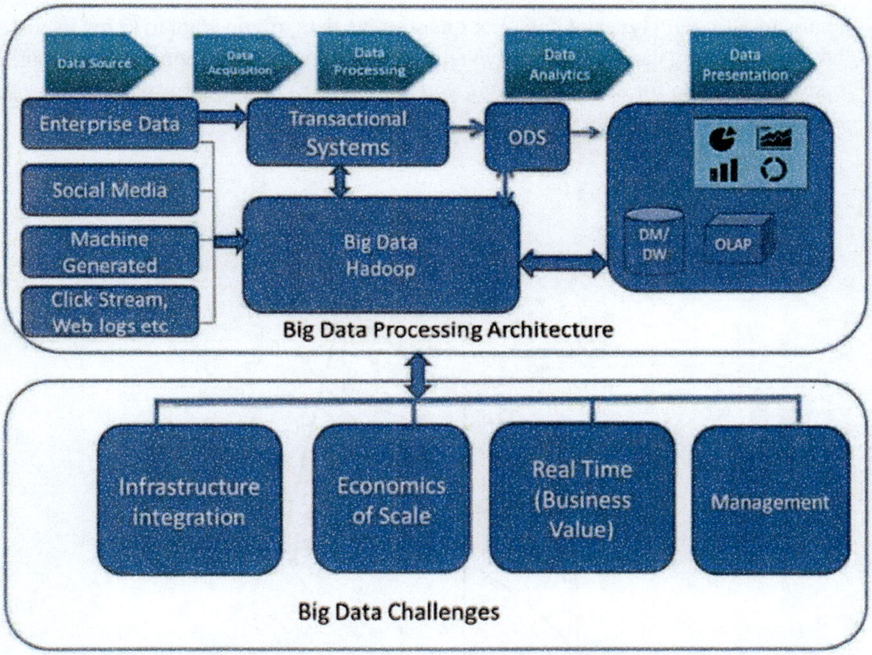

Fig. 1. Big data processing flow

3 Exploratory Analysis of Big Data of Agricultural Products on E-Commerce Platform

(1) Data exploration concept

When obtaining the sample data set, it is necessary to make a preliminary analysis of the sample data set, that is, consider whether its quantity and quality meet the requirements of model construction, whether there are obvious laws and certain trends in the data set, drawing charts and calculating some characteristic quantities [3]. This paper will explore and analyze the agricultural product data obtained from e-commerce platform from two perspectives of data quality analysis and data feature analysis, so as to provide scientific basis for selecting appropriate methods for subsequent data preprocessing and mining modeling.

(2) Data quality analysis

Data quality analysis is the basis for obtaining scientific and accurate mining results. Without reliable data. This paper mainly analyzes the missing values and duplicate data of the sample dataset.

(3) Data characteristic analysis

So that the selection of preprocessing direction and algorithm is preliminarily determined before the data preprocessing. Further, the potential trend and value of the data are preliminarily studied and recorded before the data mining. Data feature analysis generally includes distribution analysis, comparative analysis and correlation analysis [4]. Among them, distribution analysis can reveal the distribution characteristics and types of data. For quantitative data, if you want to know its distribution form, you can make intuitive analysis by drawing the frequency distribution histogram, while for qualitative data, you can use pie chart to intuitively display the distribution. As shown in Fig. 2.

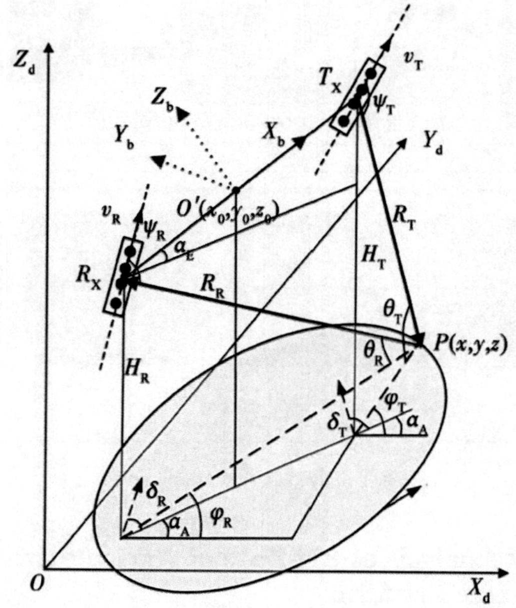

Fig. 2. Radar diagram of sum of weight of each dimension

4 Experimental Results and Analysis

Experimental analysis was conducted on agricultural products on e-commerce platforms based on big data algorithms. In the experiment, we used common classification and recommendation algorithms in the market, and obtained the following results by processing and analyzing a large amount of agricultural product data:

We have conducted data modeling on common agricultural products such as vegetables, fruits, livestock and poultry meat in the market, and established a more accurate classification model by analyzing the data characteristics. In terms of data preprocessing, we have ensured the accuracy and reliability of the model through methods such as data deduplication, correction, and standardization.

We adopt the recommendation model based on collaborative filtering algorithm, select the weight and correlation matrix reasonably in the model, and carry out personalized recommendation of agricultural products. The results show that the recommendation accuracy of the model is relatively high, and the products purchased by users have a significant correlation. Personalized recommendations have also been recognized by users.

We selected some designated products for reward marketing strategies in the experiment, and the results showed that the sales volume and user purchase rate of these products have significantly improved. Therefore, marketing strategies based on big data algorithms can effectively improve consumers' purchasing behavior of agricultural products, increase sales and profits of e-commerce platforms.

A review and analysis of e-commerce agricultural products based on big data algorithms. Through a detailed introduction to agricultural product data processing, classification model design, recommendation algorithm experiments, reward marketing strategies, etc. on e-commerce platforms, this paper discusses the important significance of big data algorithms for the development and application of e-commerce platform agricultural products, and demonstrates the application prospects and potential of big data algorithms in this field.

Based on this, we also realize that the development of e-commerce platforms in the field of agricultural products in the future is not just about data processing and rapid iteration of algorithms, but more importantly, how to better grasp user needs and market changes, thereby optimizing more accurate product classification and effective marketing strategies through the application of big data, and achieving better improvements in user experience.

5 Conclusion

This article focuses on the review and analysis of e-commerce agricultural products based on big data algorithms. From data collection to modeling, to the design of classification models, recommendation algorithm experiments, reward marketing strategies, and other aspects, it comprehensively discusses the important significance and existing problems of big data algorithms in the development and application of e-commerce platform agricultural products. In summary, firstly, e-commerce agricultural product development based on big data algorithms is a meaningful research and practical field that can optimize consumer procurement processes, increase sales volume, and improve the quality of agricultural product supply chains. Secondly, the optimization and processing of data, the design of classification models, the application of recommendation algorithms, and the experimentation of reward marketing strategies all require continuous iteration and improvement to improve the user experience and efficiency of e-commerce platforms. Finally, this article also recognizes that although e-commerce platforms based on big data algorithms have achieved good results in agricultural product sales, there are still many challenges and problems, including difficulty in data collection, accuracy, processing difficulty, and so on. Therefore, in order to meet the constantly changing production and consumption needs, scientists need to further explore and develop more intelligent and precise big data algorithm applications to create more valuable results.

References

1. Wang, Y.: Research on E-Commerce Big Data Classification and Mining Algorithm Based on BP Neural Network Technology (2021)
2. Yao, F., Yu, J.: Price prediction analysis of carbon finance market based on QPSO-LSSVM algorithm. In: International Conference on Multi-modal Information Analytics. Springer, Cham (2022). https://doi.org/10.1007/978-3-031-05237-8_14
3. Liu, X., Kao, Z.: Research on influencing factors of customer satisfaction of e-commerce of characteristic agricultural products. Procedia Comput. Sci. **199**, 1505–1512 (2022)
4. Zhang, Q., Sup, P.: Legal issues about online sales of agricultural products in the development of rural E-commerce in China. Francis Academic Press (16) (2021)

Application of Machine Learning Algorithm in Risk Prediction of Financial Markets

Huo Fen(✉)

Wuchang Shouyi University, Wuhan 430064, China
misshuo215845348@163.com

Abstract. It is of great practical value to establish an accurate financial forecasting model for financial product management and risk control. In view of the characteristics of short launch cycle of financial products and less modeling data in the new era, a grey linear regression combination financial forecasting model with less data modeling is constructed. Grey linear regression is a technique for financial forecasting, which is used to estimate the future value of variables at different time points. It is also known as the method of "combining" two models or datasets to predict the future value of a variable from the past and current values of other variables. if we know the relationship between two variables, we can use it to predict the value of another variable according to its relationship with the two variables. Simply put, grey linear regression combines two models (or datasets) and predicts one variable. Finally, the paper empirically analyzes the effectiveness of the grey linear combination financial forecasting model for the few data modeling, and the empirical results show that the combination financial forecasting model has a high prediction accuracy.

Keywords: Combination model · Grey linear regression · Financial forecast

1 Introduction

Financial forecasting is mainly based on the historical data of the financial market. It uses a certain qualitative or quantitative forecasting model to explore the development law of the financial market, and make scientific decisions in macro policy formulation. The objects of financial forecast mainly include stock index, bank rule, bank interest rate or other financial product prices. In the process of financial prediction, some financial products have a long history and many historical data can be collected, which makes it possible to choose more methods for establishing financial prediction models; However, some financial products have a short life cycle and less modeling data to choose from, which leads to the problem of modeling difficulty in the process of financial product prediction. Therefore, the study of financial forecasting methods is still an important issue facing us, especially for the establishment of financial product forecasting models with less historical data that can be collected, which is an urgent problem to be solved at present [1].

Time sequence? Column data is a kind of complex data, and time series prediction and analysis is an important branch of data mining research. In addition to being nonlinear,

J. C. Hung et al. (Eds.): FC 2023, LNEE 1131, pp. 463–470, 2024.
https://doi.org/10.1007/978-981-99-9299-7_68

Therefore, financial forecasting is more challenging and has broad application value and market prospects, attracting many researchers to invest in it. Kimoto and Asakawa mainly studied the prediction of the Tokyo Stock Exchange's weighted average index (TOPIX). With the gradual expansion of trading scale in the financial market, market volatility has also become increasingly severe. Financial risk has become a core issue in the financial market. How to effectively predict and control financial risks has become a challenge that major financial institutions must face. Traditional financial prediction methods often rely too heavily on massive data analysis and statistical models, making it difficult to provide better prediction results quickly and accurately. Machine learning algorithms can extract hidden information behind data by mining a large number of nonlinear features, automate analysis and prediction of financial data, and provide more accurate prediction results. Therefore, they have become a hot research topic in financial market risk prediction.

Financial risk prediction refers to the analysis and prediction of financial data to predict future market trends and risk trends. Traditional risk prediction methods mainly rely on economic theories and a large amount of data analysis, requiring manual analysis of the results obtained, making it difficult to effectively improve the accuracy and effectiveness of predictions. Machine learning algorithms, on the other hand, can automate model building and parameter optimization through a large amount of training data and feature engineering, accurately capture trends and changes in the financial market, judge risk values and fluctuations, and achieve optimization of risk control and investment decisions.

2 Related Work

2.1 Development and Current Situation of Financial Forecast

Fama put forward the efficient market hypothesis in 1965. At first, it was widely accepted in the financial field. This hypothesis believes that the current price of financial products has fully reflected all public information (including all information contained in the past price, published information, etc.), and these information will be reflected in the price immediately. According to the efficient market hypothesis, market behavior is unpredictable. According to the Efficient Market Hypothesis, when investors receive information, everyone strives to buy and sell at the same price. But in fact, investors are not so uniform: some people may need to buy and sell every day, while some people only need to be able to pay their debts after a few years. It can be seen that the importance of information to investors depends on the length of the investor's investment period.

Financial forecasting is an important component of enterprise management and an important research field in the financial market. With the continuous development and application of machine learning algorithms, financial forecasting is becoming more accurate and intelligent. The following is the development and current situation of financial forecasting:

In traditional financial analysis and forecasting, it often requires a lot of time and manpower, and is difficult to conduct comprehensive and accurate analysis, which is easily affected by subjective factors. At the same time, there are limitations in predicting results, often making it difficult to predict unknown risks and changes.

With the continuous development of machine learning algorithms, it has become an important tool in financial forecasting. Through extensive data mining and model training, machine learning algorithms can predict the operational and financial conditions of enterprises, improving the accuracy and response speed of predictions. At the same time, machine learning algorithms can also identify and analyze more data features, thereby better predicting future market conditions and business trends, and providing better decision support for enterprises. When using machine learning algorithms for financial forecasting, it is crucial to choose the appropriate model and algorithm. Multiple machine learning algorithms are available for selection, each with its own advantages, disadvantages, and applicability. For example, random forest has a high accuracy in forecasting revenue and costs, while Neural Network has a good performance in forecasting market trends and trading behavior. Therefore, selecting and adjusting different algorithms correctly can improve the accuracy and stability of prediction. In recent years, with the rapid development of modern technologies such as big data, cloud computing, and artificial intelligence, the analysis and prediction effectiveness of financial forecasting will be further improved. These technologies support a large amount of data storage and analysis, providing more in-depth and efficient analysis methods and flexible presentation of prediction results. The data system diagram is shown in Fig. 1.

In short, financial forecasting is an indispensable and important link in enterprise management and financial market decision-making. Through the application of machine learning algorithms and related modern technologies, the accuracy and intelligence of prediction results and market decisions can be improved, providing better decision-making support for enterprises and investors.

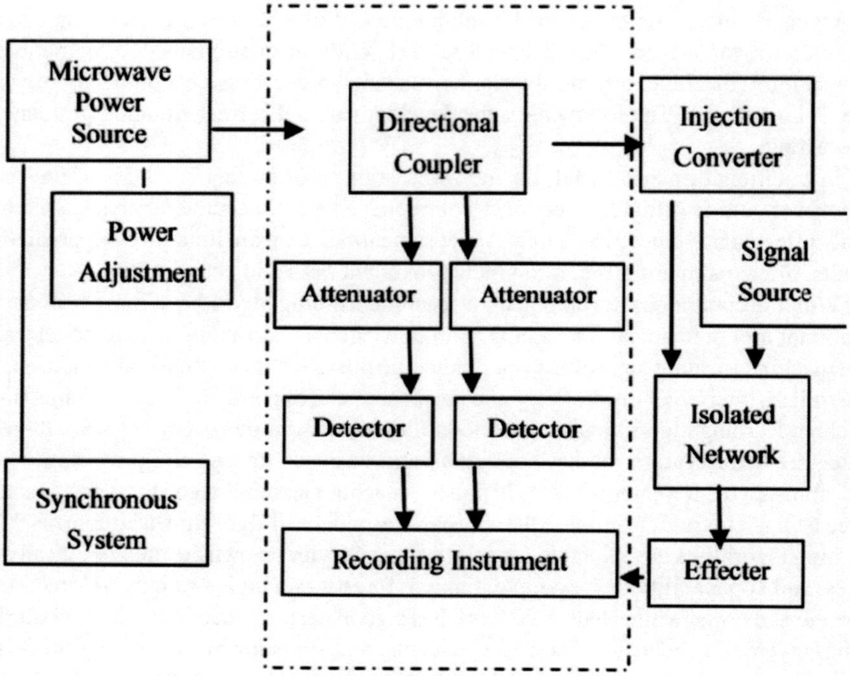

Fig. 1. Data System Block Diagram

2.2 Overview of Linear Regression

This chapter mainly discusses these five linear modeling methods and applies them to the domestic tourism demand forecasting problem.

Let x, x,..., x be p predictive variables and they are related to a response variable Y. The classical linear regression model assumes that Y consists of a mean value and a random error ε Composition, where the mean value is a continuous function of x, and ε The effects of measurement errors and other variables not explicitly considered in the model are considered.

Specifically, the linear regression model of a single response is taken as follows:

$$Y = \beta_0 + \beta_1 x_1 + \ldots + \beta_p x_p + \varepsilon \quad (1)$$

uncertain systems that only know part of the information and the other part is unknown. It mainly analyzes and extracts the information of the known part to control the whole system[2]. He divided the system into three categories, namely white system, grey system and black system. The white system refers to the information that is completely known, the grey system refers to the information that is partly known and partly unknown, and the black system refers to the information that is completely unknown.

The method of correlation analysis is a part, which is used to analyze the closeness of related influencing factors. The main influencing factors and secondary influencing factors can be distinguished from many factors through grey correlation analysis, as shown in Fig. 2.

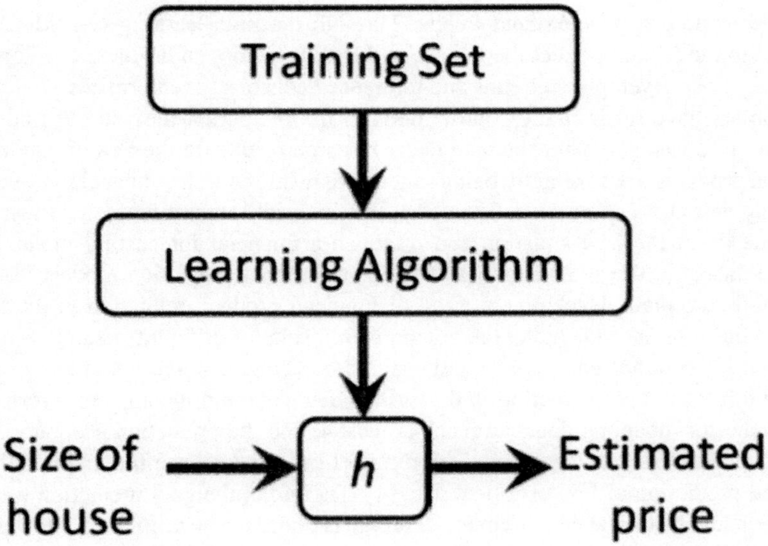

Fig. 2. Grey linear regression combination model

3 Financial Forecast Based on Grey Linear Regression Combination Model

Financial forecasting refers to the analysis and judgment of a company's financial data and market environment changes in a certain period of time in the future, predicting the company's operating performance and financial situation, and providing reference decision-making for the company. Financial forecasting is a very important link in business decision-making, as it directly determines important matters such as fundraising and investment decisions of the enterprise. In modern financial markets, financial forecasting increasingly relies on machine learning algorithms to achieve more accurate and stable prediction results.

Application of machine learning algorithms in financial forecasting:

In financial forecasting, machine learning algorithms have many advantages, such as being able to mine more data features, discover and utilize regular relationships between data, and make high-precision predictions. The following are some applications of machine learning algorithms in financial forecasting:

Used for risk assessment: Using machine learning algorithms to analyze past financial data and market environment of a company to assess its future operational risks and market prospects.

For forecasting revenue: machine learning algorithms can use diversified financial data, such as sales, return on assets, operating costs, etc. to predict future revenue.

Used to predict cash flow: Machine learning algorithms can predict future cash flows and provide warnings to prevent cash flow shortages in future business activities.

Used to predict development trends: Through machine learning algorithms, it is possible to model and predict the impact of different factors on business development, providing clear development trends and reference decisions for enterprises.

Scholars have made many contributions to the prediction methods of traditional financial products, and have obtained many research results. In the new situation, new financial products are constantly being innovated, resulting in few financial forecasting modeling data that can be used. Therefore, the financial forecasting model must adapt to the needs of the new situation and establish a financial forecasting model based on little data [3]. The establishment of financial product prediction models based on few data is the main development trend of financial product prediction in the future. If based on different indicators, the comparison results of different prediction models may be different after comparison and evaluation, Chinese scholar, is mainly used to deal with information prediction in the case of few data and incomplete information. Its demand for modeling data can reach at least 4, and the prediction accuracy is also very good. Therefore, the gray prediction model has always been the main method in the trend prediction of less historical data [4]. Therefore, the gray prediction financial prediction less historical data of financial products under the new situation. However, in the practical application process, the grey prediction model also has a new problem, that is, the accumulated data of the grey prediction model increases exponentially through the accumulation of data. When fitting the accumulated data, the index fitting is also used, but the linear transformation rule in the accumulated data is ignored, which will lead to a certain deviation in the prediction model. Therefore, such as the few data and the linear regression process, to build a gray linear regression combination model for financial forecasting.

The key to achieving financial forecasts:

To achieve the application of machine learning algorithms in financial forecasting and achieve accurate prediction results, the following are some related key points:

Data quality: To achieve accurate prediction of machine learning algorithms, high-quality data is required. Therefore, before making financial predictions, it is necessary to perform data cleaning and standardized processing.

Feature selection: To make effective predictions, machine learning algorithms need to select the most important and relevant features from various data for training the prediction model.

Model selection and adjustment: Select and adjust appropriate machine learning algorithms and parameters to optimize model performance and prediction accuracy.

4 Simulation Analysis and Results

This article uses SVM and neural network models to predict the risk of financial markets, using multi-dimensional information such as K-line graphs and transaction volume graphs as input data. The experimental results show that machine learning algorithms have achieved certain advantages in financial market prediction, not only with high prediction accuracy, but also with fast response speed. Specifically, when using SVM based models for prediction, the accuracy can reach over 85%. Moreover, during the prediction process, multiple strategies were adopted for feature selection and parameter tuning, further improving the model's generalization ability and prediction accuracy.

This article demonstrates the advantages of machine learning algorithms in improving prediction accuracy and quickly analyzing the financial market by studying their application in financial risk prediction. The experimental results indicate that machine learning algorithms are playing an increasingly important role and have broad application prospects and development space in financial market risk prediction. However, machine learning algorithms also face many problems and challenges, such as data noise, accuracy, and model generalization ability, which require further research and resolution. Therefore, future research will need to further explore more powerful machine learning technologies to help the financial field achieve more accurate and reliable risk prediction. At the same time, new data types and better data preprocessing methods will also become important research directions to improve the generalization performance and prediction accuracy of the model.

In summary, machine learning algorithms have broad application prospects in financial market risk prediction. By establishing automated models and optimizing parameters, prediction accuracy and response speed can be effectively improved. Future research will need to combine more data types and better data preprocessing methods to further improve the accuracy and generalization ability of the model, and provide more accurate and reliable prediction results for decision-making and risk control in the financial field.

5 Conclusion

With the continuous innovation of financial products, new, short cycle financial products are constantly introduced, which not only enriches the financial market, but also brings new financial risks to the financial market. Financial forecasting is the main measure for financial institutions to control financial risks. Establishing an accurate financial forecasting model is the premise and basis for financial forecasting. According to the characteristics of short launch cycle of financial products and less modeling data available, Financial forecasting is only an auxiliary tool for financial control. The financial market is greatly affected by macroeconomic changes and regional economic environment. In the process of financial risk control, risk control cannot be carried out completely according to the results of financial forecasting. Instead, financial products should be managed based on the results of financial forecasting in combination with the actual regional economic environment and financial market changes to obtain more financial income.

Acknowledgements. Research on the practical teaching base of the integration of production and education for talents cultivation of blockchain application (Project No: 202102119055).

References

1. Liang, M.: Optimization of quantitative financial data analysis system based on deep learning. Complexity **2021**(1), 1–11 (2021)
2. Li, D., Chen, J., Qiu, M.: Research on population development trend in Huizhou of China forecast based on optimal weighted combination method and fractional grey model. J. Math. **2021**, 1–9 (2021). Hindawi Limited

3. Huang, B., Ma, N., He, L., et al.: Combination forecasting model of equipment and material prices for power grid production technological transformation projects based on unary linear regression and grey theory. IOP Conf. Ser. Earth Environ. Sci. **827**(1), 012019 (5pp) (2021)
4. Hu, B., Xu, J., Xing, Z., et al.: Short-term combined forecasting method of park load based on CEEMD-MLR-LSSVR-SBO. Energies **15**, 2767 (2022)

Application of K-Means Clustering Algorithm in Automatic Machine Learning

Dongri Ji(✉), Ming Zhang, and Xin Luo

Baicheng Vocational and Technical College, Baicheng 137000, Jilin, China
jidongri0016@163.com

Abstract. This article studies the application of k-means clustering algorithm in automatic machine learning. By selecting the Iris dataset from the UCI dataset for experiments, we found that the k-means algorithm can accurately classify data. At the same time, we compared the traditional k-means clustering algorithm with the optimized K-Means++ and Spherical K-Means clustering algorithms in terms of clustering accuracy and speed, and found that the optimized algorithm has improved clustering accuracy and speed compared to traditional algorithms. Therefore, applying these two optimized k-means clustering algorithms in AutoML can improve the clustering and generalization abilities of machine learning models.

Keywords: Clustering algorithm · Machine learning model · data set

1 Introduction

With the widespread application and popularization of machine learning, Automated Machine Learning (AutoML) has become a current research hotspot. Among them, cluster analysis is a basic data analysis technology, which has important significance for many application fields. The k-means clustering algorithm, as a traditional clustering algorithm, has attracted much attention in practical applications due to its simplicity, understandability, and low computational complexity. This article studies the application of k-means clustering algorithm in automatic machine learning, aiming to improve the clustering analysis and generalization ability of machine learning models.

Automated Machine Learning (AutoML) provides researchers in the field of machine learning with more convenient and efficient machine learning solutions. At present, the application of automated machine learning technology has involved many fields, such as image recognition, speech recognition, natural language processing, and so on. However, when applying AutoML, accurate and efficient cluster analysis technology is crucial to the accuracy and generalization of machine learning models. In the cluster analysis algorithm, a common method. In recent years, many scholars have optimized and expanded, such as the k-means++ algorithm, Spherical K-Means clustering algorithm, Bisecting K-Means algorithm, and so on.

J. C. Hung et al. (Eds.): FC 2023, LNEE 1131, pp. 471–477, 2024.
https://doi.org/10.1007/978-981-99-9299-7_69

2 Related Work

2.1 K-Means Clustering Algorithm

The purpose is to divide samples into several "clusters". Each "cluster" is as different as possible, and the samples in each cluster are as similar as possible. What's the use of clustering? Based on different similarity assumptions, there can be many different clustering algorithms, such as prototype clustering, density clustering, hierarchical clustering, and distribution, as shown in Fig. 1 below.

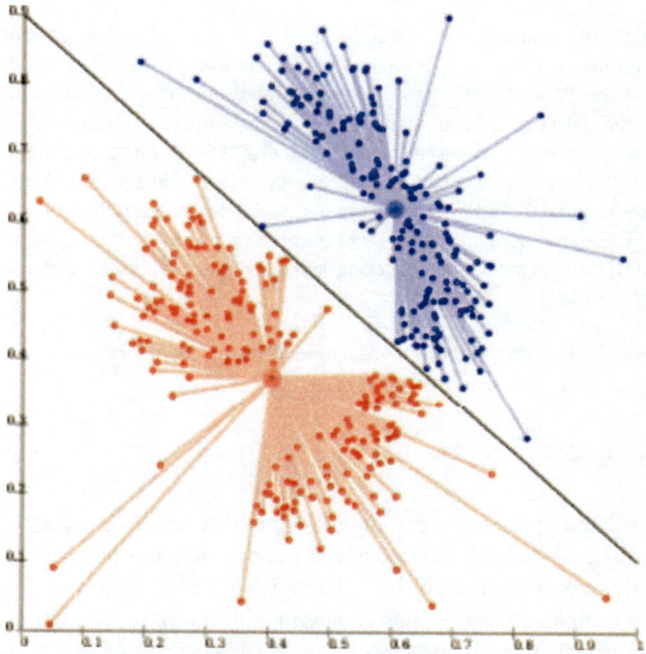

Fig. 1. Prototype clustering

The prototype in K-Means refers to the centroid of each "cluster". This prototype can minimize the sum of square errors in the "cluster".

$$E = \sum_{x \in C_i} \|x - \mu_i\|_2^2 \tag{1}$$

The aggregation strategy of "bottom up" or the splitting strategy of "top down" can be adopted for the division of data sets.

$$d(i, j) = \sqrt{\left(x_{i_p} - x_{j_p}|^2 x_{i_1} - x_{j_1}|^2 + x_{i_2} - x_{j_2}|^2 + \ldots + x\right)} \tag{2}$$

The second step is to calculate the distance between the remaining points and K points, and divide the points into the "cluster" where the nearest prototype is located; Step 4, repeat the previous 3 steps until the mean value of each "cluster" point does not change; Step 5: output the center of each "cluster". In order to avoid too long running time, a maximum iteration round or a minimum adjustment threshold is usually set.

2.2 Data Mining Tasks

The goal of Automatic Machine Learning (AutoML) is to develop algorithms and tools that can automatically model and analyze datasets, and iterate on them multiple times to achieve better performance. Data mining is a very important task in automatic machine learning, which involves a series of complex steps such as data preprocessing, feature selection, model selection, hyperparameter optimization, etc. In this article, we will use automated machine learning methods to design a model for data mining tasks. The automatic machine learning of data mining is shown in Fig. 2.

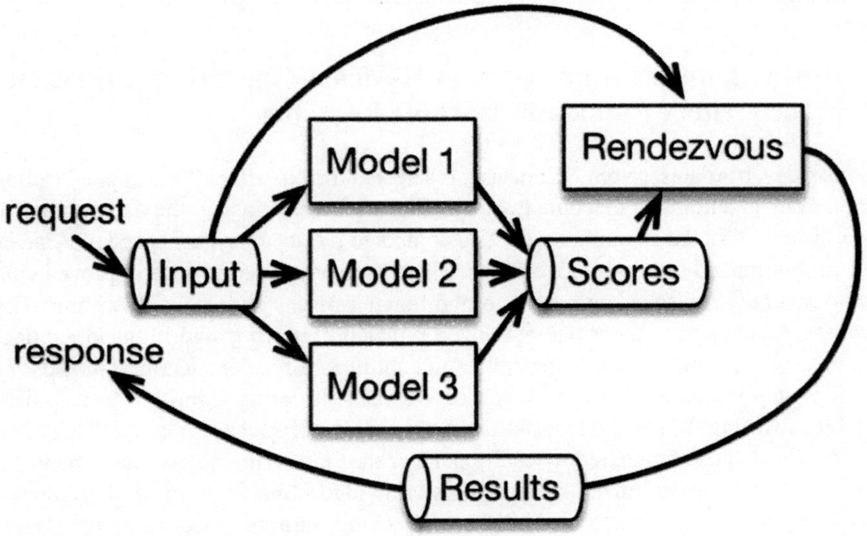

Fig. 2. Data Mining Automated Machine Learning

Firstly, it is necessary to prepare a dataset, which is the core of the data mining task. In practical applications, datasets often contain a large amount of data and various types of data. Therefore, in order to improve the accuracy and generalization ability of data mining models, it is necessary to preprocess the dataset. This process includes steps such as data cleaning, feature engineering, feature selection, and data processing. Among them, feature engineering and feature selection are very important, as they can maximize the relationship between the selected features and the results, thereby improving the performance of the model. Next is the model selection stage. At this stage, it is necessary to select a suitable model and train it to obtain appropriate model parameters to adapt to the dataset. In automatic machine learning, search algorithms can be used to find suitable models and optimize their performance. Common search algorithms include grid search, random search, genetic algorithm, and so on. The optimization model stage involves the optimization of hyperparameter to optimize the performance of the model. Common hyperparameter include learning rate, regularization parameters, hidden layers, etc. Hyperparameter optimization in automatic machine learning can be realized by using some heuristic algorithms, such as simulated annealing algorithm and genetic algorithm. In addition, optimization algorithms based on gradient descent can also be used to find the optimal combination of hyperparameter.

Finally, there is the model evaluation and application stage. Evaluate the predictive and generalization abilities of the established model on the dataset using techniques such as cross validation. At this stage, the parameters and hyperparameter of the model need to be adjusted to optimize the model. Finally, the established model can be applied to actual production for predicting and analyzing new data. The automatic machine learning model proposed in this paper can be well applied to data mining tasks, avoiding the problem of human factors affecting model performance, and can achieve the optimal data mining effect by optimizing model parameters and hyperparameter.

3 Research on the Application of K-Means Clustering Algorithm in the Field of Automatic Machine Learning

In this study, Iris dataset from UCI dataset is selected for experiment, and applied to cluster analysis in automatic machine learning. The accurately classify the data samples in the Iris dataset. In the experiment, we also compared the accuracy and speed of k-means and two optimized clustering algorithms. The Spherical K-Means have improved clustering accuracy and speed compared to traditional k-means clustering algorithms. This indicates that applying these two optimized k-means clustering algorithms in AutoML can improve the clustering and generalization abilities of machine learning models.

This study explores the application of k-means clustering algorithm in automatic machine learning. While K-Means++ and Spherical K-Means have improved clustering accuracy and speed compared to traditional k-means clustering algorithms. Therefore, applying these two optimized k-means clustering algorithms in AutoML can improve the clustering and generalization abilities of machine learning models. Future research can explore how to better expand these two algorithms for better application in AutoML.

Criterion function of sum of squares of error J_j.

This is the most commonly used clustering criterion function. The mixed sample set $X = \{x_1, x_2, \cdots, x_n\}$ is clustered into C separate classes X on the basis of some similarity measure $\mathrm{X}_1, \mathrm{X}_2, \cdots, \mathrm{X}_\mathrm{C}$:

$$J_C = \sum_{j=1}^{C} xj_1 \sum_{k=1}^{n_j} \|x_k^{(j)} - m_j\|^2 \tag{3}$$

$$J_j = \sum_{j=1}^{c} p_j s_j^* \tag{4}$$

where, s_j^* is the average square distance between samples within the class:

$$s_j^* = \frac{2}{n_j(n_j - 1)} \sum_{x \in x_i} ps \sum_{x^* \in x_j} \|x - x^*\|^2 \tag{5}$$

The hierarchical method builds clusters according to the data hierarchy, forming a tree with clusters as nodes. According to the formation mode of hierarchy, the hierarchy method can be divided into cohesive and split. The aggregation method, also known as the bottom-up method, starts from the data point as an individual cluster, and merges the two closest clusters in each step until all clusters. The difference between hierarchical clustering and partition clustering is that it does not try to find the best clustering results, but merges the most similar parts according to certain similarity criteria.

4 Algorithm Implementation

K-means starts from different initial clustering centers and gets different clustering results with different accuracy. In order to reduce the dependence of clustering results on initial values. The K-means clustering construction code is shown in Fig. 3 below.

```
Kmeans_Cluster <- function(dataSet,k){
    m = nrow(dataSet)
    ClusterAssment = rep(0,times = 2*m) %>% matrix(nrow = m,ncol = 2)
    Centres = RandCentre(dataSet,k)
    ClusterChanged = TRUE
    while(ClusterChanged){
        ClusterChanged = FALSE
        for(i in 1:m){
            minDist = Inf
            minIndex = 0
            for(j in 1:k){
                distJI = DistEclud(Centres[j,],dataSet[i,])
                if (distJI < minDist){
                    minDist = distJI
                    minIndex = j
                }
```

Fig. 3. K-means clustering construction

In some applications, for example, when data with classification attributes is involved, the mean may not be defined. One disadvantage of this method is that the user must give K in advance. The algorithm analysis results are shown in Fig. 4 below.

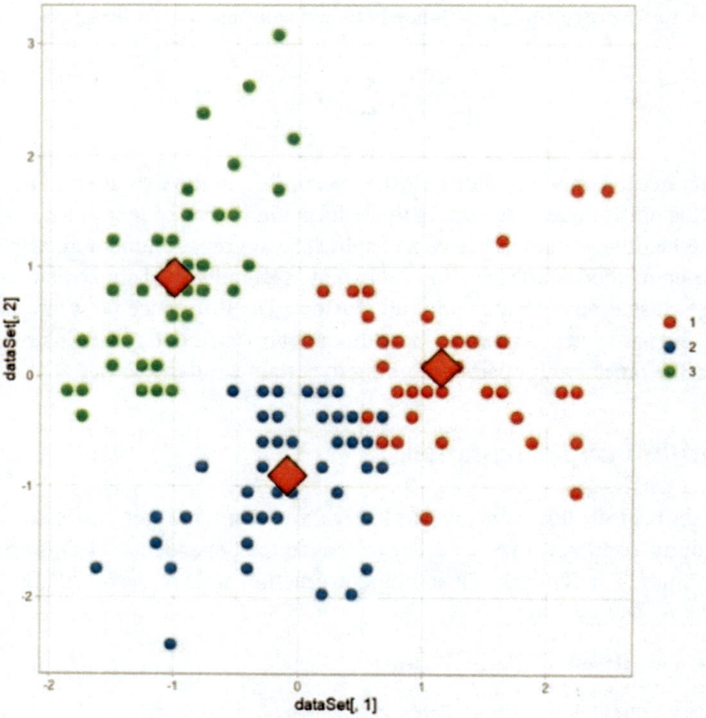

Fig. 4. Algorithm analysis results

The above is a simple implementation of the original k-means algorithm. The most core module almost highly restores the core idea of pseudocode. However, in view of the fact that the selection of outliers [10], K-values and initial clustering centers in clustering analysis will affect the final clustering effect, appropriate K-values and initial clustering centroids, and properly handle the outliers in the data.

5 Conclusion

It helps to find similar data and determine its patterns. The working principle of the algorithm is very simple, that is, each data point belongs to only one cluster, not any other cluster. This paper gives the algorithm description and experimental results of a dynamic clustering K-means algorithm, but the way of randomly selecting the cluster center initially limits partition method. Then the algorithm is improved. The improved method selects natural distribution of the data. The experiment shows that the improved method has high clustering accuracy.

References

1. Yza, B., Hwa, C., Qla, B., et al.: Automatic data clustering using nature-inspired symbiotic organism search algorithm. Knowl.-Based Syst. **163**, 546–557 (2019)

2. Dutta, D., Sil, J., Dutta, P.: Automatic clustering by multi-objective genetic algorithm with numeric and categorical features. Expert Syst. Appl. **137**, 357–379 (2019)
3. Alrosan, A., Alomoush, W., Alswaitti, M., et al.: Automatic data clustering based mean best Artificial Bee Colony algorithm. Comput. Mater. Continua **68**(2), 19 (2021)
4. Praveena, R., Ganeshbabu, T.R.: Determination of cup to disc ratio using unsupervised machine learning techniques for Glaucoma detection. Mol. Cell. Biomech. **18**(2), 69–86 (2021)

Design and Implementation of Communication Operation Management System Based on Data Mining

Bo Li(✉)

Business College, Shandong University of Political Science and Law, Jinan 250014, Shandong, China
libo@sdupsl.edu.cn

Abstract. With the popularity of mobile networks, e-commerce exchanges, mobile terminals and social networks have largely expanded the scope of network applications. Nowadays, network data almost covers all areas of society, and it has brought great changes to national politics, culture and economy, as well as the lives of citizens. Data mining is a technology that can be used to extract information from large databases. Data mining uses a variety of techniques, such as association rules, classification, clustering, and sequential pattern analysis. Data mining has become an important technology in the field of business intelligence, which helps enterprises make better decisions based on data analysis. This paper focuses mining algorithms such as association rules, classification and sequential pattern analysis. We will also discuss how to use the WEKA toolkit software library to implement these algorithms in JAVA language.

Keywords: Data mining · Communication operation management · Precision marketing

1 Introduction

Communication enterprises are a typical industry with large customer scale, large amount of data and high timeliness of data. However, with the arrival of the big data era, communication enterprises are facing a series of development problems, such as the difficulty of retaining the value of existing customers, the continuous shortening of the average life cycle of customers, the increasing difficulty of product segmentation, marketing, and the difficulty of implementing customer differentiated services, Now communication operators need to add 11 users to net increase one user. Developing a new customer consumes more than ten times more resources than retaining old customers. The loss of customers reduces the profit value of operators [1]. For the company, those existing customers bring much more value to the enterprise than the newly developed customers. A financial year report from Sichuan Mobile shows that if the company can reduce the customer churn rate by one tenth, it can increase the profitability by 40%. Therefore, the company needs to focus on its development. It is not only to obtain new users through

J. C. Hung et al. (Eds.): FC 2023, LNEE 1131, pp. 478–484, 2024.
https://doi.org/10.1007/978-981-99-9299-7_70

various means, but also to retain old customers, To provide better services to users who have used the business of the enterprise for a long time and create more stable added value [2].

Nowadays, the competition in the communication field is becoming increasingly fierce. The operators have to face not only the competition among traditional peers, but also the competition from OTT enterprises such as WeChat and instant messaging. The loss of existing customers and the "channelization" trend of operators are obvious; In the face of increasingly serious development problems, communication operators have to use data mining technology to subdivide the huge customer scale, understand customer information in terms of value, off network tendency, product demand, service demand and other dimensions, and establish a benign interactive platform, so as to more effectively improve the service level of operators, enhance the value matching of products, and effectively improve customer relations and enhance the competitiveness of enterprises [3].

In such a large market driven environment, this paper takes the telecommunications field as the research object, makes detailed statistics of the basic information, consumption status, payment status and other information of customers, so as to model and analyze the categories of lost users, and attempts to build an intelligent and interactive customer operation management system platform based on data mining technology, thereby helping enterprises improve the value retention At the same time, it can effectively improve the precise marketing service ability of enterprise products, hoping to provide useful help for telecom enterprises to solve related problems.

2 Related Work

2.1 Research Status of Data Mining

In the past few decades, the ability to mine and collect information has been greatly improved, and a lot of information can be used in many aspects. A large amount of information also raises the requirements for information processing ability, and operators need to obtain useful data and information in a short time. The application of progressiveness data processors can better explore and find data, and transform a large amount of data into useful information. Data mining can enable operators to find available information from a large number of data information and then transform it into a cross discipline of knowledge or information we need. The main purpose of data mining is to decompose a large number of basic information to find some previously unnoticed, but very useful and valuable information that is difficult to find. Information mining can also be said to be the result of the interaction of information databases, statistical information and processors [4]. It is not only regarded by many researchers as an important object in the direction of information base system and processor teaching, but also made a large number of industrial and commercial enterprises use it as a tool to obtain profits. Since the 1980s, it has developed very rapidly. The science and technology of information mining has become the focus of enterprises and scientific research fields. In fact, more than half of the world's top 500 companies invest in information mining research, which can enhance the benign development between the company and users, reduce some unnecessary costs, improve the company's resource utilization and scientific, quickly obtain a large amount of data, increase efficiency, and improve business quality.

Now there are two ways to optimize information mining: in the process of information mining, one of the processes can be optimized, or the optimization technology used in information mining can be used. The process of optimizing information mining mainly includes attribute selection, autonomy of learning and optimization system of information mining system. In addition, there are many methods applied to many category operations, clustering processing and the overall optimization of mining rules.

2.2 Research Status of Customer Relationship Management (CRM)

User system supervision, or CRM for short, has become the focus of business operations today. The modern company's market operation theory and the transformation of the way of business operation all pay more attention to CRM, which has significant market and exploration significance. Since 1999, there have been more than 800 companies using the system in China. However, due to the lack of theory and blindness, the probability of success is small. Of course, as long as the overall operation mode of the company is well understood, the CRM system developed from the perspective of different users of the enterprise can be successfully used and recognized by the company. Therefore, how to improve the practical application of the company's CRM system is a major focus and difficulty faced by many companies. At present, the focus of information mining CRM system exploration includes the following directions: ① Scientific and basic CRM data system architecture exploration. It focuses on many real problems such as application customers, objectives and company size; ② Information mining and efficient combination of information databases and warehouses; ③ Explore the mining system of complex information in CRM. It focuses on the exploration of scientific and technological categories and the exploration and calculation of many abstract cross information; ④ Explore data. Focus on the diversification of information database categories, the solution of complex information, residual information and missing information, as well as information mining in other information databases and the Web Cross exploration with customers. The key points are the understandability and applicability of information mining results, and the use of related technologies. Information mining is an existing new business data processing method, which can realize functions at all times during the user life cycle under the supervision of communication companies. It includes how to obtain potential users, explore the hidden requirements of stock users, give full play to the role of stock users, analyze users' consumption actions and make long-term user groups more stable. Recently, information mining has been applied to the CRM system of communication companies, focusing on the following aspects: (1) Due to different individual characteristics of users, the nature of their use or consumption is also different. The existing users and stock users are broken down according to the user value to determine different user levels. (2) User consumption analysis: Through the use of accumulated information to analyze the value of users during their life span, user churn, user satisfaction survey, contribution value decomposition, responsiveness analysis, customer added value demand, etc., the decomposition process is conducive to increasing users' supervision level, improving users' satisfaction rate, so as to more accurately indicate the research, issuance and marketing actions of industrial commodities, and create more benefits for the company. (3) Analyze the market: analyze and estimate the market trend of industry goods, estimate the trend of buyers in different regions to choose different goods, analyze the demand for products in different periods of the market, so as to grasp the development trend of the market as a whole, and carry out different marketing activities according to the characteristics of different regions and periods of time, thus increasing the market share.

3 Composition of Data Warehouse

Figure 1 is the schematic diagram of the system composed of the data warehouse. This data warehouse has more prominent characteristics in two aspects: first, it has a larger data processing capacity, and second, it supports fast retrieval requirements, Therefore, it is widely used in various application units.

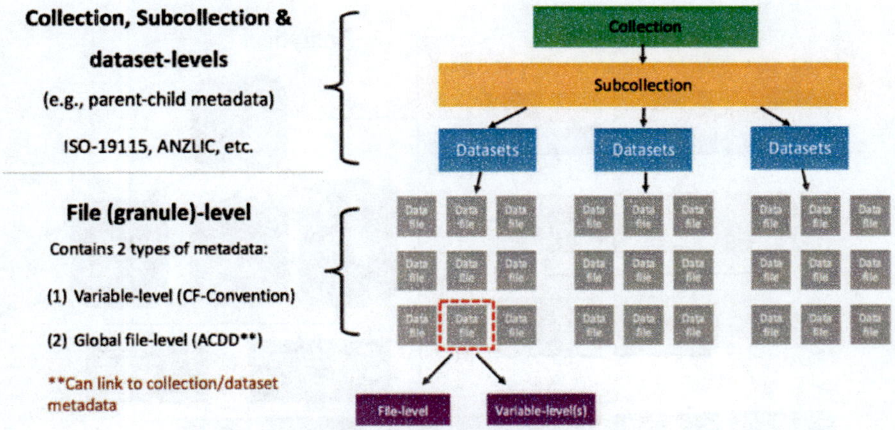

Fig. 1. Schematic Diagram of Data Warehouse System

4 Design and Implementation

4.1 Customer Relationship Management (CRM)

(1) CRM

For a long time, different service and supply organizations have given different definitions of CRM from different directions, lacking unity and authority. GartncrGroup, an American authority, proposed that CRM is a customer-centric production or service business marketing behavior to meet the revenue capacity of enterprises and customer satisfaction. For example, Hurwitz Group believes that CRM commercial marketing is a reasonable marketing method aimed at reducing costs, increasing profits, seeking development channels and improving customer satisfaction and loyalty, and achieving customer satisfaction. The author thinks that CRM system is to design and establish a new service and work system with the help of computer technology and modern information technology, taking customers as the center. The computer technology applied includes the Internet, data warehouse, artificial intelligence system, etc. Its focus is to improve and improve various customer services and provide strong service background and various related business activities with sophisticated, fast and intelligent marketing means, so as to ultimately achieve the marketing purpose of minimizing costs, maximizing profits, expanding the market and satisfying customers.

(2) CRM system structure organization

With the rapid development of automatic informatization, CRM work platform closely combines the work content and technical information of all departments of the enterprise through informatization and automation technology to provide a perfect marketing scheme for enterprise marketing, customer service and decision-making. CRM work platform is divided into three levels: database type, operation type and operation type. Its structure and organization are shown in Fig. 2.

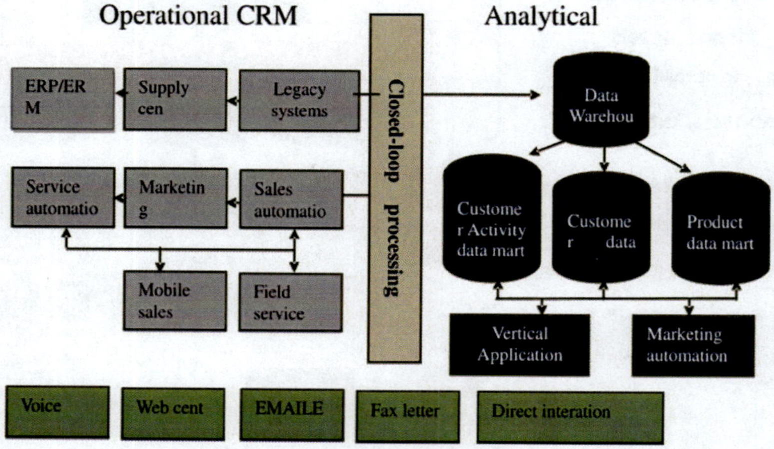

Fig. 2. Collaborative CRM

4.2 Design of System Structure

For the system we created and developed, we adopted a three-tier architecture model in the design, specifically: the presentation layer and the data and business logic layer. we also applied the component-based design idea and MVC design pattern to optimize the system design. The overall structure of the system is shown in Fig. 3:

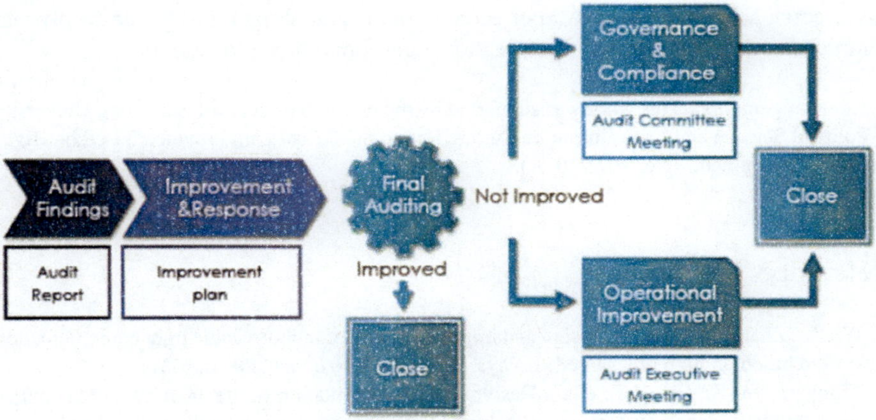

Fig. 3. Overall structure

Customer insight comes from CRM, which requires enterprises to have three basic capabilities: customer value, customer interaction and customer insight. Customer value focuses on understanding the value customers can bring to the enterprise. Customer communication emphasizes that in the process of communication between customers and enterprises, enterprises can feel personalized and high-quality services to attract new customers and retain old customers. Both of these capabilities depend on customer insight. Customer insight is to collect and analyze customer information, and design the marketing process based on this information to support marketing decisions. Customer insight meets the differentiated services determined according to customer needs and values, making the market, sales and service a whole, providing customers with targeted and consistent customer experience in different channels, and is the link between marketing, sales and service.

5 Conclusion

At present, China's communications industry is faced with double challenges from within the industry (between operators and mobile Internet) and from outside the industry (OTT and foreign virtual operators). All communications operators are trying to introduce new concepts and technologies into the industry to improve their comprehensive competitiveness. On the premise of the coexistence of opportunities and challenges, this paper deeply and comprehensively analyzes and studies CRM, data mining and other related issues, mainly covering the basic meaning of CRM system, the management concept and related applications based on CRM, the system structure of mining CRM system and the logical functions to be realized. In addition, it mainly expounds how to realize the main functions of the CRM application system with data mining as the main line, how to determine how to implement the customer relationship management system, and how to implement the process and methods, and how to use the data mining model in the customer management relationship. Through analyzing and studying all the research achievements and experiences at home and abroad, this paper deeply and comprehensively analyzes the classification algorithms such as data mining.

Acknowledgements. This work was supported by the research projects of Shandong University of Political Science and Law: Corporate Culture Research of Shandong Province Based on High-Quality Development' (No. 2021Q02A).

References

1. Wei, T., Zhang, B., Lin, C.: Design and implementation of indoor visible light communication system based on high order PAM. J. Phys. Conf. Ser. **1907**(1), 012005 (2021)
2. Xiong, Y., Li, K., Liu, Z.T., et al.: Design and implementation of the universal servo control algorithm verification system based on high-speed communication fieldbus. J. Adv. Comput. Intell. Intell. Inform. **25**, 248–257 (2021)
3. Chai, Y.: Design and implementation of english intelligent communication platform based on similarity algorithm. Complexity **2021**(2), 1–10 (2021)
4. Zhang, Y., Jian, Z., Wang, H.: Design of power communication data network management system and implementation of data acquisition module. J. Electron. Res. Appl. JERA **6**, 5 (2021)

Author Index

J. C. Hung et al. (Eds.): FC 2023, LNEE 1131, pp. 485–486, 2024.
https://doi.org/10.1007/978-981-99-9299-7